A
Dominant
Character

ALSO BY SAMANTH SUBRAMANIAN

This Divided Island: Life, Death, and the Sri Lankan War

Following Fish: Travels around the Indian Coast

A Dominant Character

*The Radical Science
and Restless Politics
of J. B. S. Haldane*

Samanth
Subramanian

W. W. NORTON & COMPANY

Independent Publishers Since 1923

For information about permission to reproduce selections from this book, write to Permissions, W. W. Norton & Company, Inc., 500 Fifth Avenue, New York, NY 10110

For information about special discounts for bulk purchases, please contact W. W. Norton Special Sales at specialsales@wwnorton.com or 800-233-4830

Manufacturing by LSC Communications, Harrisonburg

Book design by Lisa Buckley

Production manager: Anna Oler

ISBN 978-0-393-63424-2

W. W. Norton & Company, Inc.

500 Fifth Avenue, New York, N.Y. 10110

www.wwnorton.com

W. W. Norton & Company Ltd.

15 Carlisle Street, London W1D 3BS

1 2 3 4 5 6 7 8 9 0

For Padma,
the most patient recipient imaginable
of a writer's daily bulletins of torture

Suffer.

—The family motto of the Haldane clan

Pathei-mathos ("We suffer into knowledge").

—Aeschylus, *Agamemnon*

Contents

A
Dominant
Character

1.

The
Scientific
Method

1.

THE LETTER ARRIVED UNSOLICITED, like thousands of others. A retired chemist in Surrey had taken up plant genetics and set himself an immodest task: to improve the yield of his plants by a process that could then be applied by any farmer anywhere in the world. Now, in July 1948, he thought he'd cracked it. His flax plants were producing 12 or 14 seeds in each pod, instead of the usual 10—a bumper strain for flaxseed oil. "The results seem beyond doubt," he wrote in his letter, after two and a half pages of jumbled description. He had read J. B. S. Haldane's essay "Scientific Research for Amateurs." Would Haldane, as a renowned geneticist, be interested in this radical piece of amateur research?

Haldane wrote back. He nearly always did, even though he hated to be bothered by correspondence. His letters piled up around his various offices over the years: in Cambridge in the 1920s, in University College London until the 1950s, in Calcutta and Bhubaneswar thereafter. Some letters went missing, sinking under the flotsam that occupied these offices: notebooks, journals, Haldane's own papers on genetics and biometry, reprints of papers by other scientists, pamphlets, issues of the *Daily Worker*. If the letters bobbed back up to the surface, they were rescued. Haldane would first scrawl his response, often on some piece of paper on which he had been working out equations. Then his secretary typed it up. Which was just as well, because his handwriting resembled ants somersaulting through snow.

"Dear Sir," Haldane wrote, "Thank you for your letter." The chemist's results seemed striking, but Haldane needed more: fuller details of the techniques he used and the results he obtained. "You will realise that an account is useless unless it is so worded that others can repeat the work." This forms the kernel of the scientific method: that researchers elsewhere be able to replicate experiments and derive identical results. Science is held up by principles, and these principles have to be inherent in every place, not just in an amateur horticulturist's patch of Surrey earth.

Haldane never shrank from exalting the scientific method, even in casual correspondence. "Science advances by successive improvements in former theories," he wrote once to a man who sent him a hollow hypothesis about how thoroughbred racehorses inherited their coat colors. "If they are wrong"—the former theories, he meant—"the reasons for rejecting them should be stated. If they are right, this should be acknowledged." To a W. Hague of Kingswood Cottages, London, who wished to alert the world to his discovery of "a new law of nature," Haldane replied: "The test for a 'new law of nature' is this. Does it enable you to predict or control events which could not be predicted or controlled before? What is wanted . . . is a set of repeatable experiments which will go one way if it is true, and another way if it is not." The custom of accuracy in statement is essential, Haldane thought; it is, in fact, desirable to be pedantic. A scientist no doubt needed imagination to sense what nature hides, but when it came time to test and publish, Haldane considered it wise to heed Francis Bacon, to "buckle and bow the mind" to the procedures of science.

Restraint was not Haldane's style. He was a man armed with infinite provocations, and a grouch besides, his bluntness shading quickly into rude hostility. A journalist described him as a "large woolly rhinoceros of uncertain temper." Even with friends, Haldane could be pungent in his remarks if something smelled like bad science. In 1953, Hans Kalmus sent Haldane a manuscript of his new book on human genetics. Kalmus was a longtime colleague and a protégé of sorts, a Czech refugee who had, with Haldane's help, found work at University College just before the Second World War. None of these personal ties softened Haldane's assessment of the manuscript: "It ought not to be published." He listed some errors, then added: "I could go on indefinitely." The proposed book would not only harm Kalmus but the science of genetics itself. "You would be better advised, if this is possible, to go back to experimental biology," he wrote, "rather than to continue to work in human genetics."

If Haldane was merciless with others, he demanded similar rigor of himself. His career overlapped tidily with the bloom of genetics as a field of study and with the effort to discover the role of the gene in Charles Darwin's theory of evolution. Genetics grew severally studded with Haldane's contributions. He demonstrated, for the first time in mammals, the mechanism of genetic linkage, by which two genes that reside near each other on a chromosome tend also to be inherited together. (He wrote up parts of this paper while serving in the trenches during the First World War.) He mapped the genes for hemophilia and color blindness. He introduced a theory for how life began on Earth. His speculations on "ectogenesis" forecast the development of in vitro fertilization.

His most important work came in a series of 10 papers, written between 1924 and 1933, in which he subjected evolution to the unflinching stare of statistics. The papers modeled the processes of natural selection and estimated rates at which gene mutations develop and spread through a population. He was gaining the measure of life itself. The stringency of statistics delighted Haldane. Everyone should know more mathematics, he always thought. Numbers were so satisfyingly precise, equations so universal. How well they ministered to the scientific method!

Had Haldane done just this and little else, he would have been an important scientist—not as revolutionary as Einstein, perhaps, and not associated for perpetuity like Watson and Crick with a single, shining discovery, but certainly among the few who altered their field beyond recognition, pushing it forward paper by paper. This is how science progresses most of the time, after all: through the accretive power of daily work, through meat-and-potatoes research. What made Haldane one of the most famous scientists of his age, though, was not just his science but also his writing and his politics—the first clear and illuminating, the second unbending and forthright, both deeply attractive during a time of shifting, murky moralities.

In magazines and newspapers, Haldane wrote about everything. He wrote cutting opinion pieces on politics—like razor blades in print. He wrote about his own boisterous life, which was stocked with enough danger and drama for a dozen ordinary humans: his boyhood apprenticeship to his scientist father, his time in the trenches, his numerous experiments on himself, his sorties into the teeth of the Spanish Civil War, his clandestine research for the British Admiralty during the Second World War, and his emigration to India. He wrote of his views on governments and philosophies, and he wrote about history and literature. He wrote a book for children, about a magician named Mr. Leakey, and most of a science fiction novel. But mainly, he wrote columns that unpicked the convolutions of science for the inexpert reader. He preached science to the laity. Arthur C. Clarke called Haldane "the most brilliant scientific popularizer of his generation."

The breadth of these columns was staggering. They dealt with trifles like lice and the funny bone, with grave issues like lead poisoning and air raid precautions, and with grand matters like the chemistry of sex and the Milky Way. On every front of science, he seemed to know of every journal article being published, every item of research being conducted, as if scientists confided their dreams to him every morning before heading off to their laboratories. He spun his scientific lessons off the spindle of the daily world, so that no one could fail to understand them. "Start from a known fact, say a bomb explosion, a bird's song, or a cheese," he advised once. Then proceed through the science in a series of hops rather than one direct leap. His material was often filched from the week's most lurid headlines: a murder trial, the deaths of alcoholics, the monkey gland extracts administered to the players on the Wolverhampton Wanderers football team.

At first, Haldane was scornful of colleagues who wrote for the public, but he came to enjoy his role as a communicator of scientific truth. It satisfied his need to vent his opinions—of which he never ran short—as well as his belief that research ought to make its way into the public gaze. When he gave a talk—and at his most active, he gave nearly a hundred a year—the hall filled swiftly. He made for an arresting lecturer: a king-sized man in rumpled clothes, his moustache so thick and his head so large and bare that it was as if a bird had built a nest at the base of a boulder. His voice filled the room as he quoted Dante, Norse myth, and the *Bhagavad Gita* from memory, beckoning with ease his knowledge of genetics, chemistry, history, and astronomy. In Great Britain, he grew as famous as Einstein. The philosopher Isaiah Berlin, who grew up admiring Haldane, called him "one of our major intellectual emancipators." One of Haldane's acquaintances thought he was "the last man who might know all there was to be known."

Haldane's relationship with his readers was punctilious. Although he simplified the science, he was never less than exact—or at least as exact as the research of the day permitted. Questions poured in by mail, and he addressed the interesting ones at length in his columns. (Are X-rays dangerous to human beings? How do complicated tasks become easier with practice? What's the difference between reflex and instinct?) At other times, Haldane would reply by pleading that he didn't really have time to reply.

"What is the ultimate cause of Germany's retrograde mentality?" Margaret Murray, an archaeologist and a scholar of witchcraft, asked him in a letter in 1942.

Haldane spent a brave paragraph trying to respond to this riddle before huffing: "I do not propose to start a correspondence. Anyway I am pretty busy, and shall be a lot busier in the future on research in connexion with the war." Then another paragraph followed. He hated to leave a question unanswered.

Much of Haldane's writing appeared in the *Daily Worker*, the official newspaper of the British Communist Party. Haldane joined the party in 1942, having already been a fellow traveler and a dedicated Marxist for years. His work and his ideology, he felt, were in absolute concord. Haldane thought Marxism practical and transparent—scientific, really. Marxists studied history and economics the way physicists studied atoms: with objective curiosity, so that they could then predict and control events. Haldane never believed that a formula on a chalkboard or a cured guinea pig represented the climax of science. He wanted science to sweep out of the lab and into the world, to improve or perfect the way people lived. (The titles and subtitles of his books reflected this, again and again: *Science and the Future, Science and Life, Science and Everyday Life, Science and Ethics, Science and Well-Being.*) Marx, too, wanted just that kind of material change. For Haldane, Marxism was the scientific method as applied to society, and both genetics and Marxism were avenues to a more utopian civilization.

HE VISITED THE SOVIET UNION just once, for a month in the summer of 1928, in the company of his wife Charlotte. They had been invited by Nikolai Vavilov, a geneticist whose stature and networks drew many of his Western colleagues to his country. Vavilov was familiar with the men and women who staffed British science. He had worked for a while with the biologist William Bateson, who first affixed the word *genetics* upon the study of heredity; the pair had collaborated across Cambridge and the John Innes Horticultural Institution in south London, of which Bateson was the founding director. In 1928, Haldane was working part-time at John Innes as the "Officer in Charge of Genetical Investigations," balancing these duties with his role as a biochemistry reader at Cambridge. He was 35 years old, and storms of change were overrunning his life. He was newly married and beginning, slowly, to suspect that Charlotte and he could not have children. His moustache was robust but his hair was ebbing, and his stockiness was clotting into fat. He was halfway along the journey from the offhand socialism of his youth to the gritted-teeth Communism that lay ahead. He was in the midst of writing his series of 10 papers on natural selection. He was also starting to relish the sweetness of a wider fame. He wasn't yet one among the world's best-known scientists, but he was getting there in a hurry.

Like Haldane, Vavilov was born to a bourgeois family, but he possessed a most Soviet enthusiasm: to rid humanity of hunger. By rummaging through the world for hardy, productive plants and by seeking the secrets of their endurance in their genes, Vavilov thought he could culture new species to thrive in any season anywhere. His mission blessed by Lenin himself, Vavilov had set out—for Iran and Afghanistan, Canada and the United States, Western Europe and northern Africa, China and Latin America. From more than 50 countries, he gathered seeds and plants, and every time he came home, he deposited his collections in a seed bank he'd started—the world's first, its vaults lodged in an old tsarist palace in Leningrad. The bank stored a quarter of a million specimens. It was an aristocrat's bauble converted into an institute committed to feeding the poor. What more potent symbol of Soviet principle could there be?

In Moscow and Leningrad, Haldane delivered lectures on genetics, and he made an excursion to Vavilov's experimental farm in a town called Detskoe Selo, near Leningrad. He also stopped by the seed bank and noticed the incongruity of botanists conducting their noble research amidst the manor house's parquet floors and marble mantelpieces; they would no doubt have preferred laboratory benches and a decent set of sinks, he thought. Still, he grew to like Vavilov and admire his work. It promised not only to improve crop yields but also to expand the world's knowledge of agriculture and of civilization itself.

In turn, Vavilov proved a most attentive host. He was a handsome man, Charlotte noticed. His moustache was neat, his eyes good-humored, and his three-piece suits refined and elegant. He had nothing of the monkish, withdrawn quality that scientists sometimes possessed. In Moscow, Vavilov threw a rambunctious party for the Haldanes, with champagne and dancing. He arranged for them to go everywhere: to famous churches; to Lenin's tomb, on a special, private visit; to the Kremlin's museum, where they saw Ivan the Terrible's crown, with its collar of sable and its diamond-cross steeple. They went to the Bolshoi to watch *The Red Poppy*, in which a Soviet ship captain tries to rescue the employees of a cruel harbormaster in a Chinese seaport. They attended the performances of two Rimsky-Korsakov operas. In Leningrad, Charlotte happened to mention that she was fond of caviar, so Vavilov had delivered, to the Haldanes' hotel, two of the Soviet Union's greatest luxuries: a tin of caviar and a fresh loaf of white bread. That night, half asleep, Charlotte thought she heard a strange symphony of squeaks; the next morning, she found that the hotel's resident mice had climbed up onto the table and eaten their way through half the bread and even the paper in which it was wrapped.

Even in June, Leningrad was frigid. The wind lunged across the Neva River and into the Haldanes' Hotel Europa, forcing them to go to bed fully dressed. In Moscow, the couple stayed in a borrowed apartment. The main street in their neighborhood was tidy and broad, but as the streets ramified into smaller and smaller lanes, they grew progressively grubbier. "The housing situation was bad," Charlotte wrote later. "Whole families huddled in one room. The staircases of the houses were filthy, [and] cooking was done on kerosene stoves in the passages, which reeked of refuse. Although each house was supposed to have a concierge and a house committee to organize the general upkeep among the tenants, there was no real organisation. Everyone who could, passed the buck. Quarrels between tenants were incessant."

Charlotte couldn't make up her mind about the Soviet Union. She thought the Russians to be spirited people, and optimistic despite their recent cavalcade of crises. At the Red October chocolate factory, she saw women on the production line looking happy and industrious; the factory, admirably, had a crèche for their children. But she wearied of the grime and the poverty. A scientist of their acquaintance lived with his wife and three young children in a flat that was, really, just one room. She was so uncomfortable with the feeling of people being under surveillance everywhere and all the time, she would claim later, that she was relieved to leave at the end of the month.

But Haldane felt differently. In those early years, when the revolution was still warm, the Soviet Union resembled an essential experiment in itself: a state actively setting out to advance through the use of science. Lenin had believed, as Haldane did, that the chief utility of science lay in its capacity to enhance society, to improve the way people lived. The Soviet Union treasured its scientists, Haldane thought, and not without reason. At the time, Soviet scientists were still able to travel overseas, and if they qualified for a certain intellectual eminence, they were provided nearly anything they might require. "One must spare a great scientist or major specialist in whatever sphere, even if he is reactionary to the nth degree," Lenin had once told one of his commissars. For a living example of these generous dispensations, Haldane needed to look no further than Vavilov: a botanist whose research the state thought so promising and practical that it funded his expeditions, his institute, and his seed bank. In this new society, the ideas of science formed a new, privileged class unto themselves.

On the other hand, religious instruction was forbidden, and this, too, fit Haldane's sensibilities. "My practice as a scientist is atheistic," he wrote six years later. "That is to say, when I set up an experiment I assume that no god, angel, or devil is going to interfere with its course, and this assumption has been justified by such success as I have achieved in my professional career." The Soviet government exhorted its subjects to avoid superstitions or rituals; the year after Haldane visited the country, Stalin's government began to shut down thousands of churches. By way of replacement, the state suggested Marx, Engels, and Lenin; busts of these men, with their variety of beards, sprouted by the side of avenues and in public buildings. Although religious worship was not banned outright, its demotion was evident. In Red Square, the Haldanes saw, in a corner, an old icon depicting Mary and the child Jesus. Above it, in large lettering, were installed Marx's words: "Religion is the opium of the people."

Given these captivations, perhaps it was not easy for Haldane—or for anyone of his temperament who visited the country so briefly—to peer around the corner, to the persecutions and show trials that crouched in wait. After he returned from the Soviet Union, Haldane inched closer and closer to the Communists before eventually becoming one himself. For nearly two decades, British intelligence agents kept diligent watch on him, their suspicion building all the while that he was a Soviet spy. Whenever he went to the Communist Party's headquarters on King Street in London, MI5 listened in. Whenever he received envelopes that appeared suspect, MI5 peeked into them. Whenever he gave speeches, about biology or air raid shelters or Franco-held Spain, MI5 agents were in his audience, scribbling notes. What's your game, Haldane? they wondered. What kind of trouble are you trying to make?

Throughout his life, Haldane refused to trust any kind of authority—teachers, provosts, officials, bureaucrats. They knew too little, and their hidebound thinking had no science or rationality to it at all. Most of all, he mistrusted his government and the governments of other capitalist societies. He felt, as we now feel afresh in our century, that nations were held rapt by the wealthy, that they were warmongering and venal, that they placed the narrow interests of the powerful above the well-being of the powerless. They had failed to frame systems of statecraft and ethics that spoke to the modern age. "When applied science has created so many new moral problems," he once wrote, "the morality of our ancestors must in any case be drastically revised."

It is difficult now to imagine a scientist like Haldane, who recognized how deeply political his work was and who thought it his duty to speak loudly about it. Except in rare cases, scientists today wipe their public selves clean of any trace of politics or ideology. The impetus to seal science away from politics fails to realize that all science, like all art and all other aspects of life, is already cut and shaped by the society in which it is created. For Haldane, there was no hermetic boundary between the two forces. His science and his politics were wrapped tight around each other, like strands of double-helical DNA. On one occasion, though, the strands bent themselves into a knot. It would prove to be Haldane's greatest moral crisis, and it began not long after that retired chemist in Surrey wrote in about his flaxseed marvels.

2.

Diamat is a tank.
    —Czeslaw Milosz

In the summer, a pedestrian on Moscow's Prechistenka Street has to peer hard through the foliage of trees to see, in its full, wide nobility, the pistachio facade of the Central House of Scientists. It still looks like the wealthy man's mansion it used to be. The house was once owned by a textile baron, whose family sold it to a financier for half a million rubles—an unfortunately huge sum to pay in 1916, a year before the Bolshevik Revolution. After the financier fled overseas, the state annexed the house; in 1922, it was turned over to the cause of science. Here, amid ornate statuary and lush tapestries and gilt-encrusted mirrors, Soviet scientists gathered to receive their monthly rations: 40 pounds of bread, 2 pounds each of sugar and buckwheat, a pound or two of oil or butter, and some meat. The mansion became a sort of club for scientists, where they could present their research, listen to lectures and chamber concerts, and talk over cut-rate soup in the commissary.

In this House of Scientists, over one combative week in 1948, the Soviet Union abolished the gene altogether.

The Lenin All-Union Academy of Agricultural Sciences began its conference on the evening of July 31. More than 700 agronomists, researchers, and professors from across the Soviet Union attended. Every day save one, until August 7, they packed a lecture hall and listened to speeches that frequently spilled over their allotted 30 minutes apiece: a morning session from 11 a.m. to 3 p.m., a respite for dinner, then an evening session from 6 p.m. to 10 p.m. Only on August 1—a Sunday—did the scientists take a break, and on that day they climbed into buses and rode out to Gorki Leninskiye, a few miles south of the city, where the academy ran an experimental station. Through that entire week, *Pravda* covered the proceedings in all their stormy detail. After Trofim Lysenko, the academy's president, delivered his official report—edited with care by Joseph Stalin, no less—its full text was published in every central newspaper.

Lysenko was 50 years old at the time, a slab-faced man with astonishing blue eyes and a forelock of hair that hung heavy with brilliantine. He wore, out of habit, a sour look. A reporter, upon meeting Lysenko in his pea fields in the Azerbaijan in 1927, complained that Lysenko gave him the feeling of a toothache: "Stingy of words and insignificant of face is he; all one remembers is his sullen look creeping along the earth as if, at the very least, he were ready to do someone in." Perhaps he was only beset by shyness, or nervousness, or deep unhappiness, as some suspected. But he was passionate and ambitious, and as he came to dominate Soviet agronomy, he became the prophet of a new, homebrewed theory of heredity. "He is the peasants' demagogue," a visitor to the Soviet Union observed in 1944. "What he says to them, goes."

On the inaugural evening of the academy's session, Lysenko rose to speak.

The history of biology, Lysenko said, was a history of ideological battle. Darwin's theory of evolution was scientific and true, but it was not free of error, Lysenko pointed out. It was wrought too heavily in the idea of competition, in the idea that life pushed forward only by vanquishing others in the eternal struggle for resources. This was just capitalism dressed up as biology. The loose-thinking, "reactionary" geneticists of the West had debased Darwin further. They insisted that an organism's acquired characters—the ways in which a plant or an animal adapted to, or was shaped by, its environment—could not be bequeathed to its offspring. Instead, they believed that inheritance and evolution relied purely on random, unpredictable combinations of genes. But Lysenko scoffed at this notion of heredity as a game of dice, an abstract arithmetic. What happened to a gene altered the body in which it was carried, these reactionaries claimed, but what happened to the body never altered its genes. How was this not an absurdity?

Fortunately, the Soviet Union was available to rescue science from the brambles of these falsehoods. Karl Marx and Friedrich Engels, the authors of *The Communist Manifesto*, had already seeded the high laws of dialectical materialism—"diamat," an all-purpose lens to scrutinize history and nature. The conditions of the material world give rise to everything, they determine all historical change and all social reality. Soviet scientists, Lysenko claimed, had been able to pull these laws into biology. They had proved that a material change induced in an organism would modify its genes, percolate down to its offspring, reorient the very trajectory of its species. What swelling promise this held for the project of feeding the people of the Soviet Union, he effused: hardier wheat, for instance, or cattle that produce more milk. He quoted Ivan Michurin, the father of this new Soviet biology: "We cannot wait for favours from Nature; we must wrest them from her."

Lysenko's speech, by turns grandiose, cutting, and reverent, went over well; the official transcriber ensured that passages were followed by parenthetical observations such as (*Amusement*) or (*Animation, laughter*) or, at the final invocation of the two greatest teachers of all, V. I. Lenin and J. V. Stalin, (*Loud applause*). Everything was framed as an –ism, an ideology: capitalism and Communism, of course, but also diamat and idealism; Mendelism and Morganism and Weismannism, named for the reactionary biologists whom Lysenko held to be the architects of scientific lies; and Michurinism, after the Russian fruit farmer whose experiments in the early twentieth century lit the way into these revolutionary concepts of heredity. Once anything—even scientific research—became an –ism, it could be fought. Sides could be taken, traitors could be identified. Lysenko singled out several members of the audience whom he deemed to be foolish or disloyal. Their work, based on misguided principles of genetics, would soon cease to be indulged by the state's universities and research institutes, he warned. The Soviet Union had no time for their shiftless science.

Having drawn these battle lines, Lysenko ended his speech. Over the next seven days, 56 biologists, in the guise of presenting their research, defended themselves or attacked others. The conference turned into an inquisition; the speeches rang with recrimination. On occasion, Lysenko acidly heckled speakers, but mostly he sat and watched this conflagration of science and politics, and the scientists struggling to escape the flames.

THE SON OF A FARMER, Lysenko grew up working his family's fields in the Ukraine, learning to read and write only when he was 13. He enrolled in a horticultural school just as Lenin and his Bolsheviks were gathering power, so his early career—growing sugar beets; correspondence courses at the Kiev Agricultural Institute; a post at the Institute of Plant Breeding and Genetics in Odessa—neatly tracked the sharpening ambitions of the new Soviet Union. Lysenko never became a conventional, widely read academic, but this did not hinder his progress. The science his country believed it required was the practical, earthy sort that could benefit its millions. Every farmer was an agronomist in pupa; one state newspaper, *Bednota*, conscripted 23,000 "peasant scientists" to work in their "hut labs," dabbling with seeds to stimulate their development.

Lysenko launched himself into this movement. He stalked his fields in his threadbare overcoat, supervising experiments. When other scientists paid long visits to his agricultural stations, he was happy to give them the bed and sleep on the floor. *Pravda* called him "the barefoot professor." First he busied himself trying to grow peas through the Transcaucasian winter, but that project failed. Then he attempted to cultivate new varieties of grain. The Soviet Union's standard winter wheat, sown late in autumn, was often wiped out by cruel winters, and it never grew if sown in the spring. So Lysenko set about vernalizing the wheat—artificially providing it the short snap of cold required to activate its growth and then seeding it in the spring.

Early in 1929, he and his father soaked 48 kilograms of winter wheat in water, packed the wheat into sacks, and interred the sacks in a snowbank for a month. Once retrieved and sown in the spring, Lysenko claimed, the wheat not only flowered quicker but grew fabulously, granting more grain than it ever had. Further, this vernalized fertility wrote itself so deeply into this lineage of winter wheat that it persisted generation after generation. This was an ancient theory, and it had long been abandoned by farmers who never found it to deliver any real, lasting rise in yield. But on the basis of Lysenko's one experiment on half a hectare of land, Soviet officials began to sense grand, golden prospects—of tremendous harvests, bread for everyone, crops in the Arctic. In 1932, the state ordered 43,000 hectares to be planted with vernalized wheat; in 1934, 600,000 hectares; in 1937, 10 million hectares. The sheer force of politics pressed tons of vernalized seed into the soil of the steppe.

Finding himself suddenly esteemed, and flush with new titles and appointments, Lysenko turned his attention to genetics.

Through the first decades of the twentieth century, the most fundamental facts about the gene were still indistinct, but their outlines had begun to clarify. With patient experiments, scientists were peering into the clockwork of evolution, discovering how the cogs locked, how fast the wheels spun. Chance, they found, had a lot to do with it. An animal or a plant was born with an accidental genetic mutation that gave it some manner of advantage, made it better suited to its surroundings. Its descendants inherited the mutation and flourished, crowding out the ill-equipped peers of their species. The first giraffe with a longer neck was likely a freak of nature, but the neck was so useful in reaching for leaves on tall trees that the freak proved better adapted to the savanna. It prospered and multiplied and, with time, constituted an altogether new species. Fewer and fewer scientists believed—as the naturalist Jean-Baptiste Lamarck had a century earlier—that life on Earth unfolded and progressed in a preordained sequence, or that a life-form's environment imprinted itself upon genes and heredity.

Lamarck had proposed something like a Just So story: The animals of a species in the African veldt once stretched their necks to eat, those necks lengthened gradually, over successive generations, and that was How the Giraffe Got Its Neck. In the principles of this story, the Marxists of the Soviet Union saw plenty to admire. They illustrated how important the material conditions around an organism were to its development. Modifying those conditions could elevate the organism, just as revolutionizing the material structure of society could elevate a proletariat. The fable was undiluted diamat. It was also morally stirring, a call to the powerless to improve themselves through pure action, striving upward to reach the leaves that had been denied to them. It appeared to supply the Soviet Union with a blueprint for agriculture—a way to farm the future, to do rather than to just theorize prettily. "Philosophers have hitherto only *interpreted* the world in various ways," Marx had written. "The point is to *change* it."

Fired up, Lysenko organized his experiments along muddled, neo-Lamarckian lines, and he reported excellent news—always in *Pravda* or in journals and pamphlets he controlled. With his Potemkin genetics, he claimed that he could produce frost-resistant grain of any kind within two to three years and that potatoes and cotton could be vernalized just like wheat. He thought he could accurately pick out, in the very first generation of hybrid wheat, the plants whose offspring would go on to ripen quickest for decades and decades. He argued that young oaks planted in a tight cluster didn't vie with each other for sunlight and nutrients—that such bourgeois competition occurred only between species and not within a single species, and that the weaker saplings died out, in some kind of agreeable suicide, so that the stronger might survive. If wheat was cultivated in "appropriate" conditions, he suggested, it could be transformed directly into rye—the equivalent, a historian would later say, "to saying that dogs give birth to foxes when raised in the woods." In 1943, without running any tests, the government obeyed Lysenko's advice to plant sugar beets in the desiccated summer soil of central Asia. The saplings withered, but Lysenko remained undaunted. The botanical world, he bragged to *Pravda*, had now become "clay or plaster for the sculptor: We can easily sculpt from them the forms we need."

In the Soviet Union, this marriage of science and ideology proved fatal for those with different opinions. It did one's career no favors, for instance, to disagree with Lysenko when he first started to reject the classical conception of the gene, toward the middle of the 1930s. ("We deny little pieces, corpuscles of heredity," he said during a scientific congress.) The air began to thrum with words like "sabotage" and "treason." As Stalin expanded his power, such accusations spilled forth more and more easily. As in every sphere, so in genetics: Those who did not agree with Lysenko, Stalin's favored expert, found themselves in peril. Dissenters were accused of being part of "the powers of darkness." One of Lysenko's supporters called an opponent a "Trotskyite bandit"; *Pravda* denounced another's "fascist ideology." At least 22 geneticists were arrested; around 77 biologists of various kinds suffered one form of repression or another.

Lysenko became president of the Lenin All-Union Academy of Agricultural Sciences in 1938 and then director of the Institute of Genetics two years later. The appointments were overt attempts to squash the classical genetics of the man Lysenko replaced in 1940: Nikolai Vavilov. Vavilov continued to defy Lysenko's imposition of shoddy science. "We shall go to the pyre, we shall burn, but we shall not retreat from our convictions," he declared in a speech in 1939. The pyre came to him, though. The very next year, he was arrested in the middle of a field in western Ukraine, in full sight of his expedition, with no pretense. "He was taken so fast that his things were left in one of the cars," the biologist Zhores Medvedev wrote. "But late at night three men in civilian clothes came to fetch them. One of the members of the expedition started sorting out the bags piled up in the corner of the room, looking for Vavilov's. When it was located, it yielded a big sheaf of spelt, a half-wild local type of wheat collected by Vavilov."

To the military collegium that tried him, Vavilov denied all charges: He wasn't acting as a British spy, he wasn't sabotaging the Soviet Union's campaigns in agriculture, and he wasn't a member of a right-wing conspiracy. Nonetheless, he was sentenced to death and dispatched to a subterranean cell in a gulag in Saratov, where he had once taught at the local university. Vavilov died here—of malnutrition, having been allowed only mold-ridden flour and frozen cabbage—in January 1943. That sheaf of spelt in his bag, it turned out, was a new species. On his last day as a working scientist, Vavilov had made one final, modest, but ineradicable discovery.

AT THE 1948 CONFERENCE, in the House of Scientists on Prechistenka Street, academicians fell into line with alacrity. Biologist after biologist applauded Lysenko and the new Soviet genetics and then hurried on to demonstrate how their own research conformed to these radical principles. One scientist called classical genetics "a pseudoscience," another a "propaganda of obscurantism." ("Hear, hear!" an audience member yelled.) For a while, these supporters of Lysenko dominated the proceedings. "Why has no one of the adherents of formal genetics taken the floor?" a needling note, passed up to the conference chair midway through the first day, wondered. "Is it because they do not want to speak themselves, or because they are not being given a chance to speak?" After one such adherent, I. A. Rapoport, delivered a short, careful talk that merely urged reflection, he was chastised promptly.

"Is it worthy of a scientist to behave as Professor Rapoport did yesterday?" a speaker asked.

Voice from the hall: "It was ruffianism!"

"It should not be left at that," the speaker continued.

Voice from the hall: "Quite right!"

Seven others tried harder than Rapoport. B. M. Zavadovsky outlined some of his disagreements, only for hecklers to ask him, time after time, the question that was the litmus test of loyalty: "Do you agree with the inheritance of acquired characters?" I. M. Polyakov argued that only a few heritable physiological variations—the protective coloring of a butterfly, for instance—were influenced by the environment. "Can it be predicted or not?" Lysenko called out.

Polyakov tried to push forward his methodical reasoning, but was cut off again. "Can it be predicted or not?"

"It is hard for me when I am interrupted," Polyakov said.

"It is hard for me when I have to listen to wrong statements," Lysenko shot back.

Lysenko's opponents faced demands to resign their posts. Some quarrels turned personal. P. M. Zhukovsky, a geneticist, insisted that plants could not be "trained" in the manner Lysenko claimed. The science was clear: genes were lodged on chromosomes, and chromosomes were housed within cells. The genes, and not the cells themselves, were the currency of inheritance. Nothing but a mutation to a gene could produce a change that would travel into the next generation.

When Lysenko broke in, asking if anyone had even seen a chromosome, Zhukovsky announced: "I will leave this rostrum if I am interrupted."

"A certain author—I have forgotten who it was—described a maiden who blushed at the sight of a roasted capon," Lysenko said. "As soon as the word 'chromosome' is mentioned, some people blush." (*Laughter, animation.*)

"You are wrong in sticking labels on us," Zhukovsky said. "Trofim Denisovich, this year I lectured to your son. Ask him whether my lectures have spoilt him?"

"There is no need to go into family matters," Lysenko responded. "That is my business as a father. Tell us something that is more important. You complain of being ill-treated. But have you forgotten, Piotr Mikhailovich, what names I was called in your presence? Did you forget them?"

"You were never abused in public, at meetings."

"In holes and corners."

"That was the gossips," Zhukovsky protested.

On the morning of August 7, the scientists all woke to read, in *Pravda*, a letter from the chemist Yuri Zhdanov to Stalin. Zhdanov had written to Stalin in July, so it was no coincidence that *Pravda* published the text of his message on the conference's final day. Zhdanov confessed, in his letter, that he had made "a number of serious mistakes" in his recent paper on contemporary Darwinism and that he had failed to "mercilessly criticise the radical methodological defects of Mendel-Morgan genetics." He was guilty of trying to reconcile the opposing strands of genetic thought, "but in science, as in politics, principles are not subject to compromise." His criticism of Lysenko was another error. "All this is because of inexperience and immaturity. . . . I will repair my mistakes in work." It was the most public of rebukes, a dressing-down in the town square—and it had happened to Zhdanov, son of a powerful Stalin loyalist, husband-to-be of Stalin's daughter. The letter must have shredded the nerves of the biologists who had just picked apart Lysenko's theories and who now read their *Pravda* over their congealing breakfasts.

Arriving that morning at the House of Scientists, Lysenko launched into a triumphant concluding address. The party's central committee had examined his inaugural report on the state of biology and approved it, he revealed. (*Stormy applause. Ovation. All rise.*) "We recognise the chromosomes. We do not deny their existence. But we do not recognise the chromosome theory of heredity." And it simply couldn't be that random mutations drive the development of a species, Lysenko declared, for "science is the enemy of chance," and an unpredictable model of genetics was of no value to the state's farmers and planners. Lysenko was satisfied: This conference, as a whole, had turned out to be a victory for Michurin's ideas of Soviet biology over those of reactionary geneticists. He roused his audience as if he was rallying unionists at a factory. "Long live the Michurin teaching, which shows how to transform living nature for the benefit of the Soviet people!" (*Applause.*) "Glory to the great friend and protagonist of science, our leader and teacher, Comrade Stalin!" (*All rise. Prolonged applause.*)

A series of quick, startling recantations followed, by scientists who'd now sensed the permanence of the new wind and who were desperate to preserve their careers, their families, and their lives. Zhukovsky insisted he had changed his mind about gene theory the previous evening and that "there is therefore no connection between my present statement and Yuri Zhdanov's letter. The speech I made . . . was my last speech from an incorrect biological and ideological standpoint." Another geneticist, Sos Alikhanian, began, almost comically: "Comrades, it is not because I had read Yuri Andreyevich Zhdanov's letter in today's *Pravda* that I requested the chairman to allow me the floor." He would emancipate himself from his old ideas, Alikhanian promised: "We must be on this side of the scientific barricades, with our Party, with our Soviet science." Polyakov, having confirmed that he, too, had refurbished his thinking before seeing Zhdanov's letter, now said: "One must frankly say that the Michurinian trend is the high road of development of our biological science."

The dissenters had been shaken down, their opinions erased, their spirit broken. In a further matter of weeks, the party ordered universities to dismiss lecturers and professors who held views that contradicted Lysenko. Libraries destroyed their old genetics textbooks; research stations shut down; laboratories murdered every last fruit fly in the stocks of *Drosophila* that they used to run genetic experiments. Scientific institutions prominently displayed portraits of Lysenko. If you wished, you could buy a bust of the man, or you could visit a monument to him. In a state-sanctioned book of folk songs, one composition ran:

Merrily play on, accordion
With my girl friend, let me sing
Of the eternal glory of Academician Lysenko

He walks the Michurin path
With firm tread;
He protects us from being duped
By Mendelist-Morganists.

As vaunted as he was, though, Lysenko was only an instrument, the brand with which the state burned its programs upon the flanks of biology. The 1940s had been, in any case, a superheated decade, and everywhere the advances of science were bent by political purpose. At no time is there such a thing as pure research, untroubled by the politics of its age. But the middle of the twentieth century was, by any measure, extraordinarily fraught. Ideology and war harnessed for themselves the theories of race science and chemistry, of computing and communication, of medicine and aeronautics, of the gene and of the atom. In this morass, scientists struggled—and occasionally failed—to keep their footing and to keep in pursuit of the truth.

3.

Four months after the purge at the House of Scientists, the BBC broadcast a symposium on the Lysenko controversy. A translation of Lysenko's address had been published in England a few weeks earlier, to utter consternation. Two Nobel Prize winners resigned their honorary memberships of the Soviet Academy of Sciences in protest. The BBC invited four geneticists—Sydney Harland, Cyril Darlington, Ronald Fisher, and J. B. S. Haldane—to a debate, although perhaps "debate" isn't strictly the right word. Each scientist individually recited his prepared text into the microphone, without hearing what the others had to say; there were no questions or interruptions. The program ran from 6:50 p.m. to 7:25 p.m. on November 30. Haldane, now 56 and in the plump fullness of his reputation, spoke last.

The first three scientists heaped criticism on Lysenko—on his science, but also on the way he had drained his field of peers who disagreed with him. Harland reminisced, warmly and at length, about Vavilov's intellect: "Ideas sprang from his mind like a constant succession of balls of fire from a human Roman candle." Harland had visited the Soviet Union and had seen Vavilov's scrupulous work and his library of botanical collections. He had also met Lysenko, and to talk to him, he discovered, "was like trying to explain differential calculus to a man who didn't know his twelve times table." Darlington pointed out that none of Lysenko's claims of genetic jugglery were ever fulfilled. There were no new, miraculous wheat strains of incredible fertility, there was no species that had been transformed into another, and there were no crops carpeting the Arctic. "The official English translation of the Lysenko fairy tale is now available. Any gardener, any farmer, any stockbreeder, any scientist even can now judge for himself the rights and wrongs of the question," Darlington said. He was shocked not by the persecution of scientists in the Soviet Union but by the indifference of his British colleagues to these horrors. Fisher quoted ugly passages from Lysenko's address and called him a "Grand Inquisitor" stamping out the last, feeble sparks of intellectual freedom in the Soviet Union. Lysenko did not crave scientific knowledge or prosperity for poor farmers, Fisher said. "The reward he is so eagerly grasping is power: power for himself, power to threaten, power to torture, power to kill."

Haldane's short speech was a curious, un-Haldane-like affair: rickety and defensive, full of feints and distractions and muddy logic. He complained that he did not know what the other three scientists had said and that the full, 500-page transcript of the academy's bygone conference was still unavailable in English. "Till I have read a translation, I cannot judge whether the Academy's decision was right," he said. "We are like the jury in *Alice in Wonderland*, considering our verdicts before we have heard the evidence."

Nevertheless, Haldane had his views on Lysenko's work, and although he disagreed with a lot of it, he said, he found a few fundamental points to be in order. He proceeded through these, gliding over the evidence and ignoring the fallacies. His arguments were as fugitive as fog.

There was some proof to endorse Lysenko's belief that acquired characters were heritable, Haldane argued, but he cited two weak experiments by others: one that had yet to be replicated and a second, involving fruit flies and their intolerance to an atmosphere rich in carbon dioxide, that could fit several other theories. Lysenko aspired to modify the process of biological inheritance, and agents like X-rays did just that, Haldane said. But X-rays altered heredity by directly mutating genes, not by deforming the physiology of the organism. Haldane knew this and said nothing of it, even though it had been proved, beyond contention, two decades earlier. He also failed to mention that controlling the course of these mutations with a beam of X-rays, to "improve" genes in one way or another, was impossible. Dice were still being rolled. Chance still reigned over this whole enterprise.

Lysenko's denial of the gene's function, in fact, invalidated Haldane's own research. Haldane's investigations of how linked genes are inherited together, his estimations of mutation rates, his mathematics of natural selection—all of these relied on the premises that genes carry hereditary information and that they mutate in unforeseeable ways. He had even written, years earlier in the *Daily Worker*, that Lysenko's attacks "on the importance of the chromosome in heredity seem to me to be based on a misunderstanding. This would be very serious if he were dictator of Soviet genetics." Lysenko had also scorned the tools Haldane employed daily and vitally in his work. Numbers were irrelevant to biology, Lysenko once declared. "That is why we biologists do not take the slightest interest in mathematical calculations that confirm the useless statistical formulas of the Mendelists." It was an outright dismissal of the meat of Haldane's career.

Even so, on air, Haldane handled Lysenko's primary claim to biological fame—the vernalization of wheat—with delicacy. "This is a very revolutionary discovery if true," he said, before offering a thin conjecture of why it *could* be true. He didn't mention that Lysenko had run his experiment only once or that British scientists hadn't been able to replicate his results. Circular reasoning was sufficient to demonstrate that Lysenko's results were valid: "Lysenko says that these transformed wheats have proved useful in cold parts of Siberia. I find it very hard to believe that the Soviet government would back him were this false." Lysenko was right because the USSR had declared that he was right.

To the news about the Soviet state's harassment of its biologists, Haldane responded with a soft cluck of disapproval. "If, as I am told, Dr. Dubinin's laboratory in Moscow has been closed down, I am very sorry to hear it," he said, as if he had been just told that Dubinin was in bed with influenza. Then, immediately, an evasion: "But I am even more interested in London than in Moscow. In London, there is no regular practical course in plant genetics for botany students. This seems to me a serious matter which we might well put right before we start telling Moscow what it ought to do." Haldane was even peremptory about his friend Vavilov's internment and death in a gulag. "You may have been told that Vavilov . . . died in prison," Haldane said, a sentence structured to clothe itself in doubt. "According to a very anti-Lysenko article in the *Journal of Heredity*, [Vavilov] appears to have died at Magadan in the Arctic in 1942, while breeding frost-resistant plants."

But there was, by this time, not much uncertainty about Vavilov's end. For years, scraps of news had blown into the West about his disappearance. Charlotte herself, after her own journey to Moscow during the thick of the Second World War, had told her husband that Vavilov was nowhere to be seen. ("One has not heard of him for a very long time," a Russian friend told Charlotte. And then, looking at her, Charlotte thought, as a wise old hen might look at an unwanted ugly duckling suddenly thrust into her care, the friend added with meaning: "There have been many changes here, you know, since you were last in Moscow.") In 1945, the journal *Nature* was sure enough of Vavilov's death to publish an obituary, deducing correctly that he ended his days in Saratov; two years later, the *Journal of Heredity* described Vavilov as "a martyr of genetics." Journalists and eyewitnesses had written in London, Paris, and the United States of how the Soviet Union swatted down its dissidents, forcing them through show trials and then consigning them to gulags. Discomfort about Vavilov's fate ought to have coursed furiously through Haldane's radio address. Instead, he accorded Vavilov only a glancing reference, a conversational swipe, and moved on.

Haldane concluded: "I think that a number of Lysenko's views, both positive and negative, are seriously exaggerated. But so, I think, is the view that you cannot change heredity in the direction you want. . . . I do not think it will be such an easy job as Lysenko believes. But that does not mean that we can neglect his work, or that of Michurin." It was an old trick, straight from the debating union: a flimsy rebuttal disguised as an even hand and an open mind, a conclusion basking in the achievement of reaching no conclusion.

In the days that followed, Haldane received befuddled letters. A surgeon in Swansea wondered how it was that "official Russia does not know [about] the overwhelming mass of evidence for the gene theory of heredity." Whatever Lysenko's methods were, they could not lead to the madness of condemning Mendelism or of insisting on "a special brand of Russian science." An agronomist solemnly abjured his Communist allegiances. Neither Lysenko nor any other Russian scientist had succeeded in thoroughly disproving Mendel's theory of heredity, he wrote, yet it had become difficult in Moscow to even buy a textbook on Mendelism. "When I see apparent repression occurring in connection with things scientific, which I <u>do</u> understand, I begin to wonder how many things I do not pretend to understand are similarly repressed or distorted."

Haldane's sister, Naomi Mitchison, wrote to him as well—a careful, gentle letter that fingered the chinks in his defense of Lysenko. One of the other speakers had suggested that Lysenko, a second-rate man, couldn't be expected to have first-rate ideas. "I remember you saying much the same about Lysenko yourself at one time," she recalled. If acquired characters were indeed heritable, Naomi said, she would naturally be delighted. "But equally obviously, the *Drosophila* work is still valid and it doesn't help to call people names." Some of the Russian scientists who disagreed with Lysenko were sure to have been doing important, necessary work, and she had no way of knowing if they had been ousted from their positions. "This, I think, represents the point of view of a good many people who are worried, willing to suspend judgement, but also anxious not to condone the kind of thing which they condemned in the Nazi scientists. And, unless you are going to talk entirely to the converted, you have to consider them."

In 1948, Haldane was the chief intellectual in the tent of the British Communist Party, and one of the foremost geneticists on the planet. In choosing whether to endorse Lysenko or tear him down, Haldane was picking between his political fidelity and his scientific integrity. His decision to be soft on Lysenko, to be bland and ambivalent about Lysenko's so-called science instead of shaking it until its falsities tumbled out, confounded everyone who knew him. How could he have made this mistake? He, Haldane—the devotee of the scientific method, the proud unsentimentalist, the scientist whom the Nobel-winning biologist Peter Medawar called "the cleverest man I ever knew"—why did he side with the party against science and against his own colleagues? What had transpired in his life until then that led him, in this juncture, to choose the party? What happens in anyone's life, for that matter, to lead them to the choices they make?

We think instinctively of science as a realm of pure objectivity, forgetting that it is still a pursuit of humans, with all their prejudices, hopes, arrogance, and other foibles. The methods that scientists employ, and the conclusions they draw, rely heavily on their own particular histories and on the culture in which they work. As a result, the Lysenko affair is an oddly perfect way to understand Haldane. A man stepped outside his character, and in so doing, revealed that character to us. We peer through this keyhole, and we see all of Haldane: his ideas about science, the pressures of his politics, the complexity of his influences, the full and vivid sweep of his life.

2.

The Deep End

ON THE DECK OF THE HMS *SPANKER*, the boy put on his diving suit: boots with freighted soles, 20 pounds apiece; rubber overalls with another 40 pounds of weights fixed to the back and to the chest; a helmet, a bubble of copper with small portholes and valves. All told, the outfit weighed 155 pounds. Its elastic cuffs were designed to snap shut around the wrists, to keep the water out. But the boy was only 13 years old, and although he was tall for his age, his limbs were still slender. The suit hung dangerously loose upon his arms.

In the diving logs, they listed him as Jack. His full name, John Burdon Sanderson Haldane, was too long and too easily confused with that of his father, the physiologist John Scott Haldane, who was leading this expedition. At home, they called him Boy, but that wouldn't do on a scientific mission. So Jack it was.

Jack hadn't known he would be diving. Before the *Spanker* started to nose around Scotland's western coast—off Rothesay, up through the narrows of the Kyles of Bute, then near the mouths of the fjord-like Lochs Riddon and Striven—Jack was on shore, with his mother Louisa and his sister Naomi. It was still summer, the final week of August, 1906. The new term at school hadn't yet begun. Jack and Naomi, who was 8, spent the days outdoors, in the hills. They climbed into a quarry and were promptly attacked by a cloud of midges; they stumbled on a beach, where they watched barnacles close and open their shells slowly, as if they were doors on rusted hinges; they played in copses of dwarf oak and beech. Whenever they found a brook, they tried to walk upstream along its course, pretending they were explorers hunting for the source of the Nile. Then the *Spanker* arrived, and Jack was called by his father to stay on board during the weeks of experiments.

The *Spanker* was a torpedo gunboat, a long, svelte vessel with two smokestacks, temporarily ordered by the Admiralty to Scotland to help J. S. Haldane resolve a vexing question. Men had to go down more than 70 feet underwater on naval duties, in full dive regalia, while pumps fed compressed air into their helmets. Even so, they could only work for short periods of time before feeling exhausted. Sometimes they were unconscious when they were pulled back onto the deck; other times, they grew paralyzed, or their joints ached, or their hearts failed. Doctors knew why this happened. At underwater pressures, the nitrogen in the blood dissolves into the fatty portions of human tissues. When a diver then begins to rise too quickly, the nitrogen reemerges in the form of bubbles, which can lodge in an artery or a vein or set off pain or paralysis by impinging on a nerve. If the bubble lingered near your spinal cord, it forced you to double over from agony. In this stoop, you were said to have "the bends."

In theory, scientists knew how to counteract the bends. Divers simply had to return slowly and uniformly to the surface, perhaps at the rate of a few feet a minute. But no one had figured out how slow was slow enough, or what the fastest safe rate of ascent was, or whether the rate varied for fatter men. Expeditions tended to use their own, haphazard rules. On the side of caution, divers stayed in the sea longer than necessary, risking nasty weather or shifting tides; the unfortunate ones were tugged up too quickly. Even on the ground, divers with the bends could be saved by a recompression chamber, in which the pressure was quickly winched up—to push the nitrogen back into tissues—and then gradually eased, giving the gas enough time to dissolve into the blood without forming bubbles. But not every ship or construction project had a recompression chamber on hand. In 1889, when excavators were trying to tunnel under the Hudson River from Jersey City to Manhattan's Morton Street, an engineer had a chamber installed—but only after he had seen one out of every four workers die from decompression sickness.

What the Admiralty needed was a formula or a set of tables they could consult every time, everywhere, without fail. So they turned to J. S. and put the *Spanker* at his disposal.

At the Lister Institute in London, using a steel pressure chamber with a stout, circular door, J. S. had compressed and decompressed a troop of 85 goats and then two of his colleagues and had observed enough of the effects to gain the glimmers of a principle. The quickest way of ridding the body's tissues of their excess stocks of nitrogen, while still underwater, was for a diver to resurface in swift stages. If he began at a depth where the absolute pressure was 9 atmospheres— nine times the atmospheric pressure at sea level—he ought to rise suddenly to a point where the pressure was 4½ atmospheres, remain there for a calculated number of minutes, rise again quickly to halve the pressure, stay a little longer, and so on. As the diver neared daylight, the ascent should slow; he should come up in increments of 5 or 10 feet and spend more and more time at each stage. The scheme appeared to work well at the Lister Institute. But goats were goats, and a pressure chamber was a room. J. S. wanted to drop real, live men into the real, blue ocean.

Beginning off Rothesay, J. S.'s two assistants, naval officers named G. C. C. Damant and A. Y. Catto, took turns climbing into the diving suit, down the rungs of a ladder, and into the sea. Neither had gone below a depth of 138 feet before. Two pumps on deck, their wheels turned by six sailors apiece, compressed air and sent it down through a hose to a diver when he was below. One day, standing on the seabed, Catto discovered that his air hose had gotten tangled in the *Spanker*'s own lines; when it came time for him to resurface, he was still picking the coils of the lines apart, running short of air, all the carbon dioxide he'd exhaled now fuddling his mind. He needed another 20 minutes—nearly twice as long as he had planned to stay down—before he could free himself, progress upward, and emerge safely. J. S. never stopped taking notes. The dives got deeper and

J. S. Haldane in a diving suit, in the company of a German soldier.

deeper. At Loch Striven, Damant reached a depth of 210 feet—at the time, a world record. The sailors pumped furiously, turning their wheels 30 times a minute. Damant stayed on the soft, muddy bottom for 6 minutes, squinting into the gloom and taking samples of the air in his helmet, until he was summoned back up. The staged decompression took nearly 50 minutes, but when Damant stumbled back onto the *Spanker*, he felt whole and well.

To Jack, this was all a screaming adventure. By night, he slept in a hammock next to a row of rifles and pistols. During the day, he watched these brave men populate a small alley of scientific history. The *Spanker*'s crew took him up to the bridge, where he sounded the foghorn. Once a crewman, explaining to Jack the workings of the 4.7-inch gun mounted on the ship's deck, loaded it with a shell case. Jack swiveled the gun to point it at a boat some distance away.

"Oh, no, wait a minute," J. S. said.

But Jack had already fired. It was a blank, but it ripped the silence with a mighty explosion. Delighted, he ran around to the muzzle to fill his head with the acrid, metallic smell of cordite.

On the day of his dive, Jack was first seated in the recompression chamber and taught to swallow hard to clear the pressure that would build up in his ear tubes. Even so, when he started the rapid descent down to 40 feet, his ears felt fit to burst. Then he realized he had a bigger problem. The loch was streaming into his suit through the gap between his wrists and his baggy cuffs and beginning to pool around his ankles. Even in late summer, the water was frigid.

Joining Catto at the bottom, Jack tied a distance line, leading from a 50-pound weight, to his wrist, so that he could stride around the seabed. The bottom was mostly sand, with some small stones, and he spotted sponges, crabs, and starfish, with their compact bodies and long, bristly legs that reminded him of snakes' tails. "All but the bottom was the same beautiful light green, from the sky down to the dust clouds, or as one should say mud clouds, that one kicked up," he wrote later to his grandmother. "There were practically no plants but pink growths on stones, and no sound but of Catto and me."

When Jack was called up, the water in his suit had risen past his waist. He could feel it lapping around him as he moved. But he still had to go through the staged decompression, and as he waited out the minutes at each level, his suit filled even further. Up to his chest. Near his shoulders. Only the air being forced into his helmet was keeping the water from climbing any higher. By the time he was back on deck, the suit had flooded up to his neck, and he was shivering—from the cold and no doubt from terror. J. S. dosed Jack with whisky and put him to bed in Damant's bunk, where he slept for a very long time.

LATER IN HIS LIFE, Jack would proclaim himself a materialist. It was the material world that shaped a person, not the metaphysics of his consciousness or the unknowable mysteries of his mind. Jack could certainly have been a classics scholar or a newspaperman or a stock-broker. But materialism suggested that, growing up as J. S.'s son, he was never likely to become anything but a scientist.

When Jack was born—November 5, Guy Fawkes Day, 1892, ensuring that his birthdays were always greeted with fireworks—J. S. was working as a demonstrator at the Physiological Laboratory in Oxford. The position involved teaching undergraduates the rudiments of physiology, and it gave J. S. access to a laboratory and the university's splendid libraries. But unlike his colleagues, J. S. wanted to draw his research from, and inject his discoveries back into, the world outside the lab. In most British universities, students were stuffed with "facts" that had been "formed from the skeletons of dead theories," he wrote in a pamphlet published in 1890. He held up as a role model the chemist and clergyman Joseph Priestley, who had, a century earlier, been among the first to isolate oxygen. Priestley "never was at any College," J. S. wrote, "and, although engaged all day with his business of teaching and not possessed of means, yet made many noted discoveries." The questing mind could find its laboratory everywhere.

Oxygen and the mechanism of human respiration formed the central trunk of J. S.'s scientific life. It was perhaps the most practical thing a physiologist could do: study how people drew, or failed to draw, the breath that sustained them. In Dundee, in eastern Scotland, where he had taught in the mid-1880s, J. S. established the template for his work. The town's textile mills and shipyards employed thousands of poor laborers, who lived in crowded slums that were rife with disease. With a colleague and one of the town's medical officers, J. S. determined just how poor the air in these slums was. Over four nights, the team paid surprise visits to one-roomed and two-roomed houses, knocking on doors between 12:30 a.m. and 4:30 a.m., when they could be sure that all their residents were indoors and sleeping and that the windows would all be shut. The houses were crammed with people—an average of between six and seven per room, J. S. would later calculate. The scientists waved tubes and bottles through the rank air, as if they were trying to trap invisible butterflies. Later, they analyzed the concentrations of carbon dioxide, mold, and bacteria in these samples. The results were grim. A child asleep in a one-roomed house was inhaling twice the carbon dioxide, nearly seven times the bacteria, and three times the mold as a child in a house with four rooms or more. Then the scientists looked into the files of Dundee's registrar. The rate of death for children under 5 living in a one-roomed house was four times that of children living in the biggest houses, and the residents of these slums tended to die much more often of diseases like bronchitis, whooping cough, and diarrhea. The "vitiated air" of the slums was ruining the health of the people who breathed it.

Through the winter of 1885–1886, J. S. checked the air in 68 school and university classrooms. In the "naturally" ventilated rooms, the windows were shut tight to lock in the warmth of a fire. In the luckier classrooms, heated air rattled up through shafts, swept toward the ceiling, hung there until it cooled, and then sank down and out through an outlet 2 feet off the floor. The air in the naturally ventilated rooms was far fouler. In most cases, the students went from these schools to spend their nights in congested homes with still, soiled air. "We need not be surprised at the unhealthy appearance of many of these children," the scientists wrote. "They are breathing for at least fifteen hours out of the twenty-four a highly impure atmosphere." Their report contained stern recommendations for Dundee's public health authorities: more ventilation, better housing, cleaner schools. "His experience of the Dundee slums may not have made him a radical, but it kept him one," Jack would write. J. S. would always have "little patience with statements by the rich that the poor were really well off, or could be if they tried."

A few months later, troubled by richly offensive smells, the House of Commons asked J. S. to investigate the sewers below Parliament. Sewer air was considered toxic to breathe, laden with germs and gaseous toxins—and yet J. S. found less carbon dioxide below than in a bedroom occupied by three or four people, and fewer microbes than in the air of a street like Piccadilly. Returning to Dundee, he spent days in the town's sewers to confirm his findings—so long, in fact, that he could work out where he was just by the smell of any particular factory's effluent. (The discharge from one factory, he recorded, smelled like an orange grove.) J. S.'s paper overturned conventional wisdom about the air in sewers and its supposed connection to typhoid fever.

Many years after he left Dundee, J. S. visited the town with Jack. His father didn't quite remember how to get from one place to another, Jack realized, so J. S. tried to imagine himself belowground once again, in Dundee's sewers, their mazelike layout having never left his mind. When he did that, he recollected how the streets ran above his head, and then he could negotiate his way around easily.

No one who knew the Haldane clan would have thought any member would ever set foot in a sewer. The Haldanes are one of the few British families that can track their ancestry back, through the male line, to the twelfth century; the name, it is said, derives from "half-Dane," suggesting that the first of the Haldanes arrived from Scandinavia as settlers or raiders. Through the Middle Ages, and even beyond, they lived in and around a fort that controlled a pass from the hills to the plains, in the area now near Gleneagles, north of Edinburgh. They guarded these lands fiercely, fighting the hill tribes that plotted to rustle the cattle of the plains, and on occasion they ventured south to battle English invaders. The Haldanes became important landowners. They may have led strict, spare lives, obedient to their Calvinist virtue, but they were aristocrats just the same.

J. S. grew up in Edinburgh, the third son of a second wife. His father Robert, a solicitor, already had five children through his first wife, who'd died in childbirth. Then he got married again to Mary Burdon-Sanderson, from a Northumbrian family. The Haldanes lived for much of the year in Cloan, near Gleneagles, where Robert had bought a farmhouse and, to achieve the manorial look, dressed it up with a turret. "He was very devout," J. S.'s eldest brother Richard wrote, "and had fitted up a barn where he used once a fortnight to preach to a considerable audience of old-fashioned Scottish country folk who came to hear the Word of God in all its strictness." Their mother was similarly deeply religious. The Haldanes' days opened and closed with readings from the Bible, and there was church every Sunday. Even in rain, even if it pelted down so much that it threatened to inundate the family barouche, church was never skipped; the children simply squatted in the carriage, by the feet of the adults, to stay dry. In Edinburgh, Robert Haldane took his family past the houses of the sick, so that his children might toss religious tracts at the doors: brochures about the Hell-Fire Club or sermons from evangelists or illustrations of the impious being flogged. At the back door of the Haldanes' own home, a number of poor men and women appeared once a week for food or a cash dole. An implicit part of their social position, the Haldanes thought, was diligent public service, and they considered it their duty to save the lives and souls of those in the classes beneath them.

J. S. followed Richard to the Edinburgh Academy, then attended Edinburgh University for an arts degree, where he read philosophy. Even in medical school, which he started in 1879, J. S. found that his teachers girded their science with the principles of theology. This disturbed him. Already, along with Richard, J. S. had begun to react against the severe orthodoxies of their childhood. Equally, though, he shrank from the dogma of mechanism—the belief that all the processes of the natural world, even the production and survival of life itself, could be explained by physical and chemical facts. An electrical nerve impulse arrives, a chemical releases a stock of calcium ions, the ions allow one protein to bind to another, a muscle contracts, and a man runs: this is mechanism, a cool interplay of minuscule particles.

But life was no mere machine, J. S. thought, to be reduced to a grinding assembly of cogs and levers. Between the creeds of the purely physical and the purely divine, philosophers carved a vague third way that prized the vital force, and J. S. cleaved to this view, even if he couldn't explain it very clearly at all. Life has a purpose, and its purpose is to preserve and prolong itself, J. S. and his brother wrote in a dense essay in 1883. The body's individual parts, too, had adapted to fulfil their particular functions within the exquisite, coordinated whole. This balance of self-regulating entities even had a parallel in how society functioned, and within this idea, he perceived a call to action. A doctor had to intimately know how the various organs work to be able to heal a patient. In the same manner, scientists had to leave their labs, plunge into the world, and know its various constituent parts if they were to cure the ailments of the social body.

These conclusions emboldened J. S. and settled his mind. The year before he graduated from medical school, he quit the congregation of his church. (It wasn't that he didn't have faith in the divine, he explained in a patient letter to his minister. It was just that the church's doctrine, "that a man's religion depends on his belief in matters of history," presented "a scientific difficulty" for him.) At medical school, J. S. missed classes so that he could work more hours at the infirmary. When a professor failed him in a midwifery exam, he nearly decided to abandon his medical degree out of pique. He didn't need it to pursue his research, and he didn't want to spend more time memorizing the notes of lecturers whose intellects he scorned. When he did finally retake the exam and receive his MD, he left Edinburgh almost immediately for Dundee. As a parting shot, he wrote anonymous letters to *The Scotsman*, caustic dispatches about the lack of intellectual freedom, the woeful neglect of practical work, and the teachers who needed to be goaded "to live the lives of men of science, and not of overpaid drill-sergeants."

J. S. remained in Dundee for barely three years. His uncle, John Burdon-Sanderson, had become Oxford's first Wayneflete Professor of Physiology in 1882, and in 1887, J. S. joined Burdon-Sanderson's department as a demonstrator. At Oxford, he found the academic soil he had been looking for, and he planted himself firmly within it. He taught the way he had once wished his professors would teach, emphasizing the need to experiment. He developed a system to measure carbon dioxide and moisture in air: a series of test tubes containing either soda lime or fragments of pumice soaked in sulfuric acid. These captured one or the other of the gases from air blown through them, and they could then be weighed to calculate how much they had absorbed. With two other scientists, he worked out ways to determine how much heat an animal—a rabbit, say, or a rat—gave off, how much weight it lost as it starved, and how its metabolism quickened or slowed.

At Oxford also, J. S. commenced the lifelong project of using himself as a test subject. Who better? A rabbit, or even an undergraduate, couldn't convey to him accurately enough the shifting sensations felt during an experiment. J. S. had been thinking about Dundee's slums—about those shoebox rooms growing stuffy with exhaled car-

bon dioxide—and he wondered about the effects of breathing vitiated air. He had a wooden box built: 6 feet high, 3 feet wide, 4 feet
long, painted all over to make it airtight, its chinks filled with putty.
A tube bled off samples of air. J. S. and his colleague, John Lorrain
Smith, took turns in the box; while one sat within and tried just to
breathe normally, the other observed through a glass window.

J. S. occupied the box one afternoon, having already measured the
air: 0.03 percent carbon dioxide, 20.9 percent oxygen. For the first
3½ hours, he was untroubled, but then the carbon dioxide began to
climb over 4 percent, and he began to pant. At 8 p.m., when he had
been in the box for 6 hours, he fought for every breath. His head lit
up with pain. His body felt drained, as if he was exerting every one
of his muscles each time he took in air. By 9 p.m.—carbon dioxide
5.58 percent, oxygen 14.1 percent—J. S.'s breathing rate had doubled.
The walls of the box, and even his clothes, turned damp with exhaled
moisture. At some point in the evening, he had eaten a meal to keep
his strength up, but now he realized it was a mistake. Just before
10 p.m., he asked Smith to let him out, so that he could proceed to
be violently sick. Smith nipped into the chamber himself and soon
he was puffing as well. He took notes: "I attempted to strike some
matches while in the chamber, but even the phosphorus refused to
burn completely." He read off the levels: carbon dioxide 6.39 percent,
oxygen 13 percent.

J. S. spent the next day in a fug of fatigue, but even through this
he sensed the question that nagged. What was responsible for his
physiological symptoms: the accumulation of carbon dioxide or the
shortage of oxygen? So they repeated the experiment, but this time
they placed a tray of soda lime in the box to soak up the carbon dioxide. Smith shut himself in at 11:09 a.m. By lunchtime, the oxygen in
the air had fallen from 20.9 percent to just above 17 percent, and it
had become impossible to light a match. At 4 p.m., a wisp of a headache, nothing like the shattering pain J. S. had suffered. At 6:20 p.m.,
a tendency to pant. When the experiment ended at 7:15 p.m., Smith
strolled out, a little short of breath but otherwise in no discomfort.

The scientists ran other tests. They breathed air out of a bag that
got rid of the carbon dioxide as they exhaled, or they breathed an

oxygen-hydrogen mixture. As the oxygen levels dwindled into the low single digits, they turned blue, or they got mildly uncomfortable, or they slid gently into a faint. "Became quite stupid, and could not see an open book placed right in front of me," one notation ran. "Did not feel very bad." But if the carbon dioxide remained in the bag, and as its proportion in the air rose, the scientists ran into graphic distress: headaches, hard sweating, addled minds, faces so red they were dusky. Out of these gasping, breathless days emerged a sturdy conclusion: the body responded first, and most acutely, to a profusion of carbon dioxide rather than a slow drop in oxygen.

J. S. was riding the energies of a giant wave of self-experimentation. In the twentieth century, scientific instruments would become keener and more discerning, and systems of rules and experimental ethics would fall into place. But throughout the 1800s, the scientist's own mind was the best-calibrated instrument available, and the human body was, as Goethe wrote, *genaueste physikalische Apparat, den es geben kann*—the most precise physical apparatus that can be. Humphry Davy huffed nitrous oxide and recorded its effects; Johann Ritter touched the poles of an electrical battery to his eyes, tongue, and skin; chemists swallowed their own pharmacology; Max Josef von Pettenkofer drank bouillon salted with cholera bacteria. Science was pulled forward "by a common ideal of intense, even reckless, personal commitment to discovery," Richard Holmes wrote in *The Age of Wonder*, and in this it found its analogue in the age of Victorian exploration. Some men and women journeyed outward, risking their lives for glory or wealth but also to see firsthand everything that remained unknown in the world. Scientists made their expeditions inward, burrowing into their own bodies to find the secrets they held.

ON AN AFTERNOON LATE IN 1883, when J. S. was still in medical school, he was studying in front of the fireplace at home, his books open and in disarray around him on the hearthrug. His father had died six years earlier, and his mother had decided to lease their house and move elsewhere. Now a real estate agent arrived with a prospective tenant, a mother and her 20-year-old daughter. When they were shown into the dining room, J. S. struggled to his feet.

He'd just been working, he explained.

"No," the daughter remarked, "you were fast asleep."

This was how Louisa Kathleen Trotter remembered meeting her husband for the first time.

Within the Scottish class system, the Trotters and the Haldanes were near equals. The Trotters, too, owned land and dispatched their sons into public service or the military or sent them overseas. Louisa's father, Coutts Trotter, had gone to Calcutta himself for a couple of years, for the East India Company, before he fell ill and returned. Illness became his life's true vocation. He reviewed books for journals, and he administered the affairs of the Scottish Geographical Society, but he thought just as much about his "pernicious spinal complaint" or the possibility that he had cancer. "Poor dear Coutts," his mother and sisters murmured. The doctors delicately called his hypochondria "male hysteria," but they treated him for whatever he thought he had nonetheless. In Germany, at a spa, he met an Irish family, the Keatinges, and got engaged to their daughter Harriet. It all happened very quickly, one of Harriet's friends told Louisa: Coutts's assorted ailments may have been fictional, but he was "an attractive young man, and Harriet was very sympathetic with him."

The family moved around constantly for Coutts's health: in first-class carriages, with manservants and French maids in tow. To Cannes, for the Mediterranean air; to spa towns in Germany; to Brighton, to visit Coutts's favorite homeopath. For a while, the Trotters lived in Bournemouth, where a governess gave Louisa her lessons. When she learned, very slowly, to read, she was given a series of books on the histories of European countries, with titles like *The Greatness of England* or *The Greatness of Russia*. They mystified Louisa. Why wasn't England the greatest? Had she lost her greatness? Were people trying to regain it for her? Wasn't that the most important thing anyone could do?

Louisa was a born Conservative. She celebrated the Empire—its embodiment of British spirit and its virtuous attempts to elevate the inferior societies of its colonies. The men and women who spread and maintained the Empire had made the highest sacrifice of leaving their country, she thought. Their efforts deserved the gratitude of those who remained home. Once, she hotly interrupted a conversation between Coutts and his friend, who were wondering if it wasn't best that the United States be allowed to absorb Canada. "You can't do that! You can't let them go!" Louisa said. What about the glory of the Empire? "Oh, the Empire!" her father's friend said with a sneer. Louisa left the room and wept her anger in the corridor.

Perhaps these differences between herself and J. S. ruffled her, made her uncertain; she agreed to marry him only after he had twice proposed and twice been rejected. She had gotten to know the Haldanes and their political views, and she disliked, in particular, J. S.'s brother Richard for his self-satisfaction and his contempt for the ideas of anyone who disagreed with him. "Before we were engaged, after months of correspondence and some rather agitated interviews, I made John understand, as clearly as I could do, that I could not and would not alter my political faith," Louisa wrote later. J. S. held the opinions of a radical Liberal, but he wasn't interested in active politics, and she couldn't bring herself to become fascinated by his philosophies of life and physiology. But she was quick and independent and knew her mind, all appealing qualities to J. S. Already, ahead of the wedding, she was growing exasperated with the notion of being "given away," and when one of J. S.'s friends, who was performing the ceremony, sent her the text of the service in advance, she edited it furiously with a blue pencil. The wedding, in Edinburgh in December 1891, was abbreviated to just 20 minutes. She wore a gray dress with a belt that was, she thought, the color of plum juice mixed with milk. Nobody recorded what J. S. wore, but he was already distinguished by the moustache that flared below the wide Haldane nose and that would grow to resemble an unruly sea sponge. In a final subplot of defiance, Louisa entered the church through a side door.

The intellectual divergences between husband and wife persisted throughout their marriage. Over Christmas just weeks after their wedding, J. S. spent the evenings dictating to Louisa a draft of a planned book that would explain his notions of life and its vital force. He worked as if nothing else in the world mattered. When she grew bored and tried to read during his ruminative pauses, he insisted she pay attention. Later, while considering a move to Bel-

fast so that J. S. might start a lectureship at Queen's University, they
abandoned the idea. Louisa would pelt right into politics, she knew,
because she was so staunchly in favor of keeping Ireland a part of
the empire; on the other hand, J. S.'s university position might come
under threat if he became too friendly with the Home Rulers, who
thought the Irish ought to govern themselves. J. S., a progressive
for his time, believed in a form of equality and social justice, but
the threads of Louisa's ideals were more knotted and tangled. She
thought that women ought to work and vote, even if she couldn't
agree with the suffragettes' unladylike tactics. But her class prejudices were hale and firm. In Oxford, she helped establish the Imperialists' Club and, during the Boer War, the local chapter of the
Victoria League, which pledged itself to "any practical work . . .
tending to the good of the Empire as a whole." The war, in which
the British fought to maintain its grip on South Africa and built
the world's first concentration camps for Boer internees, divided the
couple, Louisa remembered:

> The Haldane family was, in the parlance of the day, "pro-
> Boer," that is to say, it greeted our disasters with "serve them
> right" or "what else did you expect?" When the troubles in
> the concentration camps were reported, John refused to listen
> to reasons why the camps were created. It seemed as if he
> were deliberately closing his mind to any reasoning on that
> matter, and could only talk about "inhumanity" and "starving
> women and children."

When Jack and Naomi were very young, they bobbed and followed in the wake of their mother's opinions. Louisa took Jack with
her when she went to meetings of the League of the Empire, where
they listened to talks on British history or passed around artifacts
from the colonies. Jack was, at the time, impressed. When he was
10, he wrote in his diary about the league, about how it was "a society for making people more able to fight for their country, and to
be useful if they emigrate to the colonies, & to let them know . . .
that the Empire isn't a lot of little colonies, but one big one, in fact

to teach them to be good citizens of the Empire." At one meeting, he noted with satisfaction, the league had expanded its membership. "We enrolled about 10," he wrote in his diary. Naomi was not quite as involved in Louisa's politics as her brother, but her childhood bore the imprints as well. When she was cast as Portia in a school production of *The Merchant of Venice*, she discovered that the prince of Morocco, vying for her hand, would be played by a brown-skinned boy—Indian, maybe, or Middle Eastern, but almost certainly from one of Louisa's beloved colonies. It didn't matter. Louisa complained to the school, objecting to even the suggestion of a mixed marriage. Another time, during an election, Naomi grew so afflicted with her mother's enthusiasm that she refused to wear one of her coats altogether because it was red, and red was a Liberal color. A brisk little campaigner, she recited Louisa's favorite political slogans, and she hand-delivered pamphlets to the neighbors.

When she was an adult, Naomi read through her childhood diaries and felt ashamed of how resoundingly she had echoed her mother's views. She also wondered why her father hadn't disputed them with his own opinions. He had always been annoyed when she rattled off a line from Louisa's stock of Conservative rhetoric; she could tell because he would frown and shake an irritated foot. (When she let off a quip about how William Gladstone, the Liberal prime minister, had been responsible for the deaths of British troops overseas, J. S. "responded with pain and anger that scared me into silence.") Perhaps he never spoke about his beliefs because he loved his wife and didn't wish to contradict her, Naomi thought. Or perhaps they had reached some private agreement, the terms of which compelled him to keep his politics to himself.

Or perhaps J. S.'s mind was so saturated with work that the rest of the world was largely absent to him. The Haldanes told fond stories of how distracted he could be. At meals, he stared in silence at the dish in front of him, shaking himself alive after many minutes to ask, "Will anyone have any of this? It looks rather bad." He bought hats that were several sizes too small for his head. Once, forgetting that Louisa had invited guests to dinner, J. S. came home late, apologized, and went to his room, intending to dress. When he didn't return for some time, Louisa investigated and discovered him in bed. Finding himself taking his clothes off, he explained, he had just figured it was time to go to sleep.

When Louisa first came to Oxford, after the Haldanes were married, she moved into J. S.'s house on 11 Crick Road, where she ran into a state of domestic catastrophe. She opened the cupboard that held his clothes, and plumes of moths rose to greet her. J. S.'s bills had remained unpaid. His papers, detailing his experiments, were waiting to be published, but first they had to be "licked into shape," as J. S. put it. Until then, they lay dismembered around his study or forgotten in the backs of drawers. The disorder followed J. S. into every one of their houses. After Jack and Naomi were born, the family shifted to a larger villa on St. Margaret's Road, where a part of the drawing room was bitten off to create a study, with a massive oak table in its center. Books and papers tumbled off the desk onto the floor and around the room, and yet J. S. appeared to be able to fish out anything he wanted. This was the earliest house that Jack and Naomi remembered well: the giant brass knob on the front door; the almond trees by the gate; the pitch-darkness of the basement in which they charred their hair and eyelashes while igniting trails of gunpowder; the bathrooms with their stained lead bathtubs. On the top of a pole, in a birdhouse, lived a crew of white fantail pigeons. Jack took to laying on the ground, dead as a log, with corn stuffed up his shorts, so that the pigeons would wriggle in to peck at the grain.

In 1906, when Louisa's mother came to live with the Haldanes, the family moved once more, into a house they had built near the River Cherwell and named after it, and which has now given way to Wolfson College. The garden, sloping eastward toward the river, included three walnut trees, two and a half centuries old; they were so bountiful that once, when Louisa tried to count how many walnuts she could drop into her basket while standing in a single spot, she gathered more than a hundred. On the wall facing the fields to the north, J. S. had imprinted his family's crest: the motto "Suffer" above an eagle with a beady eye. The house held dozens of rooms, all wired for electricity. The largest space in Cherwell, the drawing room with its grandfather clock, bookcases, and china cabinets, could have

Jack and Naomi, aged 6 and 2.

comfortably hosted a ball. The maids lived just below the rafters, in hutch-like quarters.

J. S. built himself a laboratory. He had always experimented at home, but in the manner of a man pained by the constraints of domestic architecture. In the attic of the Crick Road house, he had inhaled candle smoke to gauge its effects, and he once sealed a room off and piped fuel gas in to see how it spread. But the most complicated work still had to be done at the university lab, which was more spacious and better equipped. At Cherwell, finally, J. S. managed to bring all his work home. He extended his study into a small anteroom, where his items of apparatus were stored, and then into a bigger lab with sinks, tables, and two airtight chambers made of steel, in which a volunteer might be shut away.

In these laboratories, makeshift or elaborate, Jack grew up.

HE WAS A BEAUTIFUL CHILD, his mother thought: fair-haired and moon-faced, with a chin like a doorknob. They called him Squawks as a baby, but after that he was simply Boy. He asked a lot of questions, so in self-defense, Louisa quickly taught him to read. Before he was 5, Jack was reading aloud the newspaper reports of the meetings of the British Association for the Advancement of Science. When he came to a word he didn't recognize, he paused in caution, never proceeding until he had worked out the pronunciation in his head. Then the exercise stopped altogether, because Jack demanded an explanation of the word.

He remembered everything he heard, everything he saw in a book; his memory was capacious and precise. At night, J. S. told him stories out of Walter Scott's *Tales of a Grandfather*: Scotch accounts of audacity in battle, of the country's love of freedom, of its noble underdoggery against the English. Most evenings, after tea, Louisa sat her children down to read to them: *The Water-Babies* or Kipling or all of Tennyson's *Idylls of the King*. Around Jack, J. S. and his colleagues talked about respiration and gases. All of it soaked in. When anyone thought back to Jack's childhood, the first memory was of his precocity—and of his undaunted manner of displaying it.

Like the time Louisa sent him to run up and down Crick Road for a while, to expend his energies, whereupon he was stopped by a curious neighbor. What was he up to?

"I'm the overland mail," he replied.

"What is that?"

"It's quite true—it's in a book," Jack said, and then started to recite Kipling, one of his mother's favorite writers:

Let the robber retreat—let the tiger turn tail—
In the Name of the Empress, the Overland Mail!

"Don't you know *that*?" he concluded, in triumph.

Or the time—Haldane legend has it—he looked intently at the blood trickling out of a cut on his forehead and asked, "Is it oxyhaemoglobin or carboxyhaemoglobin?" He was not yet 4.

Or the time he fell off his father's bicycle, just as J. S. was turning a corner, and cracked his skull on the curb. When he was mending at home, his doctor and his surgeon paid him a visit. Relieved to see that Jack was alert, the surgeon turned to the doctor and suggested medication: "I should give him five drops to start with . . ."

Jack cut him off. "But you're the mechanical chap. Leave that to the chemical chap."

At the Haldanes', the stuff of science was tactile and alive. Even to two children fooling around at home, the principles of the universe showed themselves on demand. How could that be anything but endlessly exciting? They dropped shiny beads of mercury on the floor and chased them with their fingers, watching as they lolloped along or rushed like lovers into each other. Finding a dead, furry caterpillar, they skinned it, planning to use its pelt as a rug in their small wood-brick castles—except that the skin shriveled up soon after it had peeled off its body. They dipped cotton wool in methylated spirit and set it on fire. They were comrades in these capers, Jack and Naomi, so it didn't even infuriate her too much when, on occasion, he aimed his devilry at her. He filled a metaled sink, connected the basin to the house's weak line of electricity, and invited her to snatch up pennies he had dropped into the water. The current left her body ringing. He kept telling her, too, that he had a lump of radium hidden in a desk drawer. Was he fooling her? By way of proof, he took her into a cupboard, closed the door, and cut the darkness with a feeble glow. But was it radium? She was never quite sure.

Like their father, they walked plenty—long, sweaty rambles, particularly when they visited the family estate of Cloan, in Scotland, every summer. With great gravity, Jack transcribed everything he saw into his diaries, like an adventurer on a new continent. If he saw a drain leading to a septic tank, he wrote several pages about sewage and its link to typhoid. If he collected snails and mussels near a river, he specified which kinds: "I got some fairly good specimens of *Planorbis corneus, Paludina vivipara, Limnaea auriculata, & Anodonta anatina.*" (And then, as an afterthought: "We met a good few other chaps up the river.") He bought a special notebook with horizontal leaves just to record his botanical observations. Each page held five columns: Plant, Where Found, Date, Where Found, and Date, two sightings per plant. The Plant column came preprinted—an alphabetically ordered list, running from *Acer campestre* (field maple) to *Zostera nana* (a species of eelgrass)—and Jack filled in the details of where he had spotted any of them: "near Oxford," "Cloan," "Devon," "near Gloucester."

Soon after Jack was born, J. S. began to take his knowledge out of his lab and into the world, and Jack became accustomed to science in this purposeful, utilitarian mien as well. A recognition of the Industrial Revolution's malign effects was whipping across Britain. Factories and mines were proving dangerous places to work: dim and poorly ventilated and prone to accidents. Laborers lived in deprivation, in slums of the kind J. S. had seen in Dundee. Cycles of economic depression and competition from industries overseas kept wages low. Decades ago, the miseries of the working classes might have been ignored, but no longer. Voting rights still hadn't reached the poorest 40 percent of Britain's men, but trade unions had been legalized, and the pulse of labor was quickening. In 1893, nearly 30½ million working days were lost in strikes and factory lockouts. Increasingly, unions, voters, and socialists were demanding that the state intervene more to hasten social and economic reform. Within these anxieties, J. S.'s skills promised to diagnose and alleviate some of the nagging symptoms of industrialization.

Just as he had spent days below Dundee, traipsing through its sewers, J. S. now turned to the mines. Seams of coal and metal around Britain fed the country's industrial appetites, but they were primed for catastrophic explosions: an ignition of dry coal dust or the foolhardy use of dynamite could send a blast rippling through the tunnels. When an accident of this kind occurred, death claimed its victims in sheaves: 290 in Cilfynydd, 178 at Salford, 164 at Seaham, 159 in Ferndale, 146 at Risca. Puzzlingly, however, most of the miners weren't dying in the inferno itself. Their bodies weren't being brought up charred by flames or smashed by collapsing beams. These tragedies came laced with scientific mystery: What was killing the men?

The only way to find out, J. S. realized, was to visit the mines himself—preferably in the immediate aftermath of an explosion. Whenever he received a telegram from the Home Office, he set out for the train station. He wore heavy boots and mining overalls, and he carried a safety helmet, a small leather case with the words "London Fever Hospital" inscribed in red on its lid, and a Gladstone bag full of oxygen tanks, valves, tubes, and some clothes. Sometimes he took along a cage of mice—test subjects that would warn him of unhealthy air. On his journeys, he was doubtless an intimidating figure: craggy brow, rowdy moustache, grubby overalls, and the glacier-blue eyes daring you to step in and share his compartment.

If the mine's shaft hadn't been obliterated, Haldane descended into the earth. Sometimes he went as deep as a mile, where the heat was so fierce that the men had worked in bathing drawers and, during their breaks, emptied their boots of pooled perspiration. In the tunnels, he collected samples of air, sucking through a rubber tube to pull the gases into his corked bottles. Aboveground, he accompanied local doctors to their postmortems of dead miners. He brought his samples of air and dust, vials of blood, and autopsied organs back to Oxford, where he set about unraveling the nature of these deaths.

Initially, J. S. hypothesized that the men were being killed in one of two ways. Miners in the immediate vicinity of an explosion were burned to death or crushed by collapsing tunnels, he thought, and those farther away were suffocated, the explosion having consumed all the oxygen in the air. In the postmortems, though, J. S. repeatedly noticed that the men without a mark on their bodies had pink fingernails—not a sign of the leaden-blue stains that should have appeared with oxygen deprivation. The workers who retrieved these bodies had also reported that the lamps beside them had still been burning with a steady flame. Had the air been divested suddenly of its oxygen, the lamps should have been snuffed out as quickly as the lives of their owners.

The blood tapped from the jugular veins of the dead miners, on the other hand, gave J. S. a different idea. Its color was a deep, carmine red, of the kind that appears when carbon monoxide displaces oxygen from its bond with hemoglobin. Some of the blood samples, upon testing, were nearly 80 percent saturated by carbon monoxide. J. S. had already heard stories of rescue parties that ventured into mines after explosions—of how they gradually felt their legs growing rubbery, their sight dimming, and their minds overcome by drowsiness. If he went into the tunnels himself and stayed too long, his memory was apt to grow comically blurred; once, out of a mine, he sent a telegram home saying "I am all right"—and then continued to send the same message in several more telegrams, until his faculties cleared enough to recall all the previous telegrams. A deficit of oxygen would have caused similar effects, J. S. knew, but the rescuers' lamps and his own had burned the whole time they were belowground. The symptoms didn't match an excess of carbon dioxide, he also knew, from the time he had spent in his wooden box in Oxford breathing the gas in. Among the chemical constituents of "afterdamp"—the gases that were produced in mines following an explosion—it was, clearly, carbon monoxide that was killing most of the miners, sometimes even as they ran in the direction of rescue. They couldn't smell it or taste it, but still the poison worked its slow, stealthy way into the blood.

At his lab, J. S. ran experiments to make certain. In a small chamber, its air containing 3.6 percent carbon monoxide by volume, he placed a mouse; it fell over, unconscious, in 15 seconds, suffered convulsions 5 seconds later, and ceased to move entirely in 1¼ minutes. When the air was 1.8 percent carbon monoxide, the mouse lived longer: 3 minutes. But now, having come to his old problem of not knowing what the mouse was experiencing, J. S. inserted himself into the experiment. He rigged up a mouthpiece so he could breathe the same mixture as his mouse, and well after it was dead, he continued to inhale air tinged with carbon monoxide. He varied the percentages—from 0.021 to 0.5—and tracked how long it took for his pulse to race, for his breathing to become ragged, for his legs to quiver. Periodically, he took samples of his blood, which grew pinker and pinker in dilution—a sure sign that the carbon monoxide was latching onto his hemoglobin. Then, putting down his mouthpiece, he tried running up and down a flight of stairs, simulating the exertions of a miner sprinting away from a blast. It made him giddy, his heart shuddered, and his vision grew clouded. Sometimes, when carbon monoxide levels were particularly high, he needed hours to recover. Reeling home from the university, he was once stopped by policemen who suspected him to be extravagantly drunk.

"I know how you feel, ma'am," Louisa's housekeeper commiserated. "My husband's just the same on a Friday night."

A tiny taint of carbon monoxide was enough to make a person feel unwell, J. S. concluded, and any concentration greater than 0.2 percent induced helplessness, defeated faculties, and eventual death. The afterdamp mixture frequently contained more than 3 percent carbon monoxide. Even when this was cut with regular air, the gas could be lethal. Most dangerously, the miners had no way of telling what was in the air around them—no smell like the foul-egg odor of hydrogen sulfide, no physical tell like the smothered flame of a lamp in an atmosphere of vanishing oxygen. But here, too, a thought struck J. S. Long before carbon monoxide started to trouble a man, it would render a mouse breathless and its limbs dysfunctional. If miners carried cages of mice or small birds like canaries that tumbled off their perch after a few sniffs of carbon monoxide, they would have an early-warning system right at hand—a ready signal to retreat into cleaner air.

Assisted by his brother's rising career as a Liberal politician, J. S. became the government's best-beloved expert on the health of miners, the quality of air, and the intricacies of breathing. He was asked to ride the Tube from King's Cross Station, through tunnels filled with noxious clouds of gas thrown off by the coal-burning trains; mid-journey, to the bafflement of other passengers, J. S. leaned out of the window and captured samples of air in his bottles. He was consulted on the right way to ventilate submarines. (Once, on a new vessel, a young officer pointed to a cage of canaries and told him, with great solemnity: "Oh, not just pets, sir! I am told they give valuable information. When the atmosphere gets dangerously polluted, they begin to sing very loudly and flap their wings.") He was appointed one of London's three gas referees, responsible for maintaining the quality of the fuel gas used for light and heat. He was called on to advise divers how to sink and resurface, which led to the expedition aboard the HMS *Spanker*.

In 1902, J. S. was paired with a mine inspector to report on the health of tin miners in Cornwall, who were dying young in alarming numbers. Their lungs were wasting away, or they suffered from a hookworm disease of the gut called ankylostomiasis. That year and the next, J. S. traveled to Cornwall frequently, and he sent letters back to Jack about the mines, about how unsparing their conditions were on the men who worked in them. He wrote clinically, but Jack could not have missed the currents of empathy that coursed through the letters.

"My dear Boydie," J. S. began in a letter to Jack a week before Christmas in 1902, when he'd been visiting a mine every day. He had met men infected with ankylostomiasis and men whose lungs were caked with rock dust. In one mine, there weren't even enough pails for the miners to relieve themselves. A second mine was filthy and poorly ventilated, and pockets of the tunnels were filled with the fumes of exploded dynamite. Even in the heart of winter, the heat was brutal. J. S.'s clothes were soaked through in 3 minutes.

Another day, another mine: this one near Land's End, on a day so blustery that the spume on the waves looked like white horses galloping toward the shore. The mine's workings ran under the sea for a mile, and it took 45 minutes to get all the way down. They got bewilderingly lost. He was, J. S. thought, going to dream about it all night:

It was very hot at the bottom. We then walked out along a level under the sea for more than a mile. It got hotter & hotter, but was a little cooler at the furthest point. There was almost no air current. I took off all the clothes I could, but couldn't leave them behind as we weren't coming back the same way. We then climbed to a level below, & this was still hotter—about 89 degrees—& the air steamy and stagnant. I took my own temperature, & it was 103 degrees, & I was getting rather uncomfortable. We then tried to get to a lower level in order to get out, & we climbed down the lode on the bars of timber but got into a cul de sac & had to get up again, as the manager couldn't find the way. At least he found a way through various clefts & holes, & we got down at last, & out along the level to the shaft, where the air was much cooler— about the same as a very hot summer day at Oxford. We telephoned up to set the man engine going (it had stopped

for the day) but no one heard for some time. At least we got
up & were very glad to be above ground again. I was never in
such a . . . badly ventilated mine before.

Naomi read the letter as well. Her father wrote to Jack as if they
were fellow scientists, she thought, not without envy.

As little children, both Jack and Naomi were avid apprentices in
J. S.'s lab. They washed his bottles, and he explained to them how
his gas analysis apparatus worked. Sometimes, when J. S. shut himself into one of his airtight chambers, he gave Naomi instructions
on what she had to do if he keeled over unconscious: unlatch the
door, first pull him out, and only then begin artificial respiration.
When J. S. devised a method to calculate hemoglobin levels from a
single drop of blood, he tested it on his children and then asked them
to round up others. On the road, Naomi ran into another girl from her
school. "You come in here," she said. "My father wants your blood."
After the terrified child had been hurried along by her nurse, Naomi
stalked back in. "No-one seems ready to have their finger pricked."

But as they grew up, it became evident that their father would
enlist only Jack as a full partner. For all his liberalism, J. S. was still
a man of his century, assuming without question that the world of
science—particularly science done his way, scrabbling around in
sewers and mines—was not made for women. It wasn't that Naomi
wasn't keen. Her favorite funny word, as a child, was "physiology,"
and she read all the newspaper reports about mine accidents as well
as the lurid bits of the *British Medical Journal*. But Naomi's education always remained outside the radius of J. S.'s attentions, and she
feared that to try to learn more, in a difficult but interesting subject
like mathematics, for instance, would be thought silly or extravagant.
So she contented herself, very early, with the excitement that merely
spilled into her life from the work of her father and brother, although
she still felt prickles of regret. In Scotland, during J. S.'s diving experiments, she watched as Jack was summoned aboard the *Spanker* by
their father. "He says he will take Boy down," Naomi wrote in her
diary. "I wish I were a boy."

When Jack was very young, he sent letters to J. S. during his travels, eager to know what he was doing. He called J. S. "Uffer," after his boyhood mangling of the word "father." "Dear Uffer," he wrote, when he was 4, "What peart of the mountins wear you in? How did you like it? . . . How was it like at the labretree?" Or: "Dear Uffer: Are there any intrsting things if there are plees tel me about them." In his letters to others, he recounted his own scientific exploits and observations. "Dear Grannie . . . We have an experiment in progress, with Uncle Richard's chemicals. I know the crystallising one will succeed!" he wrote at the age of 7. "PS. Uffer's home tomorrow. I have finished Huxley's phisiology and am reading it over again, it is very interesting." Or: "Dear Grandpapa . . . The garden is getting on, we have planted some horse chestnuts & limes & transplanted the yew. 'Marco Polo' is one of the most interesting books I ever read; it is the last place I expected to find out about the aloes for Naomi's sucked finger in." From a holiday in Dorset: "Dear Grannie . . . Apropo of Aunt Bay, I am sorry to say we could not get her around to [the] Old Harry [rocks], though we could to his children, to them we have to go through a tunnel into Swanage Bay, where there is a most splendid archway through the most prominent point; at this place there are innumerable quintillions of limpets of all sizes which make tremendous burrows, about an inch or so. There are also some sea anemones that keep tentacles always spread out." The letters hold few references to friends or ice cream or the fripperies of life; they strain to be mature and serious. ("With love to Aunt Nan & Granpapa I remain, yrs sincerely, Jack," he signed off in one letter, sent when he was 9, the handwriting already resembling the strangled scrawl he would employ as an adult.) It was as if he could not wait to grow up, could not wait to become the kind of man his father already was.

Thus, when J. S. embarked on a new breathing study in which participants had to be encased up to their neck in a coffin-like box, Jack was among the volunteers. When J. S. was asked to ride a submarine on a trial run, and when he fretted that he couldn't take along one of his regular assistants on confidential Admiralty business, he roped Jack in. (Jack had to tend to the soda lime during J. S.'s experiments. Did he know the formula for soda lime? He did, Jack said, his animation in barest restraint: It was a mixture of $Ca(OH)_2$ and NaOH.

"Well, that would simplify matters," J. S. said, "but remember, you mustn't even *look* as if you knew anything about it.") When J. S. had to rate the effectiveness of killing plague rats with sulfur dioxide on ships, he took his son along onto the SS *Bavaria*. Jack remembered, even years later, the game he played with the sailors, to see who could hold their breath, dash into the hold, and collect the most dead rats in a single swoop.

In Staffordshire, Jack joined J. S. on a tour of an old coal pit. They climbed aboard a large bucket and were let slowly down the shaft until they hit the tunnels. After some distance, the ceiling dipped, so they crawled on all fours until they reached an enlarged chamber, about 8 feet high. Someone raised his safety lamp to the ceiling; the flame blazed into a lapis blue before going out with a pop.

Still crouching, J. S. asked Jack to make himself useful. Could he stand up and recite Mark Antony's funeral oration from *Julius Caesar*?

"Friends, Romans, countrymen," Jack began, "lend me your ears," but already his mind had started to come unmoored, and as he stumbled through another two sentences, his head grew lighter, his legs flimsier. "The good is oft interred with their bones," and by now he was panting, "so let it be with Caesar." Midway through this fifth line, just as he was bringing up the noble Brutus, Jack collapsed. The lesson was unforgettable: The methane in mines, being lighter than ordinary air, pooled near the ceiling, so it was safer to breathe with the nose close to the floor.

A textbook could have told Jack this as well. What he gained from his firsthand experience of his father's career was not just a command of scientific principles but a sense of life itself. He met people— miners, soldiers, sailors, railwaymen—he would never otherwise have encountered as the son of an Oxford academic, and he learned of the hardships of their trades and the misfortunes of their classes. He grew up as a child with few illusions. There was a world outside the university, and it failed to treat its denizens well. The government was often unequal to the task of solving their problems: what made the residents of slums ill, how to keep naval divers safe, why miners died. Specialists like his father were indispensable. Their scientific methods were clear-eyed and fearless, always prepared to upturn the old, unquestioned ways of doing things. They delivered order out of chaos. The physical bravado of J. S.'s work felt thrilling, of course, to a young boy. But it was the intellectual adventure that fired Jack's spirit. If you tried to think with rigor and patience, if you ran experiments, made accurate measurements, and gathered evidence, if logic carried you from one idea to another, then nature gave you answers. And those answers then led to further questions and further answers. This was what Jack liked about scientific research: You never quite knew where it would take you next.

IF J. S.'S WORK LEANED into the twentieth century, the family home at Cloan, where the Haldanes spent their holidays, appeared to cling to the nineteenth. The estate, once a farmhouse, had been augmented again and again, as if to cast in stone the status of the Haldane clan. A spiral staircase ran, like a malformed spine, up the new tower; the bedrooms were somberr, the drawing room was chilly, and the gloomy corridors were lined with monochrome reproductions of old masters. The family always dressed for dinner. The grounds ran to yew hedges and croquet lawns and a walled garden in which the bushes bristled with gooseberries and raspberries. Every August, on the moors nearby, the Haldanes held a grouse shoot.

The trammels of class, which slackened so readily in a mine tunnel or a submarine, were unbendable at Cloan. Mary Haldane, J. S.'s mother, lived at the estate until she died at the age of 100, and she maintained a full staff of servants, many of them men and women whose parents had themselves cooked or cleaned for the Haldanes. Jack and Naomi grasped quickly the nature of relations between the family and the domestics. Respect was essential, but so was the distance of formality; even as babies, the children were always Master Jack and Miss Naomi. Some of the servants' skills were worth admiring—ironing or polishing silver—but the chores of others, such as the emptying and scrubbing of chamber pots, were considered just inferior work. None of the adults ventured below stairs, into the staff's domain; no one deigned to be familiar or friendly. In every way, the children understood, one must not descend. "It was a funny business, really," Naomi wrote later. "Here were people living in the same house, walking through the same rooms, but thought of differently."

Once a week, though, the hemispheres of Cloan came together. On Sunday evening, upon the thundering of the gong in the hall, the servants trooped into the drawing room to join the Haldanes in prayer. The members of the family sat in their chairs, and the servants on long plush benches carried in for the purpose. Everyone on the staff was present, from the butler and housekeeper, to the chauffeur who had induced the car horn to play the opening notes of *Lead, Kindly Light*, to the youngest scullery maid. J. S.'s sister played dolorous hymns on the piano. The eldest male Haldane present then read a chapter of scripture and improvised a prayer. The company stood to sing during the hymns, sat for the lesson, and knelt to pray—the servants on the floor, the family on footstools covered in chintz.

At Cloan, even the sincere practice of religion was inflected by a consciousness of class. It was important for the Haldanes to perform their diligence, to set an example to those beneath their status, and to inspire them to moral rectitude. Every Sunday morning, the family walked to church: past the railway station, over the glen, and up onto the ridge. As they neared, they were joined on the path by more people, the women wearing black gloves and the men in their Sunday suits, "to whom we talked a little, though with a certain awareness that they were not quite our social class," Naomi later remembered. Upon the villages nearby, the Haldanes bestowed their good works, dropping in on the sick or feeding the needy. Even to employ someone was considered a moral act, a provision of livelihood and security to those who had neither. As soon as the children were able, they were sent to the poorhouse to read to its residents from the Bible, and they accompanied the grown-ups on their visits to other homes in the villages. "One sat dangling one's legs on one side of a black-leaded cooking stove trying to be good," Naomi wrote. "Perhaps at the end one got a bit of shortbread."

The unflagging religiosity at Cloan was distinctly different from the looser attitude to faith at home in Oxford. J. S. and Louisa tended toward agnosticism, of the kind that the biologist Thomas Henry Huxley described when he coined the word in 1876. "It is wrong,"

Huxley wrote, "for a man to say he is certain of the objective truth of a proposition unless he can produce evidence which logically justifies that certainty." Evidence! The word was a clarion to a man who attached a laboratory to his house, and to his wife, who had shorn down her own wedding to the barest acceptable duration. There was nothing of the church and its articles of blind faith that J. S. or Louisa desired. When they first moved to Oxford, in fact, they had failed for so long to turn up at their church that a clergyman materialized on their doorstep at Crick Road, wondering if they were in any spiritual difficulty. Neither Jack nor Naomi was baptized. When they went to services at all, it was at their father's college on Sunday evening, where they fidgeted and grew bored.

Jack read and absorbed the Bible just as he read and absorbed everything else at hand. The Bible contained stories, after all, accounts of high romance and heroism similar to Walter Scott's *Tales of a Grandfather.* They entertained him; they even moved him to write poetry. At the age of 11, he scribbled down his first long poem, a 22-line recital of the selection of David to greatness and kingship. But these early fascinations with the narratives of religion were always destined to fade. His parents' interests lay elsewhere, and as Jack's education in science progressed, he found alternative employment for his imagination and his intelligence along avenues that ran in very explicit contradiction to the lessons of the Bible.

You know, you're wrong about all of your children being murderers. Now, I've studied the Mendelian laws of inheritance and their experiments with sweet peas. According to their findings—and they've been pretty conclusive—only one out of four of your children will be a murderer. The thing for you to do would be to just have three children. No, no, that might not work. The first one might be the bad one. I'll have to look that up.

　　—Gilbert Wynant, to his sister, in *The Thin Man*

In 1901, when Jack was 8 years old, his father took him to the Oxford University Junior Scientific Club, where the biologist Arthur

Darbishire was giving a lecture on Mendel's laws of inheritance. The explanatory power of these laws was only just being appreciated, and they were transforming the state of science in the Western world.

Gregor Mendel, the Augustinian friar, had conducted his experiments on pea plants five decades earlier in his abbey in Brünn, now a town called Brno in the Czech Republic. First in a kitchen garden and then in a two-room glass hothouse, he had planted 34 true-breeding strains of peas—strains that always produced offspring whose every trait matched that of the parent. If the parent plant's seed had a wrinkled texture, and if its flower was purple and its pea pods were yellow, then its offspring's seeds, flowers, and pea pods, too, would possess the same qualities. Mendel tracked seven traits, each of which came in one of two variants. If seed shape was a trait, round and wrinkled were its variants, its alleles. Then, to see what their offspring looked like, he began to cross these true-breeding types. Bending over one strain of pea plants—with difficulty, because he was already, by this time, on his way to barrel-like rotundity—Mendel used tweezers to pluck the pollen-filled anther from their flowers. These flowers, now effectively rendered female, then received careful dabs of pollen collected with a camel's hair brush from the anthers of a second strain of plants. Sworn though he was to celibacy, Mendel was impelling his plants, otherwise rooted in their pots, to engage in a form of long-distance copulation.

For seven years, Mendel engineered his hybrids, noting what happened when a tall plant was crossed with a short, or when a plant with wrinkled seeds was crossed with a round-seeded one. In the first generation of offspring, every new plant inherited its key trait from only one of its two true-breeding parents. A tall crossed with a short always produced tall progeny, and a round-seeded crossed with a wrinkled always produced round-seeded plants. In each pair of the seven traits Mendel was observing, one allele was dominant and one was recessive. Tallness was dominant; so was roundness of seeds and the purpleness of flowers. A true-breeding parent stamped its dominant trait on every one of its offspring.

When he crossed the first-generation hybrids with each other, though, the recessive traits seemed to resurface in the second generation and in a relatively neat ratio: one out of every four occasions. So in one run of second-generation plants, when Mendel patiently counted 7,324 seeds from 253 hybrid plants, 5,474 of these were "round or roundish" and 1,850 were wrinkled. The other traits, too, in other second-generation batches, divided themselves up on average in the same way—3:1, dominant to recessive. Mendel watched over at least 10,000 plants, and he sorted and counted 40,000 flowers and 300,000 peas.

In 1866, Mendel tidied up his results—perhaps a little artificially, eager to fit all his data into his 3:1 ratio—and published them in a journal called *Proceedings of the Brünn Natural History Society*. He didn't speculate too elaborately on the factor—*der Elemente*—that carried a trait from parent to offspring. He desired only the logic of heredity—the arithmetic behind the dispersal of *die Merkmale*, the traits.

Until the turn of the twentieth century, Mendel's paper remained unrecognized. It wasn't immediately obvious at all how his results plugged into the era's grand biological puzzle: Darwin's theory of evolution. Darwin may not have noted the mentions of Mendel in the books he owned, but Mendel certainly read the German edition of *On the Origin of Species* in 1863, just as he was finishing his experiments. He marked up parts of it in pencil, and terms from these paragraphs appear in the final two sections of his 1866 paper. In another paper, published in 1869, Mendel mentioned "the spirit of Darwinian doctrine." But he never realized that his principles could help solve a problem that nagged at Darwin's theory.

As plants and animals reproduce, the offspring vary from their parents. Darwin had proposed that sometimes an accumulation of slight variations makes an organism better suited to its environment than its parents or its peers. It thrives, it finds more mates or more food, or it lives longer and better. Its own descendants, carrying this robustness—or "fitness," the biological term—down the lineage, evolve eventually into a wholly new species. But for natural selection to work in this manner, the apparatus of heredity had to allow both for variations to appear and for those variations to be fixed and passed onward.

Darwin had only an unsatisfactory view of how this happened. Every part of the body formed tiny particles called gemmules, he thought, and these gemmules held blueprints for their respective organs and limbs. During reproduction, the information in these gemmules was collected and distilled into the sex cells, so that the sperm and the egg both had a compendium of instructions within them: where an eye should go, what a foot should look like, how a stomach should work. In the embryo—and here Darwin, having no theory of his own, yielded to the prevailing idea of his time—the information from the gemmules of both parents was blended together to create a new organism. Darwin didn't particularly like this paint-pot theory of heredity. It was, he wrote in a letter to Huxley, "a very rash and crude hypothesis." If a chance variation appeared in one organism, it would just be blended away over the centuries, diluted and diluted again with each successive generation, until it was finally undetectable, like a drop of orange paint in a gallon of ink. The paint-pot theory was no help in securing a variation or in privileging it in inheritance. In fact, it was the opposite; it was a way to quash variation, a way to purée a species into homogeneity.

Even in Darwin's time, his gemmules proved a frail construct. When Francis Galton, a scientist who was Darwin's half-cousin, transfused blood from one rabbit to another, he ought to have been ferrying gemmules over as well—and yet the rabbit receiving the blood showed no signs of inheriting its donor's characteristics. (Darwin, who had first applauded the design of Galton's experiment, reacted defensively to its result. He had "not said one word about the blood" in his theory, he wrote in *Nature*, and gemmule transmission didn't depend on blood at all.) Darwin died in 1882, just as a German biologist named August Weismann was further dismantling gemmule theory. In a series of experiments, Weismann lopped off the tails of mice for generation after generation and then bred them with each other, but baby mice were always born with full, jaunty tails. Why weren't the gemmules from the tail stumps carrying instructions to build abbreviated tails? It could only be, Weismann reasoned, because there were no gemmules anywhere at all. Every drop of hereditary information must be held in the sex cells, the sperm and the egg, where they could remain oblivious to the unhappy mutilations of other parts of the body. Then, in 1900, three other European scientists converged on Mendel, recognizing how his paper explained so much about inheritance: how it was packed into discrete units, for instance, rather than into a kind of miscible fluid; how these units could be shuffled around to produce and fix variation; how a random variant, a "mutant," might prove to be the fittest in its environment. Within a decade, these units came to be called "genes" and the study of their function "genetics."

It is too easy to assume today that the manifestation of these links between Mendel and Darwin should have tipped the science of heredity into a moment of universal clarity and light—that alternative prejudices would be abandoned, as if some cosmic conductor had hollered "All change!" upon every other train of thought. In fact, though, science advances not with the linearity of a locomotive but with the muddled gait of a drunk: with zags and zigs, with stumbles and hesitations and backslides, even if the overall direction of progress is forward. Just as Mendel's paper did not instantly reveal its solutions to the troubles with the Darwinian model, the solutions themselves, once they appeared, still failed to validate the model in everyone's eyes.

Darwin had rocked religion back on its heels, but he never knocked it down altogether. Many of the Christian faithful rejected his theories outright, but others—scientists among them—tried to stitch together their views of the scientific and the divine. They had one ready-made philosophy at hand: the natural theology of William Paley, a clergyman from the previous century, who sensed "an intelligent, designing author" behind the complexity of the natural world. But Darwinian natural selection allowed for no designer, only for the haphazard emergence of weaker and fitter variants. This was vexing for both theological and moral reasons. Not only was the necessity of God fading altogether, but organisms and species were living and dying as if by the spin of a roulette wheel. Darwin's model was one of inching increments, often producing variants that were neither better nor worse off; it was laborious and wasteful, almost uncaring of the sanctity of life. So while Darwin had turned out a fine carriage in evolution, the dissenting scientists seemed to think he had made a mistake in yoking it to a cart horse like natural selection. A different, cleverer beast was required.

Several were trotted out, and each earned its disciples. Lamarck's ideas were revived and refashioned. They offered a moral satisfaction: that as plants and animals—and human beings—evolved, they bettered themselves with every generation, drawing ever closer to perfection. Darwin himself was unable to let Lamarck go entirely. In edition after edition of *Origin of the Species*, he allowed for the transmission of a limited set of acquired characters, speculating that some physical changes in an organism could alter its gemmules. He also flirted with the idea that a life-form improves as it evolves, that "all corporeal and mental endowments will tend to progress towards perfection." The novelist Samuel Butler suggested that a species could direct its evolution and that even the simplest life-form had "a little dose of reason and judgement" with which to aim itself toward its fittest variations. Neo-Lamarckism rescued life from the whims of pure chance. It gave organisms control—and it gave mankind, possessed of more Butlerian reason and judgment than any other species, the most power of all. If humanity was determined to slip out of the supervision of the Creator, it ought, at least, to be able to throw a bridle over its own fortunes.

Some scientists favored a theory called orthogenesis, in which a nebulous inner force drives an organism's evolution toward higher planes of fitness and ability. (A German compound word was, of course, conceived to describe this perfecting principle: *Vervollkommnungskraft.*) Unlike the unpredictable temper of natural selection, orthogenesis displayed intent, and it was neat. Other scientists endorsed mutationism, which imagined new species appearing via sudden leaps in evolution and not through Darwin's tedious, piecemeal variations, full of switchbacks and dead ends. A German geneticist, Richard Goldschmidt, would call these abrupt creations "hopeful monsters." In refusing to believe that the development of life could be inelegant and purposeless, these theorists were clinging to the final vestiges of a world ordained by religion, a world of order and consequence, a world still compatible with the existence of a Creator. If evolution was directional, then there was, some-where, a director.

At the beginning of the twentieth century, these alternative ideas were scuffling with Darwinian natural selection in their ambition to lay bare the machinery of evolution. Darwinism had its apostles, par-ticularly in Germany and America, but its critics were so numerous and spoke so loudly that it struggled to find wider favor. The German botanist Eberhard Dennert wrote complacently that science stood, at the time, at the death bed of Darwinism "while its friends are solicitous only to secure for it a decent burial." Only after a handful of young scientists—Jack Haldane among them—bound Mendelian genetics to natural selection did the full, unclouded light of Darwin-ism shine through.

WHEN JACK WAS 6 YEARS OLD, he enrolled as a day scholar in the Oxford Preparatory School, now known as the Dragon School but then just as frequently called Lynam's, after its headmaster. For his time, C. C. Lynam was a teacher of progressive methods, nothing like the stuffed shirts elsewhere who stewed their pupils in prepa-ration for public school and Oxbridge. "I confess to not attaching much importance to outward politeness," he told his wards. "I hate to be called 'Sir' every half-minute; I prefer to be called 'Skipper.'" The Skipper dressed like an old sea salt, remembered the poet John Betjeman, who studied at Lynam's after Haldane. On some evenings, with all the lights switched off and his back to the fire, he told the boarders stories from his life. The Skipper believed that his students would discover their own minds and that they otherwise ought to be supervised minimally, free to ride their bicycles around the country-side or mess about in canoes on the river. He wanted only, he said, to show them "the falseness of all the gods of society: gold, sham reli-gion, and sham patriotism." For a while, Naomi attended Lynam's as well, the only girl among a clamor of boys. "He would ask a chosen few to tea with hot buttered toast and anchovy paste, as well as still-warm fudge, and we had what appeared to be immensely intellectual conversation, touching many subjects," she wrote. "Presumably the real thing he did was to treat us as equals."

Jack joined Lynam's as the school's youngest boy, and when he was bullied, he was quick to anger and turn combative. It helped that as he grew older, he became bigger and more powerful than the other boys in his form. If he thought he was in the right, he refused to apologize or change his opinion; in such situations, a quarrel or even an outright scrap was nobler than a compromise. He fought slowly, stolidly, and though he was not always fleet enough to avoid the blows directed at him, he took his beatings and waited to lay his own, devastating fists upon the other boy. He never lost his taste for a row, even when he was clearly the predator and not prey. "He must remember that there is no manliness, only cowardice, in attacking weaker people than himself," the Skipper wrote to Jack's parents in a midterm report in 1902. And then the next year, again: "I hope he has learnt never under any provocation to hurt weaklings. A nation very possibly has to do it but certainly not an individual in dealing with school fellows."

Jack studied Latin, English, French, and mathematics, and his intelligence asserted itself instantly. He stood second in his first form, even though he was at least a year younger than every other pupil; only in mathematics, ironically, did he start poorly. He placed fourth in English and third in French. His Latin teacher, after conferring with Louisa, had thought it best that he start the language the following year, but when she was distributing the first exam paper, Jack held his hand out for one.

"You can't do this, Jack," she said.

"I can," he said. "I know you did not *teach* it to me, but I heard you all the same."

That paper, Louisa was told later, "was quite up to the average of the form." In his Latin exams that year, he scored 45 out of 50 and then 72 out of 100.

As long as Jack wasn't bored, he remained at the top of his class. If his interest flagged, or if it leaped upon a different subject, the voltage of his performance dimmed. "An able mathematician, but often curls up and does nothing," a teacher remarked in 1902. Later that year, in another report, he repeated himself: "Is an able mathematician but occasionally falls to the level of the lowest boy in form." The next year: "Good at the harder work, despises and therefore fails at the easier sums."

Other teachers complained similarly about Jack's temperament. He was "much too aware of his own cleverness." He could be pleasant and friendly, but he "occasionally wants suppressing both by boys and masters." He wasted his hours, spent them extravagantly on his own thoughts. He was, at moments, something of a nuisance: "I should be grateful if he would behave better in class." One teacher remarked, "His mind like his body is powerful, but neither at present under his control." He was never one for gym: too clumsy and slow and too uninterested. His handwriting tormented anyone who read his work. "He must aim at brevity of expression and clear enunciation—he often knows but cannot state what he knows," the Skipper wrote, and added that he ought to "avoid the prefix 'Well' to all his sentences!"

Nevertheless, Jack improved through his years at the school, defeating not the difficulty of his subjects but the volatility of his own focus. His report cards yielded first to grudging approval and then to warm praise. By Easter 1904, he had become a "delightful boy" with a "splendid memory and knowledge of facts." He stood first in mathematics, second in English and Latin, and third in French. For some years, with Naomi's help, he set for the school an informal exam that resembled a general knowledge quiz, full of questions on current events and tricky riddles. (What was *vers libre*: a bookworm or a political party slogan that meant "Towards freedom"? Was Virgil really the mother of Our Lord, as depicted in the painting "The Virgil and Child"?) By his last year at Lynam's, Jack even approached the status of a model student.

In the baking summer of 1905, Jack took the King's Scholarships exams for Eton, the most prestigious of England's schools. The Skipper, after reading through the exam's question papers from previous years, bet the headmaster of another Oxford school that Jack would not only win a scholarship but take first place. The Skipper's brother, a teacher at his school, took Jack to Eton for four days of exam writing. They stayed in rooms that felt like sweatboxes, taking boats onto the river in the evenings just to cool down. Each day Jack toiled through question papers that he thought were stiff and beastly long, purposely too difficult to finish in the time given. So he worked backward, starting with the hard questions at the end of the papers, which promised him more marks, and skipping the simpler, lower-marked questions at the beginning. When he entered Eton in September 1905, Jack was the top-ranked King's Scholar. The Skipper won his bet.

Eton plunged Jack into misery. He had never lived away from home before and had always been mildly spoiled; now he was whisked from his family and into a school where he was fresh and anonymous. He slept, in his first year, in one of several cubicles in a dormitory, with only a curtain for privacy, shared tin baths, and lights out at ten. Once more, he was among the youngest boys in the school, and like all new boys, he was summoned to run errands for senior students, having to drop anything he was doing to make toast or brew tea or copy out passages. The bullying was merciless, even in his very first days, when one of his arms, which he had broken back in Oxford, was still in a sling. Jack didn't behave like the other boys, he knew; he was outspoken and clever and so "probably a nuisance," he reflected many years later in an autobiography that remained partially written and unpublished. "At any rate, the senior boys in college did not like me. On one occasion, I was caned by them every night for a week. At least once on the soles of my feet." He was laughed at for wanting to go to bed early. He hated being forced to play sports. "I was ill (surgical shock) & v. badly treated, as well as lonely," he lamented in his diary. "Poor me!" One of his teachers was concerned enough to write to Louisa: "I was rather unhappy about his last half, fearing that the other boys were tending to 'rag' him more than was quite fair. As you may imagine, it is a difficult thing to interfere in just the right way." Jack was, during his early years at Eton, precisely the kind of persecuted underdog he would defend throughout his life.

Eton pressed religion upon its students, much more of it than Jack was used to at Oxford: 20 minutes of chapel every weekday morning, 10 minutes of prayer every evening, and 2 hours of services on Sundays. Until he learned to get through them on pure reflex, Jack considered the Sunday services dreadful and annoying. Eventually, he developed a tepid fondness for the rituals of the Anglican Church, but he also built up a complete immunity to the faith. "Religion, none, but no convictions of atheism," he wrote, assessing himself, in telegraphic half-sentences, in 1910. He was still young; those convictions were yet to sprout. "Am really a Buddhist—Karma, but with belief in dominance of law, and perhaps illusion of personal identity."

Most disquieting of all, for a boy from a sheltered home now tripping into his teenage years in unfamiliar surroundings, were the currents of homosexual coercion that swept through the studies and corridors of Eton. Sex itself—the accepted kind of sex, between a man and a woman—was only rarely spoken of, a habit that, in the minds of the pupils, forever smothered the activity in layers of moral guilt. Homosexuality, though, was the truly nameless vice of the English public school, to be ignored except in the face of the most egregious complaints. The year Jack joined Eton, the school gained a new headmaster, who removed from the captaincy of a house a senior student whose reputation for buggery had turned too embarrassing to ignore. But nearly every other episode slid entirely past the notice of the masters.

Sometimes the relationships were consensual, boy drawn to boy in a tumult of hormones and sexual doubt. More often, they consisted of nothing less than forced sodomy. "Where there was much disparity of age, the younger boy was not always a free agent," Jack wrote. "The Eton Society, or 'Pop,' included the most distinguished and popular athletes. The shapely youths who were alleged to assuage the desires of this august body, often in return for presents, were known as 'Pop bitches.'" Students carried these experiences like calluses upon their memories of Eton. Years after he left the school, the biologist Julian Huxley, who was five years ahead of Jack, met an Old Etonian who had belonged to a house with "rampant and horrible" sexual coercion. "Luckily, I was large and ugly," he told Huxley, "but the pretty little boys . . . ," and here he broke off, simply unable to finish.

Jack never participated, and was never compelled to participate, in these activities, he maintained. "Morals: Sexual business quite chaste. Some lust after women (never enough to upset me the least). V. little after boys," he wrote in his staccato self-evaluation. But what else could he say? Even if he had endured incidents of molestation, it was difficult to drop these stories into his correspondence home or even into a quarter-finished memoir many decades later. When he wrote to his mother, his letters throbbed with some large, unspoken pain that may have just been homesickness, but might also have been acute distress. "When will any of you come down here, as I do want to see someone," he once asked plaintively. The sentiment recurred in letter after letter: "Could Uffer come down the Saturday after next, I should be absolutely free from 11.30 to 1.30 . . . or any Tuesday, Thursday or Saturday." "It will be most awfully nice seeing you again. Do stay till tea." "It is rot not seeing any of the family for such a time." "Come down soon . . . I am rather sick with people and things in general." "I should like Uffer down on Saturday very much indeed." "Dear Uffer, When are you coming down?"

Two months after Jack joined Eton, his father's uncle died in Oxford, and Jack went home for a couple of nights for the funeral. Her son looked so acutely unhappy that Louisa suggested he leave Eton altogether. "He did not want that 'exactly,' he just cried and said that he 'couldn't bear the place,'" she wrote later. She asked J. S. to talk to Jack, "but he could not get anything more out of him." When he returned for the holidays later that winter, "he begged to be taken away," Naomi remembered. Jack must have confided more to his sister than to Louisa or J. S. "When he told me . . . what they had done to him, I tried to get my parents to take his side, but they apparently paid no attention. How could they have taken a son away from Eton and sent him, say, to Radley?"

In his dejection, Jack unraveled somewhat; the hard-won progress of his final terms at Lynam's flaked away, and his report cards brimmed again with worries and remonstrations from his teachers. His classics tutor salted his reports with dissatisfaction. "A queer, untidy youth," he called Jack in 1906, and then the next year: "A failure this [term] and, what is worse, I didn't see where to begin a cure. I tried chasing and acquiescence and waiting; then an appeal to the discomfort of failure; but I couldn't find the ground in his character to lay a foundation. He has the irresponsibility of an infant." The King's Scholars wore gowns, looking, as John le Carré described the pupils at a fictive school modeled on Eton, "like crows, or black angels come for the burying." Jack's gown always seemed to be rumpled or askew and his appearance bedraggled. He was forever late for classes and with submitting work. His unpunctuality was, he realized, his besetting sin; he wished he had two or three additional hours in his day. "I can always sleep 11 hours and only get 7 or 8," he wrote. "Cannot always get to sleep in evenings, and always sleepy in mornings."

Jack meant well, his teachers knew. He was "most appreciative of true feeling and good workmanship; also as always most affectionate and anxious to please," a master noticed. He wanted to work, even though the boys were quick to denounce and rag as a "sap" a student who paid any effort at all to his lessons. And he was—everyone admitted this fiercely, transparently brilliant. Once, during a spot of extempore German translation, he deduced his way to the right verbs and nouns, leaving his teacher astonished; the other boys in the division, who had all studied German much longer, fared poorly in comparison. "He knows far more mathematics and has a better idea of how to apply them than any boy I remember to have met at Eton," a tutor wrote. But Jack's attentions were inconstant. He would begin an essay, find himself bored by it midway, abandon it for another assignment, and drop that for a third project, lighting upon a task only as long as his interests were stoked.

He often fell ill: fevers and colds, harrowing coughs, and booming headaches. His first term, a report noted, was packed with "complaints and misunderstandings and tears and despondencies." His teachers grew concerned that he was straining himself, that he verged on some form of nervous fatigue; one master observed an increasing hesitation in speech, so pronounced on occasion that Jack was almost stammering. He could be careless about his health, not stopping to wear warmer clothes before he passed from heated rooms into the chilly air. Eton, parked just by the Thames, upriver from London, always felt cold and wretchedly damp. Banks of fog lolloped onto the football fields, Jack wrote to his grandmother, and the boys stood around in the clouds, amused at how thoroughly they enveloped them:

They are sometimes so thick that after a hard kick at football
the ball has to be searched for. Some come in waves, the fog
being now 10, now 2½ or 3½ ft high, and people's heads and
shoulders now and again appearing like ghosts in Dante.
But when your head is under you can see things (to a certain
extent) in the fog but not above it. This is caused by a difference of refractive indices (ask Uffer).

A couple of years in, when Jack began to take more classes in the
sciences, he revived. He had to argue himself into this additional
work. The school had to find time to accommodate it in a way that
didn't cut into his other subjects or his rowing, the only form of exercise he seemed to enjoy. His father was concerned that he would over-
exert himself. "Now don't say, 'That's impossible, but look here,'" Jack
wrote hotly to J. S. in May 1907, pleading to do extra chemistry—or
"stinks," as the boys called it. "The total work would only be 2 hours
more than what I did last half. Surely it's better to do too much stinks
than too much classics. . . . Surely you'll admit it's better to have too
much of a good thing than more than too much of a less good?" His
mother worried that he had thrown his syllabus out of balance, load-
ing it with science and neglecting languages, history, and the classics,
the prime meat of Eton's education. "I will now answer some of your
accusations," Jack wrote to her, attempting to account for his time.
"You say I am doing nothing but science. I am doing 5 hrs a week of
history (not to mention essays & work for the Rosebery prize.) Also
Latin & Greek. I am trying also to get through some English literature. . . . I get a game of football or a drill every day. I have also been
rowing & swimming lately."

Jack liked his science teachers, and they liked him. For biology,
he had M. D. Hill—"Piggy" to the boys, a nickname that had been
handed down out of the mists of the past, but that derived, they could
only guess, from his protruding, snout-like nose. Hill was a little
man, a fussy but inspiring teacher. He was an affirmed Darwinian,
and as a result, he taught less physiology and more taxonomy, the
categorization of the natural world that laid the bedrock of evolutionary theory. "My brother Aldous also began biology with him," Julian

Huxley wrote, "and thus laid the basis for his constant fascination with science." When Jack was still studying mainly Latin and Greek but itching to spend more time with biology, Hill gave him short assignments outside of regular class hours: to draw, for example, the curving spinal column of a rabbit, with its delicate Y-pronged lower vertebrae. "Beautiful objects," Jack thought of these bones, and presented with such ebullience, Hill forgave his frustrating unpunctuality and his tardiness with written work. He would let Jack off school early, Hill told him, if he helped ready specimens for the class. "I remember preparing the membranous labyrinth of a dogfish and the limbs of a *Cypris* (a crustacean about the size of a pin's head). I wonder whether these are still in existence," Haldane wrote, in a letter to the *Eton College Chronicle* in 1958, after hearing of Hill's death that year. Hill regarded his student with affection as well. "[H]e works with remarkable insight and always has some question to ask or remark to make to the point, where the others have seen nothing peculiar," he wrote about Jack in a report. "To work with him is always refreshing and I shall miss him greatly, more than I can say."

The really first-rate teacher, Jack thought, was the Rev. Thomas Porter, whom he had for stinks. Porter was a ruddy-faced eccentric, a jovial man so incapable of keeping order and so sympathetic toward loafers that some of his pupils suspected he must have been defrocked years ago. If he had ever preached at all, they thought, he must have talked about the uses of asbestos in the heat of Hell or about the gaseous composition of the universe on the first day of Creation. Porter was an impresario of chemistry. He built working models of erupting volcanoes; he claimed to have revived dead cats with the help of a galvanic battery; he floated bubbles of hydrogen toward the roof of his classroom and then exploded them with the smoldering tip of a long bamboo pole. He boiled scientific laws down into short sentences, which pupils then had to learn and repeat verbatim at high speed and without drawing breath—a memory trick that seemed, magically, to work. Frequently, he passed around a glass vial of clear liquid, urging his students not to lose their hold on it. It was a sample of the world's most explosive substance, he said, and dropping the vial would risk striking not just Eton College but nearby Windsor

Bridge off the map. Inevitably, the vial slipped and splintered on the floor, whereupon it was discovered to have held only water. Porter, nevertheless, bellowed: "Down on your knees, boys, and thank God for your deliverance!"

No one at Eton told stories like Porter. A doggerel had it:

So sure his touch, that fame affirms
A boy exists who caught a
Disease from only hearing germs
Described by Dr. Porter.

Behind all this showmanship lay an enthusiasm for science and an insistence on precision in the lab—both qualities that Jack had seen in his father. Porter considered Jack to be sharp and gifted, but he wished he could drill care and order into the boy. "It will be some time before he learns that, in experimental work, nothing can atone for a want of attention to details: that a gas won't go down a tube against water pressure if it can get out through a loose India rubber joint. That some caution is necessary in smelling poisonous gases and other trifling details of this kind," Porter reported. He was, himself, engrossed by the practical aspects of chemistry and physics. He had his own photographic darkroom at Eton, and he was among the first people in the country to work with the newly discovered roentgen rays—or X-rays, as they came to be called. He read about experiments in journals and if possible tried to conduct them or tweak them in his laboratory; he knew, as Jack learned, that the scientific method hinged on the ability to replicate both procedures and results. Porter pushed his students to think and analyze for themselves and not to swallow whole every word of their textbooks. "It never crossed Porter's mind for a moment," a student later remembered, "that someday one would have to pass an examination." The examinations weren't important; the thinking was.

So Jack thought, and not only about science. He thought about the frictions in British society—about how, for instance, the teachers at Eton simply assumed that every one of their students was rich. He thought about religion and concluded that it was in decay. He thought about patriotism and deemed it irrational. He thought about the unemployed men from Manchester who came to Eton in protest, trying to get an allotment of land from Windsor Park. They "speech-ified" in the schoolyard, Jack wrote home: "They were led by a long-haired man who is in favour of various things, such as tariff reform, and a return (in some respects, such as all landowners being soldiers) to the feudal system, and who quoted obscure parts of Leviticus with great effect, and a rustic-looking man in knickerbockers who wishes to beat the railway lines into ploughshares, etc.!" He thought about other poor or oppressed or subjugated people. "We must foster 'a large and liberal discontent' with the lot of our fellows, if we wish progress," he wrote, quoting the anti-imperialist poet William Watson. "I think discontent, of one kind and another, is the only cause of progress."

In his house at Eton, Jack joined a debating club, an informal affair in which, after the harangues back and forth had ceased, the boys wrote their arguments into a ledger, each in his own hand, each following the custom of deprecating himself and commending the next speaker. The entries were all in mock-formal third person. Mr. Haldane, lauded for his "ringing eloquence" and "broad humour," argued once for the motion that it was better to have loved and lost than never to have loved at all: "For love to be worth having, it must be dangerous and potentially unsuccessful." On another occasion, he insisted that a state-owned railway system was fated for failure and bankruptcy. (He had clearly been allotted a stance contrary to his beliefs, because he described the end of his speech thus: "After a few more anti-socialist lies, he sat down.") In one debate, he contended that life was too short: "He thought that science and care could prolong our lives to about 140, when we might, perhaps, be willing to die naturally."

This was no mere debating point; it was something he had genuinely come to believe. "Man has now to look to his own unaided efforts for progress, but he ought to be a god, or something very like it, if he exists 10^6 years hence," he wrote in his diary. "On the other hand, for all we know, he may be in 100. . . . Still, we may live to see diseases abolished, which would be worth having lived to see." Science was, Jack decided, the only human activity that made for definite, continuous progress, and he wanted to be a part of it. "I intend to do some good to mankind, and having this ideal before me, can make it overcome other cravings. I do not expect any special reward, here or hereafter. I look for ambition to be its own reward. . . . I am probably right in saying that I am boundlessly, quite boundlessly ambitious."

His sharpening opinions brought him trouble. In 1911, not long before he left Eton, he announced in his diary that there had been a row. He had read good reviews of a book called *The Nature of Man*, by the biologist Élie Metchnikoff, but having bought it, thought it mostly dull; half of it, at least, muddled through aspects of religious philosophy, and Metchnikoff discussed how evil emerged from fundamental "disharmonies" in human reproduction and in the species' relationship to its environment. Still, he loaned it out to some friends, one of whom was "owl enough to leave it in the reading room (he is owl enough for anything)." The book was reported to A. W. Whitworth, the classics master, who thought that Jack was corrupting his classmates with radical literature. "Whitworth has a natural and laudable objection to rationalism in religion or ethics, or to talking openly on sexual topics," Jack wrote in his diary, drizzling sarcasm. "I have not. . . . Whitworth . . . tried to get me taken away. Uffer much amused and rather bored. Mother furious with Whitw." Jack sent a cocky letter to Louisa, urging her not to tear up to Eton to present the master with a piece of her mind. Unless he was sacked, he said, he would stay on. "Anyway don't come down tomorrow. You may have your grand row with Whitworth at the end of next half. I rather think Uffer would like one too."

Slowly, as happened at Lynam's, Jack found his bearings at Eton. Perhaps he only required time, or perhaps he needed tutors who could be happier with him. But it was also as if the habit of thinking—about science, but also about everything else he saw and read—clarified him, made him surer of himself. Jack had always been brainy, but now he was relying on his wits and his rationalism to build himself planks to stand on, positions he could take. In turn, these lent him confidence and insights into himself. Ideas made the man, so they were to be worn always on the sleeve. If he knew what he thought, he knew who he was.

Jack (middle), as captain of the school at Eton, at the annual Fourth of June festivities, 1911.

In his last year at Eton, Jack became Captain of the School and Captain of the Boats; a winner of the biology prize and a Coronation Medal; a member of the library committee; and the boy who delivered the students' address to King George V during his visit to the school. (Jack mentioned this proudly in his "obituary," the summary of his Eton career that he inscribed into the traditional ledger. In his handwriting, it looked as if he had written "Presented actress to George V.") He also won a mathematics scholarship to New College in Oxford. Naomi and her mother went to Eton in the summer of 1911 to watch Jack act in a play and saw him in full school regalia, complete with a carnation in his buttonhole. He was, Naomi wrote, now the wielder of a blue-ribboned cane, licensed to administer the kind of punitive licks he had himself received. "I think Jack may well have enjoyed an occasional ritual beating," she wrote, but "I doubt if he did it much."

Somehow, he had managed to fit better—but not completely—

into Eton, like a head jammed into a hat two sizes too small. His masters seemed to know it. "I trust that at Oxford he will not become too much of a recluse," one teacher told the Haldanes. "He must work hard but at the same time make friends and enter into the social life which is so valuable."

In the late nineteenth century and early in the twentieth, Eton suffered from the reputation of existing solely to groom rich men's sons to be rich men. Pedagogically, to use one Old Boy's phrase, it was a scholastic sausage machine. Little of intellectual value was transacted within most of its classrooms. "We are often told that they taught us nothing at Eton," Herbert Plumer, the field marshal, remarked. "That may be so, but I think they taught it very well." It was possible to escape being educated at all, Jack thought, but he had nonetheless contrived to learn a good deal. He could read Latin, Greek, French, and German, and he had enough chemistry to take part in research and enough biology to conduct his own. He knew no economics, but he could outline the Provisions of Oxford and the League of Schmalkalden, name the Hungarian prime minister, and talk about the relationship between the Reichstag and Bundesrat. He had lots of ideas, he wrote, "probably more than anyone else at Eton." His science masters had been profoundly influential. But the most fruitful part of Jack's education was dispensed to him not by Eton but by his father and by the projects and experiments that engaged him far outside his school.

IT WAS NEVER J. S.'S PLAN for Jack to remain just a willing test subject, sealed into diving suits or dragged down mine shafts. His father gave him more and more work of consequence—the mathematics of his respiration research, in large part—which yielded Jack's first collaboration on a paper. When Jack was at Eton, J. S. and an Oxford colleague were conducting studies on hemoglobin's affinities for oxygen and carbon monoxide, and it fell to Jack to work the numbers. J. S. presented some of this ongoing research to the Physiological Society in London in the spring of 1911, and he wrote to Jack's teacher seeking permission for his son to join him. "Jack's part is a very important one as he has evolved an equation which has thrown a flood of light into what was a very dark region," J. S. wrote. "I should like him to be present if possible, partly to fortify me against possible attacks from people who know the higher 'mathematics'!" The paper was published in June 1912 in the *Journal of Physiology*, its pages fuming with graphs and equations. The byline of three scientists was footnoted: "The former two authors are responsible for the experimental results; the latter for the mathematical analysis."

At Cherwell, Naomi started keeping guinea pigs, idly observing their routines and trying not to grow overly fond of them as pets. After a couple of years, Jack proposed that they recruit the guinea pigs into Mendelian tests. A condominium of hutches was constructed next to a hedge, by the side of a path that ran past the tennis court and down to the boathouse. They named their male guinea pigs, many of them after the scientists of the time, and when Jack was away at Eton, Naomi fed the guinea pigs, cleaned out their apartments, and took notes on the generations of progeny that were produced. (Once, curiosity overwhelmed her, and she milked one of the animals and licked up a drop. It tasted foul.) They recorded the color of the guinea pigs' coats, the length of their fur, and the whorled patterns that sometimes ran through the fur. If a creature was born with an extra toe, they followed its offspring with attention to see how the toe appeared or failed to appear. They were watching, as Mendel had with his peas, for signs of the steady, law-abiding shuffle of dominant and recessive traits.

But difficulties emerged. For one, the guinea pigs didn't breed as rapidly as Jack and Naomi would have liked; for another, the animals' traits were not clear and definite enough to supply any firm conclusions. (When Jack and Naomi, on one occasion, bought a specimen with a pink tint, they thought they had stumbled on an uncommon variation. But the pink faded so completely it might well have been painted on.) A run of an experiment was also lost when the neighbor's fox terrier sneaked past the Haldanes' front gate and raised hell outside the cages, causing several guinea pigs to keel over from fear. "It must have been a most bitter disappointment to you," the dog's owner, Cedric Davidson, wrote to Haldane decades later, recalling the incident. "You came up to see us & told us not to worry, that you yourself were quite satisfied that your theory was correct. . . . I thought then and still think that you lied . . . most nobly."

The broader problem was one that scientists elsewhere were facing as well as they ran Mendel's experiments in different plant species. If you took dihybrids—plants of the same species that differed only in two distinct, observable traits—and crossed them, a reasonable application of Mendel's laws suggested that the traits would shake out among the offspring in a ratio of 9:3:3:1. Nine out of 16 new plants ought to have the dominant versions of both traits, and only one out of 16 would have the recessive versions of both. The arithmetic, as laid out in a diagram called a Punnett square, foretold this with cast-iron logic.

The ratio failed to materialize, though. The numbers came out skewed, more heavily weighted in some combinations of traits than the Punnett square predicted. In one of Reginald Punnett's own experiments, for example, his team gained dozens more sweet pea plants with purple flowers and long pollen grains than they expected and dozens more with red flowers and round grains as well. But in the alternate combinations—the purple-flowered and round-grained or the red-flowered and long-grained—their tallies fell short. It was as if the genes weren't independent of each other—as if the purple-flower allele was linked in some persistent way to the long grain allele, as was the red-flower allele to the round-grain allele. Evidence of such linkages turned up repeatedly in experiments, but the reason for these linkages was, for a time, obscure.

One of the earliest papers to apply Mendel's principles to vertebrates was published in 1904 by Arthur Darbishire, the lecturer who had introduced an 8-year-old Jack to genetics. Darbishire cross-bred specimens of the Japanese waltzing mouse and the albino mouse, noting how the colors of coats and eyes were distributed through the new generations of mice. Maybe Mendel's theory simply didn't apply to these mice, Darbishire thought. But Jack, reading through this paper during his last year at Eton, slogged faithfully through the 23 pages of data that Darbishire had appended, and he spotted something the biologist had missed: evidence of genetic linkage, the first to be deduced among any vertebrate.

In 1911, Jack presented Darbishire's reanalyzed data in a seminar at Oxford University, but when he wrote to Punnett, wishing to
publish his conclusions as a paper, Punnett suggested that Jack run
his own experiments and obtain his own results. So Jack and Naomi
returned to the complex of hutches by the river, this time breeding—
for the sake of speed and convenience—not guinea pigs but mice. For
help, Jack roped in a friend at Oxford, Alexander Sprunt.

Even as mice begat mice behind Cherwell, geneticists started to
understand why linkage came about. In a laboratory—an exaggerated
closet, really—at Columbia University, a scientist named Thomas
Morgan was studying fruit flies, breeding their larvae in old milk
bottles stuffed with overripe fruit and feeding the adults on rotting
bananas suspended from the ceiling. Bottles crowded the laboratory's
eight benches. The smell of the fruit was bad enough, and Morgan
also tended to squash the stray fly he deemed unnecessary, leaving
it to decompose right where it was executed. But the fruit fly's great
advantage is that it doesn't hang about; it hurtles from birth to death
within a couple of months, pausing only to produce its hundreds-
strong batches of babies.

In the descendants of his flies, Morgan could follow a range of
physically distinct traits as they were handed down through many
generations, and he saw, once more, the visible evidence of linkage. Nearly always, for instance, it was his male flies that had white
eyes and his sable-colored flies that had miniature wings. Morgan
dismissed the argument that some alleles attracted each other and
repelled others, as if they were the poles of magnets. Rather, he said
in a letter to *Science* in 1911, "the original materials" behind these
traits—the genes, although he didn't use the word—must simply lie
close to each other on the chromosome. That was why they tended
to be linked, or coupled—passed down together intact, as if within
the same package. "We find coupling in certain characters, and little or no evidence at all of coupling in other characters; the difference depending on the linear distance apart of the chromosomal
materials that represent the factors," Morgan wrote. The nearer the
genes were to each other, the more likely they were to be passed
down in tandem, and so the more likely their traits—maleness and

white eyes, say, in the case of fruit flies—were to appear as inci-
dences of linkage.

Morgan's proposal had a clarifying effect. Naomi read about it,
she remembered, "in great gulps, curled upon the school-room sofa
and seeing how much it explained." (Jack was so struck by the theory
that in a 1919 paper, he would recommend naming the measure of
distance between chromosome positions after the scientist: the cen-
timorgan, a unit that has passed into the nomenclature of genetics.)
In their experiments with mice, Jack and Naomi watched for signs of
linkage between two traits: albinism and pink eyes. They had to be
meticulous, tracking every animal's characteristics to keep their data
clean and robust. Sometimes a mouse died, or it was injured during
an exercise run on the croquet lawn by an ambushing owl or hawk.
Badly mutilated mice were killed in their father's lab, placed under a
bell jar and gassed to sleep. But if they were pregnant females, a post-
mortem still had to be done; it was essential to know the coat colors
of the litter that would never be born.

The results were published in the *Journal of Genetics* in 1915:
authors J. B. S. Haldane, A. D. Sprunt, and N. M. Haldane, this
being Naomi's singular byline on a scientific paper. In the frequencies
with which albinism and pink eyes appeared in their mice, they wrote,
they had recorded "a close enough agreement" to the ratios forecast
by linkage. But the paper was barely two pages long, and it wore a
parenthetical proviso: "Preliminary Communication." It had become
necessary to publish prematurely, they explained. Sprunt had already
died in a field hospital in March that year, of wounds sustained in
action near Neuve Chapelle, in France. And Jack was attached to the
Black Watch regiment, to do what he had never dreamed of doing: to
go to war for his country.

JACK HAD COME HOME to Oxford in 1911 to attend New College
on the mathematics scholarship he had won. He was fresh from his
captaincy at Eton, and the praise of his teachers still rang in his ears;
he popped with confidence in the way only a privileged, ambitious
young man who has been assured of his vast promise can. He read
mathematics for the first year, dabbling on the side in zoology. Then,

realizing he was learning the sciences on his own in any case, out of sheer interest—or perhaps eager to play up to his fast-forming image as a maverick—he switched to the Greats, a course of ancient history and philosophy. He liked the process of sifting evidence and writing essays, and at Eton, in any case, it was only the classics master he had abhorred, not the classics themselves. "The subjects studied had little relation to modern life, so thought on them was free," he wrote later. "The successful Greats Man, with his high capacity for abstraction, makes an excellent civil servant, prepared to report as unemotionally on the massacre of millions of African natives as on the constitution of the Channel Islands." A love for the classics stayed with Jack throughout his life, and quotations from Plato and Aristotle would sometimes spring up in his papers, like check dams arresting the torrent of equations and technical language. He graduated from Oxford with a first in the Greats and another in mathematics. As he grew fond of pointing out, he never got a degree in the sciences at all.

His Oxford years were thick with levity and blitheness—qualities that were crushed by the war and that Jack and even England struggled to recapture through his lifetime. One of his acquaintances remembered him, at the time, as "a great shaggy bear, and matching his wit with the bawdiness of his repertory of songs." He acted in plays, some of them written by Naomi. He was never more than a tankard of ale away from reciting poetry by the yard: Housman, Milton, Shelley, and Shakespeare, word-perfect and readily unreeled. He was a member of the University Officers' Training Corps, and in 1914 he joined the signallers' unit to learn telegraphy. He rowed for his college and even enjoyed it; he was, by now, so burly and heavy that the oar custom-built for him cracked on occasion. Training for a race entailed a measure of dietary discipline, but the night of its completion, by way of reward, promised rounds of roaring drinking. Some of the young men additionally procured "a bit of skirt," as Naomi called it. Not Jack, she thought. "Jack didn't go in for the tobacconist's assistant type who was more or less available," she wrote. "In fact up to 1914 his sexual experience, like mine, was practically non-existent, except for erotic verse and guinea pig watching."

One night in June 1914, Jack was sleeping on a patch of heather near the town of Camberley; he didn't explain, when he wrote about this incident later, why he was out in the open, far from Oxford, in the middle of the night, and he didn't remark upon the dreamlike oddness of what happened next. "About 1 a.m. an Austrian commercial traveller named Sobotka arrived on a motor bicycle attempting to interest me in various new radio devices and also announcing the murder of Archduke Franz Ferdinand and the probability of a war between Austria-Hungary and Serbia," he wrote. "We had not envisaged the possibility that the Angel of Death should arrive on a motor cycle, not only announcing the death of many of us signallers but the death of the culture into which we had been born."

In the beginning, the war seemed like just a crease in the cloth of life, nothing more. It was even exciting in a mildly perverse way. Most of the young men in the Haldanes' circle hurried into khaki and into training, hoping to see at least a little action before the whole affair folded up in months. Instead of taking a six-week walking holiday across Europe as planned, Jack joined the family regiment, the very Scotch Black Watch, along with some of his cousins. As a commissioned second lieutenant, he was sent to Nigg, north of Inverness, held in reserve with the 3rd Battalion. Even then, it wasn't difficult to believe this something of a lark. But then, unexpectedly, people began to be killed: family and friends and colleagues, slain in a war that refused to be as brief and meek as expected.

Through the rest of 1914 and the early winter of the following year, Jack remained in Nigg. Ostensibly, the reserves had a job: to guard the entrance of the Cromarty Firth from an assault by land and to protect the town of Invergordon, with its strategic oil reserves, from a raid by sea. Neither attack, really, was likely. Cruisers and torpedo boats prowled within the Firth, and searchlights bounced off the water. Jack and his platoon of 60 men, and all the other reserves at Nigg, trained and shivered, snatched at rumors, and waited to be dealt their walk-on parts in the larger theaters of war.

There was always work to be done. "E.g. today," Jack wrote to his father:

7 a.m. – 8 a.m. Drilling platoon & get NCOs to do so, correcting their mistakes

9.30 – 1.00 Route march, during which I try to instruct my NCOs in map-reading etc, & get a few general facts e.g. that the sun is south at noon into the heads of the stupider recruits, also teaching them to keep their eyes open.

3.30 – 4.30 Fire control & range taking . . .

In the evening, we generally practice signalling, but today is rather a slack one for me . . .

They practiced digging trenches, their walls pocked with what looked like rabbit holes but were in fact cubbies for cartridges. Once a week, in rotation, they spent a night in a trench; another night of the week, they were on picket, sleeping in their uniforms and slipping out for a recce an hour before dawn. Jack liked his men, he told his sister, and if any war remained for them to fight, they would fare very well: "I think within 3 months they will be as good as any Germans who are left."

The weather grew frigid, the wind blowing savagely off the snow-clothed slopes of Ben Wyvis, eastward and down toward the sea, goosing the men of the Black Watch in their regimental kilts so that they were forever warm about the waist and cold around the knees. The rain soaked through the tents, puddled around the camp, and gave the recruits enteric infections. One soldier committed suicide, cutting his throat after lights out, making, as Jack recounted, a beastly mess of his tent mates' blankets and gear with his blood. In batches, men kept shipping out: 90 in a week, 100 the next, so that it seemed forever imminent that Jack himself would leave. He didn't know where he would go. At one point, he thought he would be sent first to Bombay, and at another, he reckoned that it was likely to be Schleswig-Holstein. "To judge from letters, the battle of Ypres doesn't seem to be bad fun in parts although very violent indeed in others," he told his mother. The days flapped past, and there was no news—"Still here"; "Nothing special doing here"; "Here I am, & likely to stick on for some days"—until suddenly there was. Late in January, Jack was posted to join the 1st Battalion in France.

He arrived, at the front near Givenchy, into a universe of mud. Mud clogged the innards and consumed the joints of the rifles, many of them Lee-Enfields left over from the Boer War. Mud snagged shoes, sucking them off feet as their occupants tramped or ran. Mud worked its way into hair and between toes and "too far into some parts of me to wash out," Jack wrote to his sister. Mud on the ground made it impossible to dig new trenches, so that soldiers had to shelter instead behind low wooden defenses called breastworks. When the ground was drier, the men tried to get ahead of the weather and sink their trenches, but without the right drainage or enough wooden duckboards, these quarters, too, turned into rivers of mud when it rained. It rained often. In the spring of 1915, the 1st Battalion of the Black Watch fought both the Germans and the waterlogged earth of eastern France.

Near the front, the war autographed its presence upon the landscape. At first, Jack saw houses that had merely lost their windows or were missing tiles off their roofs. Gradually, the rafters grew more and more exposed, like the ribs of people whose chests had melted away. Nearer still to the firing line, whole houses were smashed down to the ground, and not incinerated, as he had expected. Right behind the line, most of the intact buildings were barricaded for use by the troops, while the others had their woodwork stripped and burnt. In these buildings, with their confusion of mortar, plaster, rubble, and old wallpaper, Jack felt a heavy air of squalor and ugliness. But their billets, in villages well behind the line, were surprisingly comfortable, running to straw mattresses and perhaps even sheets. This was coal mining country, and the communities supplied the collieries with workers; the bigger villages had railway stations, and the soldiers heard the blessed, regular sounds of trains, with all their connotations of civilization and the settled urban lives they'd left behind. Jack once lodged in the house of a miner, a delightful man who poured him beer and discussed coal mines in France and Britain. If Jack had stayed any longer, he thought, with a pang for his old academic habits, he would have taught his host English and his youngest daughter arithmetic.

On the line, with a platoon of 50 men under his command, Jack was enveloped by the grind of battle. This was not the kind of war he had read about, the kind in which men like Caesar or von Wallenstein punctuated long periods of inaction with brief, frenzied spells in which they hurled arrows or cannonballs at the enemy. Here the two sides sniped at each other nearly all day and loosed off shells and bombs and flares from time to time; at night, in their stuttering sleep, they still heard the *pup-pup-pup* of machine guns. The Germans' rifle fire, Jack realized, was mostly unaimed and so not especially unpleasant. By the time he heard the *whiwww* of a bullet passing him, he was already out of danger; it annoyed him, in a way, that so comic a sound could have so tragic an effect. The shelling from the other side was erratic as well, although Jack learned quickly that when a direct hit did land near him, when he was seeing the men around him fall and when his breastworks were collapsing, the wisest thing to do was to sit as still as possible.

Every day, there were reminders of the frailty of life. As the weather warmed that spring, the corpses that had strewn the fields since the previous winter began to come out of cold storage, the smell of the dead filling the nostrils of the living. Men were extinguished in random, pointless ways. A sergeant in the 1st Battalion, practicing his bomb throwing, killed himself and an accompanying major when his fuse exploded prematurely. Exchanges of shells and bombs turned into heated arguments, each side insisting, quite against logic, on having the last word; in a squabble of this kind, after several hair's-breadth escapes, Jack saw a bomb kill four of his company. Once, by the light of a flare, the Germans spotted Jack in front of his breastworks and let fly. He managed, somehow, to drop down into the nearest hole. The fusillade intended for him slew an adjacent sapper instead.

Not long after he joined the battalion, Jack's commanding officers figured that since he knew something about explosives—or science, at any rate—he should be taught the use of the trench mortar and given charge of the hand-thrown bombs. The bombs—Bethune bombs, named after a French town not far away—weighed about a pound apiece, resembling four thick black disks stacked on one another, the fuse poking out of the top. Each bomb made a crater about 8 inches deep and 4 feet across, breaking into several small pieces as it exploded. The trench mortars, their sleek cylindrical bodies 15 inches long and their muzzles 3½ inches in diameter, had to be stuffed full of gun cotton, aimed at a 45-degree angle into the sky, and fired. They didn't always work; indeed, their contents were only slightly less liable to blow up at their origin as at their destination. Jack tinkered with his mortars for hours, almost as if he were back at a lab bench. "Somehow high explosives seem a far more remote and ridiculous method of deciding a question of right and wrong than spears," he wrote to a friend, "and the contrasts of our life are so glaring as to appeal irresistibly to one's sense of humour."

Jack discovered he was good at this work; unexpectedly, he even came to enjoy it. The month of April 1915 was one of the happiest of his life, and being a bomb officer, he concluded, was just the most ripping job he could hope for. He happened to be in the midst of a defensive phase of the Black Watch's campaign, a phase suited to saboteurs like himself, who could sneak about the line and lob their explosives. It demanded coolness rather than dash; he wasn't sure he would like the attack quite as much. On the nights when he was sent out for some recon and light bombing, he got close enough to the Boche positions to hear their soldiers humming themselves to sleep. "One night I was on my feet till 7 a.m. The next night I got some sleep sitting up & then got wet up to the waist coming back from bombing," he wrote to Louisa. "I came back along a burn, not fancying flares & machine gun fire on me. They didn't turn it on, as it happened, but might have done, & even rifle fire is a little troublesome, though I don't duck now."

These sorties galvanized Jack beyond belief, and he began to affect a swagger toward danger as it swiped at him. Some part of this was the battalion's own attitude: its men pretending to hold death in contempt, charging so fearlessly at the enemy that the Germans, it was said, called these kilted warriors *"Die Damen aus der Hölle,"* or "The Ladies from Hell." But Jack also practiced his own rituals of daring, as if by tempting fate loudly—catcalling her, jeering at her— she might turn away in disgust and seek her victims elsewhere. The best people at bombing, in any case, as he explained to his father, "seem to be the reckless kind, who are always breaking rules and things in peace time." When he taught other soldiers bombing, he frequently tamped his pipe down with a detonator. When flares burst above him, he fixed his attention on their colors, trying to deduce the metal being ignited. ("How my father would enjoy this!" Jack once thought.) In the throes of swapping fire, he often stuck a lit fuse into a handful of gun cotton and shied it at the enemy, heedless of the risk of blowing his arm off at the shoulder. Bravery and bravado felt just like isotopes of each other. Courage, he wrote to Louisa much later, lies in "taking a novel risk, which you are not ordered to take, and enjoying it." But she had always spotted in Jack's audacity a component of willful blindness. Much like his father, she realized, Jack "would walk along a knife edge of rock or timber in a mine, quite unaffected by the yawning gulf beside him, provided it was dark and he could not see the danger."

Jack flitted from company to company, setting up his mortar and bringing it to roaring life, but then that invited a rain of shelling and rifle fire in response, so the commanders of his companies grew to dislike him actively. But the other men adored him, which must have felt novel for Jack after the way he had been bullied and teased at school. It was precisely his aura of bluff valor that they admired. They invented names for him like "Bombo" or "The Rajah of Bomb." Douglas Haig, then a general and later a field marshal, called Jack "the bravest and dirtiest soldier in my army." A friend of Louisa's, writing from a French field hospital, recounted how he met a wounded colleague of Jack's from the 1st Battalion. "He informed me that Jack was the greatest man in the world," the friend wrote, going on to describe how the soldier and his captain had fabricated for themselves a cozy two-person dugout on the front and had looked forward, throughout a day of fighting, to returning to the dugout for an unhindered night's sleep. "But alas! what should they find on arriving but the whole dug-out occupied by every description of bomb & grenade, on top of which Jack was peacefully sleeping." Bombo was nerveless. One March evening, the colonel summoned all the officers and men and praised Jack lavishly, calling him an example to them all. "It is possible that, for the first and last time on record, I blushed,"

Jack wrote to his sister. He was no born leader of men, he knew, but at least he had won the reputation of being "a proper wild man" and the authority that went with it.

The trench was a democratic place, too, at least as Jack saw it. Several of the wellborn and wealthy had been commissioned as officers, it was true, and they supervised many men from the working classes just as they did back home. But the hierarchies were now strictly of rank, not of social class; a private with a public school education got no special treatment or undue privileges. In any case, in the midst of incoming fire or in a crowded billet or an even more cramped trench, it was difficult to hold to these distinctions of class. The officers around Jack insisted on a plain informality in their interactions with their soldiers. "One of the last recorded remarks of Colonel Grant-Duff," Jack told his sister, "was when [the mess sergeant] Wallace had told him off (not seeing who he was in the dark) for asking for marmalade when there was none, and the Colonel said: 'Don't apologize, Sgt. Major Wallace, treat me just like you do the others.'" It warmed Jack's egalitarian spirit.

Jack's fondness for the army—for its comradeship; for how it prized him; for how he, as a young man, gained a sense of purpose and a way of being—was understandable. What mystified even him, although only in retrospect, was his taste for the violence and the pleasure he took in killing an enemy. In his letters, he referred to the troops opposite him with detachment. "I certainly hit one German with the mortar yesterday as he was heard to scream": placid, content, cool. "I was well aware that I might die in these flat, featureless fields, and that a huge waste of human values was going on there," he wrote in an essay titled "Some Reflections on Non-Violence" in 1959. "Nevertheless I found the experience intensely enjoyable, which most of my comrades did not." It was war as the great poets had sung it, he wrote: an affair of glory and romance, in which similarly armed men were pitted against each other. It was a war of righteousness, of the Hindu concept called *dharma*. "I was supported, as it were, on a great wave of *dharma*."

He may well have suffered a deeper disquiet. In the trenches, he wrote short poems in a ruled notebook, the front sections of which held lecture notes from his training at Nigg. The poems spoke of "unhallowed desolation," of "these poor victims . . . half buried in the mud of their damnation," of "the sweet smell of corpses on the breeze." In one, titled "Destruction of Richebourg L'Avoué," he wrote of the town's defenders, hurled

> to darkness and confusion and the grave;
> And overhead great black clouds densely curled
> Hide from the sun the anguish of the world.

Later, Haldane tried to reckon with the gratification he milked from the war. "I believe that these sentiments are fairly genuine, for they were borne out by my dreams," he wrote in his unpublished autobiography. "In my pleasant war dreams, I am in the trenches; in my unpleasant ones, I turn up late on parade or cannot find my tie or Sam Browne belt." He knew he felt an enhanced sense of life when he was in moderate danger. That was it: a simple, primal thrill. But it had to be acknowledged. "I think war is a monstrous evil, and yet admit that I enjoy it. This is an internal contradiction in my mind. . . . If we do not admit the existence of these contradictions, we shall merely invent rationalizations to cover them up, including the loftiest moral reasons for involving our country in war."

THROUGHOUT HIS TIME IN GIVENCHY, Jack sent Naomi instructions about their mice and asked after the results of their breeding experiments. He wrote up the final sections of their brief manuscript and sent it off to the *Journal of Genetics*, becoming, as he would later observe, the only Black Watch officer to publish a scientific paper from the trenches. When he mislaid his literature, he asked his father for more: "Send me a copy of your conclusions & the *Lancet* article. I lost the others during some battle when I couldn't be looking after such things." Then, toward the end of April, science insinuated itself into the war and returned for a while to the center of Jack's life.

In Ypres, Belgium, on a spring evening, the Germans decided the wind was just right, pushed their infantry back, and popped the valves on 5,730 metal cylinders. A cloud of green gas ambled westward. "It reached the parapet, paused, gathered itself like a wave, and ponderously lapped over into the trenches," a Canadian soldier, noticing the queer fog from well behind the French and Algerian positions, would recall later. The French, preparing for supper, watched the gas with interest. "Then passive curiosity turned to active torment—a burning sensation in the head, red-hot needles in the lungs, the throat seized as by a strangler. Many fell and died on the spot." The Germans heard the yells of their enemies and their panicked, useless attempts to fire every one of their weapons into the mist. After 15 minutes, the guns began to fall silent; another 15 minutes, and they heard only a stray shot. The German infantry picked up its heels and pushed cautiously forward. Willi Siebert, a German soldier at Ypres, later wrote an account of the attack for his son:

> What we saw was total death. Nothing was alive.
> All of the animals had come out of their holes to die. Dead rabbits, moles, and rats and mice were everywhere. The smell of the gas was still in the air. It hung on the few bushes which were left.
> When we got to the French lines, the trenches were empty but in a half mile the bodies of French soldiers were everywhere. It was unbelievable. Then we saw there were some English. You could see where men had clawed at their faces, and throats, trying to get breath.

Five thousand French and Algerian soldiers were killed, most within the first 10 minutes of drawing the fumes into their bodies.

The Allies had suspected that an attack of this kind was coming, but even the Germans had not known precisely how the gas would behave or how effective it would be. Twice more, in quick succession, the Germans emptied canisters of gas into the Allied positions. Forced to find a way to keep his men breathing, Lord Kitchener, the secretary of state for war, sent for J. S., England's preeminent expert on respiration. Four days after the first attack, J. S. and another scientist left for the front: a train to Dover, a boat to Dunkirk, and then a car through the night to general headquarters in Saint-Omer.

The gas was chlorine, J. S. confirmed. He saw the brass buttons, newly greened, on the uniforms of dead men. He saw the insides of their lungs, stripped of their linings and so choked with watery mucus that they seemed to have liquefied. The survivors breathed with wheezes and crackles, their blood so starved of oxygen that their faces were flushed with blue. Some soldiers had lived by obeying their officers. "Piss on your handkerchiefs and tie them over your faces!" they had yelled, suspecting the gas to be chlorine and remembering that ammonia could absorb chlorine by reacting with it. But urine-soaked kerchiefs were not a permanent solution. J. S. sent his report to Lord Kitchener and returned home to develop a mask-like respirator. Cherwell soon reeked of chlorine. J. S. deployed himself as a test subject once again, hacking and retching at such volume that he could be heard well outside his laboratory.

Within a week, he had a stopgap device: a gauze pad dampened with sodium thiosulfate and glycerine, to be tied over a soldier's nose and mouth. It would have to do until a better respirator could be manufactured. Returning to Saint-Omer in May, J. S. installed a glass-walled cabinet in one room of a hospital, and the army briefly recalled three veterans of J. S.'s experiments out of active duty so that he could shut them into his chamber and try his mask on them. Thus, for a few days, Jack was reunited with his father, to take up his familiar duties of guinea piggery and scientific assistance. Man by man, they took turns in the chamber, the respirator bound around their faces, standing still or turning a wheel until their breathing roughened and cracked. Outdoors, they sprinted 50-yard lengths, making sure soldiers could move quickly and work hard while wearing their masks. Inhaled through the gauze, the chlorine didn't incapacitate them, but it had its effects. Some of J. S.'s volunteers were temporarily confined to their beds, and Jack's lungs felt so clogged that he was unable to run for a month afterward.

He was still panting and coughing when he reached Richebourg-Saint-Vaast on the afternoon of May 9 to rejoin his brigade. The uproar engulfed him. Up ahead, batteries of field artillery loosed screaming shells into the German trenches; then the soldiers crouched under the returned favor of bombardment, hoping to live long enough to open fire once more. The din reminded Jack of a never-ending chain of thunderclaps. He passed first-aid posts, dressing stations, and small houses. On the side of the road, he saw a dead Indian soldier, who had fallen in the midst of such commotion that no one had yet been able to remove his body. The battle scorched his senses, but he felt a strange exaltation. As fast as he was able to trot, his chlorinated lungs rasping all the while, he hurried toward his troops.

Had he managed a fleeter pace, he would have been right where the shell landed; as it was, even outside the crater, he was spun around and knocked over. His first inkling of something amiss was that the earth was so close to his face. "I've been hit by a shell at last—how funny!" he thought, and lifted himself up. His memory felt tottery, but he tried to calculate if the crater had been made by a 5.9-inch shell or an 8-inch one. He was covered in mud, and his head and chest ached; his chin was so scarred and peppered by gravel that he would never again manage to shave it close. He didn't really register that he was hurt, though, until other soldiers pointed to the blood saturating his right sleeve. Then he walked to a house nearby, where a two-room medical station had been set up. A doctor dressed the shrapnel wounds in his arm and the left side of his torso. Afterward, Jack climbed the stairs to the next floor and sat for a while with other field staff, watching the battle through his own daze and through the haze of thick, oily smoke that hung over the countryside.

Not far away, a large farm had lost its roof, and its rafters were on fire. The German trenches were barely visible, a jagged and discontinuous line chopped into the ground. Using his binoculars, Jack saw the Allied troops break into a tear toward the Germans under the cover of shelling. Then, through that cannonade of noise, he heard a new, repeated sound, like distant applause, and the charging soldiers began to fall—so quickly and in such synchronicity that he initially thought they must have been commanded to lie flat. They were being cut down by machine-gun fire. He had never realized how quickly people could be killed. A few soldiers rose and tried again to advance, but none of them made it to the German trenches. Jack saw only one man in inglorious retreat, pelting back in wild fear, zigging and zagging to try to evade the shells falling around him. He nearly reached home, but close to the Allied parapet, he was overtaken by a sudden, small cloud of smoke, and when it cleared, Jack couldn't spot him anywhere.

Feeling faint now, he came downstairs, and after the shelling weakened, he walked a mile and a half, in stuttering rain, until a passing car picked him up and took him to another dressing station at Le Touret. All the ambulances setting out for the hospital in Béthune were full, so a Black Watch officer flagged down a car and helped Jack in. The driver happened to be the Prince of Wales—the future Edward VIII, piloting his own automobile around the front. "Oh it's you," the prince said, having met Jack the previous year in Oxford. "Yes, I'll take him, he looks fit for nothing but hospital." At Béthune, a doctor swabbed hydrogen peroxide into the gash on his arm; the splinter in his side, the staff found, had pierced Jack's haversack, slowed down in its passage through his volume of Anatole France, and finally come to rest just beneath his skin. The ward trembled with the groans of the injured and the delirious, and the racket of shelling still blew through the windows, but Jack fell instantly asleep.

A PERIOD OF CONVALESCENCE AT CHERWELL, then back out to Nigg, in Scotland, where a new bombing school needed instructors. The British army's No. 1 hand grenade, resembling an enlarged spark plug, was still in use, and Jack gave lectures on its anatomy: why its core of picric acid exploded, how a cane stick was used to throw it, and where to tie the streamer of cloth so that it landed nose-first. He asked his students to attach detonators to fuses with their teeth; not everyone jumped to obey him. Despite his graze with death, Jack had brought with him the bombast he had worn on the battlefield. If the grenade wasn't faulty, he reasoned, then only carelessness or panic could cause an accident. Those could be conquered. He didn't understand the psychology of the man who, having lit his bomb, held on to it, paralytic with terror and transfixed by the hissing fuse the way birds are by snakes. At Nigg, Jack carried matches and gelignite powder in his pockets. In demonstrations, a fellow officer would light a bomb and throw it to Jack, who then had to toss it sideways, out of harm's way. "Provided you are a good judge of time, it is no more dangerous than crossing the road among motor traffic," he wrote, "but it is more impressive to onlookers." None of his men, he pointed out, ever suffered anything more than a scratch during their training.

Jack's antics may have been his way of slaking his desire for action. The work was too simple, and his weekends in Inverness were spent getting drunk—once so disgracefully that he nearly got court-martialed. Nigg bored him; he longed to return to the front. (Years later, to his chagrin, he learned that he had been sent to Scotland, rather than back to France, because a War Office general was a friend of his uncle's and wanted to keep Jack safe.) After nine months, he was transferred to a post in intelligence in Edinburgh, which he recognized as "a particularly silly administrative job." It was October 1916 before he was sent out again, joining the 2nd Black Watch in Mesopotamia and entrenching himself against a line of Ottoman troops.

The Black Watch was helping to hold a front stretching from the Tigris River to Suwaikiyeh Marsh, southeast of Baghdad and not far from the town of Kut al-Amara, which the Ottomans had just snatched from the British. Three battalions of Indian infantrymen—and a host of Indian muleteers and camel drivers—waited for orders from the Black Watch. Jack was placed in charge of the 2nd's snipers, and as he had with his trench mortar command in France, he readjusted his duty until it became a freewheeling experiment. Discovering that the Ottomans shot wildly and badly at any provocation, he told his men to hoist up helmets on sticks so they could draw fire and thus gauge the enemy's positions. He tuned the telescopic sights on rifles, giving his snipers clearer views. His marksmen became so effective that they were allowed to rove up and down a dozen miles of that front, inserting themselves into the action whenever Jack saw fit. "I am glad to have thought of a game that exercizes [sic] at once that vast brain and large body, and makes a most deserving and high-minded officer happy in slaying a proportion of less thoughtful humanity," Andrew Wauchope, who led the 2nd, wrote in a letter. "The men think a lot of Haldane." This expanse of the country held few civilians, and no batteries of artillery scattered mass death into the trenches. The battle pitted gunner against gunner, man against man. It felt like a purer war.

In Mesopotamia, Jack felt at home. A photograph shows him inspecting a trench: tall and broad and as thin as he would ever be, a pith helmet on his head and just inches of skin between his kilt and his hiked-up socks. He isn't close enough to the camera to display his moustache: full of vigor, if somewhat disheveled at its extremes. He liked the brazen sun, even though it turned him pink and made his skin blaze with prickly heat. At night, jackals howled among the camel thorns and myrtles; by day, he watched spiny-tailed lizards, an extraordinary kind of armored caterpillar, and beetles the size of large buttons. There were no mosquitoes, but there were plenty of rats and mice, doing the work of earthworms by burrowing and turning up the soil. ("I should like to Mendel them," he wrote to Naomi.) He didn't mind that the food was dreadful: bully beef and a kind of biscuit so famously indigestible that it merited an entry in an artillery officer's comic poem, "The Mesopotamian Alphabet":

B is the biscuit that's made in Delhi.
It breaks your teeth and bruises your belly,
And grinds your intestines into a jelly,
In the land of Mesopotamia.

The pace of the fighting was slower here, so he had time to read, play interminable hands of bridge, and learn some Hindustani. "I can tell people to fetch things, & ask where places are, & things like that," he told his sister. He wrote maundering, affectionate letters to her, asking her if she had read O. Henry, discussing Hegel and poetry and Jane Eyre, telling her that he thought officers' wives "are perhaps the principal argument for class war." To Naomi and to his father, he described the scrubbed clarity of the Mesopotamian sky: the broad band of light that beamed from the dimmed evening sun, pointing like a finger toward the planets; the lucent moon and the odd shadows it threw, especially of men wearing kilts; the play of constellations, in which the dog barked at the bull by night and the

snake curled past the scorpion early in the morning. "Astrology and
star worship would be almost inevitable in a country like this," Jack
wrote, "and it would take a lot of persuasion to make a man suppose
that the behaviour of the stars was quite irrelevant to his destinies."

Jack himself was wholly unpersuaded. "It had been determined
by the fates," he wrote sardonically, from hospital, "that I was to get
an accidental bomb wound on 22 ' 2 ' 17." At noon on that date, at
his regimental depot, an adjutant dropped an armed grenade, which
burst a few yards from Jack. Miraculously, he caught no splinters
from that blast. Later that evening, though, kismet tried again. A
hangar near the depot went up in flames, and the bombs and petrol were stored so close to each other that Jack witnessed a prank of
chemistry: whenever a petrol flare set off a bomb, the blast sucked
up the oxygen and extinguished the fire, but even the stray, glowing spark was enough to ignite the petrol all over again. One bomb
exploded so violently that Jack struggled for breath and supposed his
throat had been ripped away. There was another explosion, and then
a third, even as Jack ran for his life. In his tent, feeling a moistness in
his leg, he removed his gumboot and came upon a wound above his
ankle. His ears rang and buzzed.

A further trial was in store. In a dressing station, his comrade
in the next bed lit a cigarette, which set the tent above their heads
burning, as if, he wrote, "someone had thoughtfully soaked [it] with
oil." Their frantic calls fetched them assistance in time. But what
rankled most was that a chaplain, having asked Jack for his religion
and received a curt "None" in response, put him down as Church of
England. It is not obvious, Jack wrote later, "that military discipline
is improved when wounded men are insulted by attributing to them
opinions to which some of them object strongly."

For a week, Jack rested in a hospital in Amara, then was shipped
to another in Basra, where he lay near a man with a colostomy and a
man with no hand. When his wound refused to heal, the army sent
him to recuperate in Pune, giving him his first long draught of India.

IN ALL, JACK SPENT nearly a year and a half in India; later he would
remember it as the place where he lost most of his hair. From Pune,

as the summer rolled over the subcontinent, he went to Shimla, in the Himalayas, where he stayed for six months, researching and revising the British army's guidebooks to militarily sensitive areas. For a while after that, he moved to Delhi and then to the cantonment town of Mhow, where he ran a bombing school. He contracted jaundice there—because, he was sure, his witless commanding officer insisted on feeding the men meat and no vegetables in the hot weather. Once again, Jack was invalided to the Himalayas. In India, his twitchy curiosity came alive. He learned to speak and write some Urdu; he roamed the sandstone courtyards of Mughal palaces; he went to the Kumbh Mela near Allahabad, the giant Hindu pilgrimage that occurs once every 12 years; he watched meetings of the legislative council in Delhi, where the Indian leader Mohammad Ali Jinnah scored laughs and verbal victories but not an inch of impact on policy. India provided an opulence of experience and a potent political education.

Why, Jack wondered, were the British so profoundly uninterested in the country they held? Most colonists made no effort to learn an Indian language or to understand the religions of the land. During the Kumbh Mela, he believed, he must have been the only Englishman in an ocean of a million and a half pilgrims. The British despised Indians and were unaffected by their condition—a state of being that Jack thought both ludicrous and uncivilized. The Englishmen around him conversed at any length only to Indians who spoke English—who had served as officers in the army or who had university degrees. For a while in Shimla, he wrote to his mother, he joined a "gang of British and Indian people who occasionally meet for a friendly chat on not too controversial subjects." It was a mixed group—Sikhs, Muslims, Hindus—but they were all rich or successful, and even with them, he suspected they were interpreting his politeness as servility. In any case, he could only feel truly close to another human being when they could both presume to be equal to the other. In the relationship between colonizer and subject, that kind of equality was impossible. He knew the gulf was maintained deliberately, and he knew the British regarded India, above all, as an economic resource. But to his mind, in the absence of a genuine British passion for India, any possible intellectual rationalization for the imperial project evaporated. The Raj then became an exercise in exploitation, a machine to perpetuate iniquity. "We cannot wonder at any amount of discontent," he had written in Eton, in an essay describing the manner of England's rule of Ireland. The same, he was now learning, was true of India.

Jack was not immune to the strains of colonial thought—to preconceptions about what constituted "civilization" and "progress." He was glad that Indians were being made officers in the army, he remarked in a letter home, because it would give the martial races some education. "The Punjaub [*sic*], where half the Indian army come from, is still in a rather primitive state as regards literature. 60 percent of the books published are poetry." But he felt easily and tremendously moved by India, by the grace of its Islamic architecture and the emotional fervor of Hinduism, which crept over the pilgrims in Allahabad "like gusts of wind over ripe wheat." Under the British, the Indian genius had turned sterile, but it was impossible not to wish for its return to life and strength.

In July 1918, Jack sailed to England, via Aden and Egypt. He was to be trained in intelligence work and sent back to India, but before he could even leave London, the war ended, and he was duly returned to a civilian's existence.

IF HALDANE REFLECTED ON the war's effects on him, he never wrote of it anywhere. He formed plenty of broader conclusions, however, and these he aired. During his father's experiments with respirators, he had witnessed the ineptness of what he liked to call the "ruling class"—how, for instance, some unknown worthy sourced caustic soda instead of bicarbonate of soda, so that the young women assembling the first batch of useless respirators did so with raw, bleeding fingers; or how, after J. S. and his volunteers enfeebled their lungs with chlorine, the government awarded a Military Cross to the "young officer who used to open the door of the motor-car of the medical General who occasionally visited the experiments." The men born into the ruling class appeared to be insular know-nothings who didn't value their scientists enough.

"The ignorance of highly-placed persons" was accompanied by cowardice. Haldane came to look with contempt on those wielding temporal or ecclesiastical power who stayed safe but sent others into battle. Years afterward, he compared them to the Duke of Plaza Toro, the character in Gilbert and Sullivan's *Gondoliers* who, "In enterprise of any martial kind / when there was any fighting / he led his regiment from behind / (he found it less exciting)." Clergymen, he noticed, had tended to become army chaplains during the war, ranking alongside commissioned officers but running "the irreducible minimum of risk. . . . In my war experience, I never saw a chaplain display courage." It all showed how remote and weak the structures of the state and the church had become. When the revolution came, he wrote to his mother from the front, his lips practically smacking with satisfaction, the people would "strangle the last Duke in the guts of the last parson." From his fulminations about authority, he exempted only the officers who served with him. He admired their aspirations to courage and their willingness to die alongside their men. But if he had entered the trenches with any vestige of automatic regard for figures of the old establishment, nothing remained when the war finished.

The war was also Haldane's first immersive exposure, as an adult, to people outside his own, patrician class. In Mesopotamia in particular, at least half the officers around him had begun as privates, rather than being commissioned directly from the ranks of the aristocratic or the wealthy, and several officers came from a working-class background. Other British troops across various fronts cherished this diversity as well: "so many fellows together from all walks of life," as one soldier recorded, "some rich, some poor, never were there quarrels." Fighting a war was, in a capitalist world, the one true socialized activity for men, Haldane wrote later: "The soldier is working with comrades for a great cause (or so at least he believes). In peace time, he is working for his own profit or someone else's." Haldane never forgot that warming glow of camaraderie.

About the war's shocks to his psychology, though, Haldane was silent. Even his mother and sister realized they had to speculate at the state of his soul. He had thought, when it all began, that the

hostilities would simmer only briefly, but they had kept at a roll-
ing boil for four devastating years. Individuals turned brutal; states
turned amoral. "It has brought to light an almost incredible phenomenon: the civilized nations know and understand one another so little
that one can turn against the other with hate and loathing," Sigmund
Freud wrote in 1915. The intensity of that hatred, and the scale of
consequent death, proved bewildering and disillusioning, especially
for the men who had to manufacture that death.

Naomi, who knew Haldane better than anyone during those
years, was sure his personality had changed in some unknowable but
disturbing way. Indeed, given his experiences, it was impossible for
him *not* to have changed, she wrote in an unpublished manuscript:

> Jack had been killing people for several years, apart from the
> breaks which followed their attempts to kill him. That does
> something to shift one's personality on its base, as happens
> with a knock on the head or a bad trip with some powerful
> drug. Return is slow. Nothing any more looks innocent. It is
> no use thinking it was simply the war. In this last war [the
> Second World War] you could fly over a city, drop bombs on
> what you knew must be families clinging to their children,
> whom you were about to suck into a fire storm. But you
> had not seen them or heard them. You came back mission
> completed into the real world. . . . In trench warfare there
> was no coming back of that kind. You had to live among
> the smell and sight and thought of close death all the time;
> even a few miles back it was still there. Most of the . . . non-
> commissioned officers in the Black Watch were killed. In
> this kind of war they had to be leaders. But Bomber Haldane
> would creep out alone through the mud holes, throw his
> bombs and see bits of the men he had killed—arms and legs
> up in the air, perhaps hear shrieks of those he had mangled.

Once, nearly half a century after the war, Naomi saw her brother
letting a horsefly feed from the flesh of his hand, watching it patiently,
almost kindly. She asked him if the blood he was giving up was, in

any way, penance for the blood he had shed. "He disclaimed it so fiercely that it may have been true."

But Naomi was guessing, too. Her brother didn't talk to her about his time at the front; a distance crept between them, and their relationship strained and buckled. Maybe the experience of the war condensed his thinking, clarified his character, and liberated him to live boldly. Or maybe his sense of self was so robust that the great unhinging of the world brought no personal disillusionment or dejection. At least once, it occurred to Haldane that it was remarkable he hadn't died in battle. In 1961, he told Robert Graves: "It has always seemed to me plausible that I did not buy a return ticket over the irremeable stream, but have imagined events since 1915. They are now becoming rather outrageous, so perhaps I shall wake up." He had run the grandest risk of all, laying his very existence on the line. The rest of life held no terror to compare.

3.

Synthesis

THE REAL TROUBLE WITH HYDROCHLORIC ACID was that it was so damnably difficult to drink. A suitably strong pour would melt Haldane's teeth and burn through his throat. He tried a dilute solution, one part acid in 100 parts water, but after a pint of it, his stomach rebelled. In any case, a pint wasn't enough. At that concentration, he would need to wash down a gallon and a half of the stuff before the acid would begin to spread through his blood, as he wanted.

What happened when the blood acidified or when it turned alkaline? J. S. had learned that respiration was governed by carbon dioxide levels in the body, but he wasn't sure of the way this worked—of the way the body knew it had to breathe slower or quicker. Scientists suspected that the pH of the blood—its acidity or alkalinity – entered into it. Late in 1919, when Haldane joined New College in Oxford, he was set to the task by his father. It felt like a revival of the prewar, even pre-Eton, life, complete with self-experimentation. He could have acidified a rabbit, of course, but it was never easy to be sure how it was feeling.

Haldane learned to analyze his blood for gases. His lab skills had always been choppy, and he needed three months to gain accuracy with his methods. (Hitting a vein was simple, but in trying to pierce his radial artery, next to his pulse near the base of his thumb, he constantly pricked a skein of nerves; it triggered lancing pains in his palm, of the sort, he thought, that victims of crucifixion must have once experienced.) With the theory, he was, as ever, a master. When the body is laden with carbon dioxide, the blood floods with hydrogen ions from carbonic acid. It was these ions, Haldane and his father

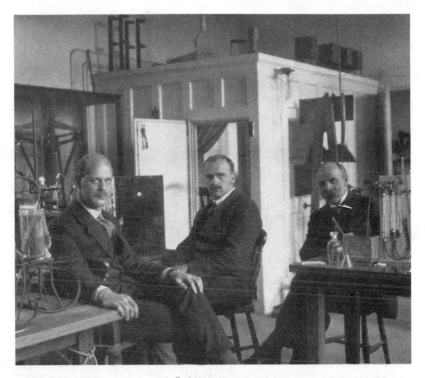

J. B. S. Haldane (middle) and J. S. Haldane (right) with their colleague H. W. Davies (left), c. 1920.

thought, that signalled the body to breathe faster—to dump the carbon dioxide and replace it with oxygen. If that correlation was valid, artificially turning the blood more acid must inspire quicker breaths; turning it alkaline must slow respiration down.

To make a man more alkaline was easy. Just after a large breakfast, Haldane and his colleague H. W. Davies ate a light snack of sodium bicarbonate. Their breathing grew leisurely. When they tested their blood, they determined an increase in its capacity to carry carbon dioxide, and the pressure of the gas in their lungs built as well, since they were exhaling it at such a sedate rate. By turning their blood less acid and drawing down the number of hydrogen ions in it, the scientists had meddled with the body's ability to gauge its carbon dioxide levels.

For the reverse experiment, to smuggle acid into the blood, Haldane tried ingesting magnesium chloride and strontium chloride.

The strontium chloride fuddled his senses. On it, he dreamed once that he was reading a life of Christ, written and illustrated by Edward Lear, but the only detail he could remember upon awakening was the moustache of Pontius Pilate. Finally, he worked out that the most sensible way to become more acidic was to ingest ammonium chloride. On six occasions, he drank solutions of the salt, its portions ranging in weight from 5 to 55 grams. The blood's carbon dioxide capacity and the gas's pressure in his lungs both fell. His body, reacting to all the hydrogen ions, worked hard to expel carbon dioxide and take in oxygen, and Haldane noticed "very marked air-hunger"— deeper and more rapid breaths, as if he was gasping after being held underwater. Twice, he remained short of breath for a week before his kidneys managed to hustle all the acid from his system.

Haldane and his colleagues wrote two papers on their results, and when they were published in the *Journal of Physiology*, scientists in Heidelberg read them. They had been studying babies with tetany, a disease often accompanied by rickets, in which the limbs and windpipe cramped involuntarily; the infants battled so hard for breath that their faces turned blue. In adults, tetany sometimes struck patients who had been treated with sodium bicarbonate or who had thrown up the acidic contents of their stomachs, so the scientists, E. Freudenberg and P. György, speculated that restoring the babies' acid levels might clear the symptoms of their tetany. "Unfortunately, one could hardly try to cure a dying baby by shutting it up in a room full of carbonic acid, and still less would one give it hydrochloric acid to drink," Haldane wrote later. But doses of ammonium chloride soothed the tetany in several babies within hours. This immediate practical application of his work delighted Haldane and endorsed his belief that experimenting on himself provided the kind of intimate knowledge that was unobtainable in any other way: "You cannot be a good human physiologist unless you regard your own body, and that of your colleagues, with the same sort of respect with which you regard the starry sky, and yet as something to be used and, if need be, used up."

Through his brief fellowship at Oxford and then in his early years as a reader in biochemistry at Cambridge beginning in 1923, Haldane kept up a regimen of self-experimentation. His trust in its value aside, it also locked neatly into the image he desired for himself. Haldane's entrance into a regular adulthood—insofar as anything about Haldane was regular—had been delayed by the whirl of the war; but now, in his 30s, the various streams of his experiences pooled within the basin of his character. His daring, which had won him so much adulation on the battlefield, expanded into flamboyance. His brilliance tipped sometimes into bombast and his plainspokenness into a brusque manner. His unorthodoxy, learned in part from his father, generated a love for the contrarian, for provocation and shock. "I am all for making people's flesh creep," he said in a speech years later. "It's great fun." Haldane relished the role of the odd genius, and he played it with knowing brio.

His presence was vivid and memorable. Julian Huxley, a friend from Eton, was at Oxford in the early 1920s and saw plenty of him:

> He dropped in whenever he liked—which was usually at tea-time—and devoured plates of biscuits, protesting that he couldn't eat a crumb, while reciting Shelley and Milton and any other poet you chose, by the yard. He had a fantastic memory and knowledge of the classics, and enjoyed displaying them. Once he went on reciting Homer so long that I had to escort him, spouting Greek all the time, downstairs to the front door. When at last the flow stopped: 'What a rotten memory I have got,' he said, and lumbered off.

In Cambridge, Haldane was regarded as a walking encyclopedia, but also as an academic guide of uncommon quality. James Murray Luck, a Canadian biochemist studying then at the university, wrote about the informal lectures Haldane delivered in his rooms in Trinity College. "How big do you think my liver is?" he would ask. The students, knowing that Haldane weighed around 100 kilograms, ventured estimates. "How much blood do you suppose I have?" Again, conference and guesswork. "How may one determine the blood volume?" And so on, question trailing question, answer begetting answer.

Haldane's demolition of social niceties was partially inspired by his view of university life, its gentility so distant from the horror of the recent war and the other realities of the world. The men around him seemed satisfied with "cultured" conversation, so he dropped his own thoughts into these exchanges like boulders onto porcelain: descriptions of the ways a man could be shredded by explosives, or stories about the color of entrails. Luck remembered how, in the midst of further experiments with acidosis, Haldane went for a dip in the River Cam and encountered a senior, newly knighted colleague and his wife, who were taking their guests out in a punt.

"I am now excreting the most acid urine that has ever been excreted," Haldane hollered, swimming under and around the boat.

"Yes, yes," the colleague murmured, rubbing his forehead in distress.

Aldous Huxley, Julian's younger brother, fed his imagination on the meatiest characters around him, and he couldn't resist the temptation that Haldane offered. In his first novel, *Crome Yellow*, he satirized a segment of the Bloomsbury circle; Haldane thought it catty. His fourth novel, *Point Counter Point*, published in 1928, would contain one phosphorus-obsessed Lord Edward Tantamount, who read admirable papers on biology and kept busy through the night in the laboratory built into his rambling house: J. S. personified. For *Antic Hay*, published in 1923, Huxley swiped Haldane himself.

The cynical, confused intellectuals of *Antic Hay*'s postwar London included Shearwater, a physiologist of ponderous build, with a big, domed forehead and a bushy brown moustache: "Wherever he was, Shearwater always seemed to take up the space that two or three ordinary people would normally occupy." He was apt to talk about the function of the kidneys or to sniff the air and declare: "Too much carbon dioxide and ammonia in there." In the throes of writing a paper, Shearwater began a paragraph with "The hydrogen ion concentration in the blood," an undisguised tilt of the head toward Haldane's two papers on acidosis. To measure the effects of profuse sweating on a human being, Shearwater set up a stationary bicycle in a heated chamber. Then he hopped on. Every time his assistant Lancing peeped into the chamber,

Shearwater was always at his post on the saddle of the nightmare bicycle, pedalling, pedalling. The water trickled over the brake. And Shearwater sweated. Great drops of sweat came oozing out from under his hair, ran down over his forehead, hung beaded on his eyebrows, ran into his eyes, down his nose, along his cheeks, fell like raindrops. His thick bull-neck was wet; his whole naked body, his arms and legs streamed and shone. The sweat poured off him. . . . Another time, Lancing reflected, they'd make the box airtight and see the effect of a little carbon dioxide poisoning on top of excessive sweating. It might be very interesting, but today they were concerned with sweating only. After seeing that the thermometer was steady, that the ventilators were properly open, the water was still trickling over the brake, Lancing would tap at the window. And Shearwater, who kept his eyes fixed straight before him, as he pedalled slowly and unremittingly along his nightmare road, would turn his head at the sound.

In his own, unpublished memoir, Haldane called *Antic Hay* "extremely amusing," but the fact that he wished to distance himself from Shearwater even decades after the novel's publication suggested that he was paying the compliment through gnashed teeth. What truly stung him was Shearwater's ignorance of love and his neglect of his unsatisfied, unfaithful wife. "I was not married when the book was written," Haldane clarified. "Shearwater is further endowed with a hopeless passion for a not really inaccessible lady. This is also contrary to fact." These felt like the weak rebuttals of a man who knew that an uncomfortable truth lay beneath the inaccuracies. Broad as the blade of Huxley's caricature was, he had somehow jabbed it through a chink in Haldane's armor of self-assurance and nicked a nerve.

HALDANE NEVER EXPLAINED WHY he finished a degree in the classics, went away to fight in the war, and returned to sink into the sciences. He did think that the biologist was the most romantic figure of his day, charged with a higher purpose as well as an unavoidable responsibility. Year by year, during this slice of the twentieth century, scientists were shaking nature loose of her secrets: the proton, the behavior of radiation, the tricks of relativity, the control of chemical reactions, the functions of nerves and glands and organs and hormones. Within this surge of discovery, biologists were circling the mysteries of life itself, and through the 1920s and 1930s, Haldane was at the vanguard of his field, publishing dozens of papers on physiology, biochemistry, and genetics. This would be his most significant, productive spell of research, helping to burnish Darwinian natural selection until it outshone other models of evolution.

Haldane's facility with numbers provided the scaffolding for nearly all of this work. A sturdy equation could hold a deep and undiluted truth, embracing a universe of specific cases. "If you're faced with a difficulty or a controversy in science," he liked to say, "an ounce of algebra is worth a ton of verbal argument." At Cambridge, for example, in considering the mechanisms of enzymes, Haldane reached for the quantitative. He hadn't previously thought of enzymes "for five consecutive minutes," he admitted, and he didn't have the leisure to begin learning experimental techniques. But the science of enzymes was still so basic and the manner of their action so hazy that Haldane's mathematics had a clarifying effect.

By the mid-1920s, biochemists had agreed that enzymes acted on their subjects—their substrates—by forming a brief chemical union with them. This was a chemical reaction, so it had to possess a rate and direction of action, a system of kinetics that governed how rapidly it rolled through each of its stages. Two scientists, Leonor Michaelis and Maud Menten, studied the enzyme invertase and how it catalyzed the transformation of the complex sugar sucrose into the simpler sugars glucose and fructose. They assumed that invertase and sucrose formed an intermediate product first, and they provided a formula: a way to link the rate of the reaction with the concentrations of enzyme and substrate. But Haldane spotted a caveat in the Michaelis-Menten model. It assumed that the reaction from substrate to intermediate was in equilibrium, proceeding in forward as well as reverse directions at equal velocities, and that the rate of formation of the final products was slow compared with the equilibrium velocities—something that was true of the invertase-sucrose example but not, Haldane saw, of every enzyme reaction. With a colleague, George Briggs, he assembled a more general version of the equation, and the Briggs-Haldane equation continues to influence the design of enzyme experiments today.

The form of Haldane's work on enzyme reactions characterized nearly all his research in his 30s and beyond. He did no lab work himself; Michaelis and Menten had been the ones to conduct and monitor the original reactions between invertase and sucrose. Rather, Haldane seized on the experimental data of other scientists, and with an uncanny eye, he recognized in the data a problem, or a pattern, or an opportunity. He built equations: basic, to begin with, but then with complication after complication layered upon them, the variables tweaked to mimic the conditions of the real world—the conditions that *other* experimentalists had observed in *their* data. What happens if the products of an enzyme reaction themselves react with the enzyme? What happens if two substrates compete for an enzyme, or if the concentration of the substrate is so high that it inhibits the activity of the enzyme, or if the enzyme is destroyed during the reaction? Haldane considered all these cases, and others, and provided mathematical ways to resolve them. His equations weren't elegant; he ground them out and shoved them into the world in an unrefined state. (Ronald Fisher, Haldane's great contemporary in evolutionary biology and a paragon of mathematical polish, once referred disdainfully to "all that algebra that Jack seems to find necessary.") But if his equations lacked beauty, they were still clear, effective, and full of insight. Haldane was a synthesizer. From upon a hill, he surveyed the terrain of biological knowledge, noticing how different areas were connected—or might be connected better.

As it turned out, at the time, a synthesizer was just what genetics needed.

Ever since the rediscovery of Mendel's theories, in 1900, a rift had opened in biology. Mendelians, like William Bateson, Thomas Morgan, and Reginald Punnett, believed that evolution drew all its energy from mutations in genetic particles. These mutations could produce small as well as large changes, but the effects were always discontinuous: Mendel's peas were either wrinkled or smooth; there was no gradation of textures in between. Moreover, a mutation arose spontaneously, through some as-yet-indistinct process, affecting an organism's fitness beneficially or adversely, so that it either prospered or struggled in relation to its environment. Natural selection rewarded an increase in fitness: the population grew fitter, and a species evolved in a single adaptive jump.

Darwinians, on the other hand, harked back to the small variations that *On the Origin of Species* had first proposed and to the slow trudge of evolution that Darwin had envisioned. In his theory, variations could only ever be small—dilutions that grew potent over many generations. Here natural selection acted not as a sieve of the fit (as the Mendelians believed) but almost as a force of creation: slight variations that might prove beneficial could accrete in the population and lead to a superior fitness. The birth of a new, fitter species was thus an impalpable process. By the early 1900s, the appeal of the blending idea—the paint-pot theory—had worn away. But the notion persisted that genetic material changed in gradual, imperceptible ways, so that a range of variations accumulated within a population. For Darwinians like the statistician Karl Pearson, the variations in a single trait could be mapped onto a smooth bell curve. Men, for instance, were rarely very short or very tall, and their heights mostly occupied the belly of the curve, right in the middle. (For Mendelians, the diversity of heights represented mere "fluctuations," owing mostly to environmental factors—like diet, in the case of height—and so impossible to inherit.) "*Natura non facit saltum*," Darwin had written: "Nature does not make jumps." Darwinians held that motto close; Mendelians rejected it. The rift—and the consequent inability to explain how evolution worked—was endangering the very theory of natural selection. A limerick, its author lost to time, ran as follows:

Karl Pearson is a biometrician
And this, I think, is his position:
Bateson and Co.,
Hope they may go
To monosyllabic perdition.

Viewed from the twenty-first century, the Mendelians and Darwinians appear closer to each other than they thought they were at the time. Both camps trusted natural selection; their differences concerned only how it gave rise to new species. Speculative compromises were broached. Bateson suggested at least once that the bell curve of variation could be squared with Mendelian principles, and Pearson eventually came to accept Mendel and the premise of discontinuous variation. Nevertheless, the scientists seemed unable to knit their ideas tightly together. They were too wedded to their methodologies, the Mendelians to their experiments with fruit flies and primroses, the Darwinians to their recitations of statistics and biometrics. The debate churned biology. In 1922, delivering a morose speech to the American Association for the Advancement of Science, Bateson dwelt on the perplexities around evolution. "When students of other sciences ask us what is now currently believed about the origin of species we have no clear answer to give. Faith has given place to agnosticism," he said. He worried that in all this uncertainty, the "enemies of science"—Lamarckians or the disciples of orthogenesis—would spot their chance and push forward their views, obscuring natural selection altogether. But a path would clear, Bateson predicted—a dab of comfort. The theories would come together: "That synthesis will follow . . . we do not and cannot doubt."

In fact, by the time Bateson made these remarks, the modern synthesis—as it came to be known—was already sprouting. Four years earlier, Ronald Fisher had published a paper showing how numerous discrete changes in genes—Mendelian changes—could yield a smooth, Darwinian curve of continuous variation. Mendel had suggested this himself, and others had solved special cases of this principle, but Fisher explained how it might apply as a general rule. (His manuscript was turned down by a team of referees that included Pearson, on behalf of the Royal Society in London; Fisher submitted it to the Royal Society of Edinburgh and, out of pique, detested Pearson thereafter.) Mendelians had assumed, as the monk himself had, that genetic factors worked as binaries: Peas had one allele for a wrinkled coat and another allele for a smooth coat, and the distribution of these in the offspring, as well as their dominant and recessive qualities, directed the new plant's peas to be either smooth or wrinkled. But Fisher reasoned that in a trait such as height—the Darwinians' pet subject—a host of genes, each supplying a small tweak to the trait, could produce a range of heights, the bell curve that was the Pearsonian ideal. Not only did Mendelian inheritance explain continuous variation, Fisher argued, but it was the best of *all* possible explanations. The union of Mendelians and Darwinians had begun.

HALDANE'S FIRST MAJOR PAPER in evolutionary biology after the war was published in 1922 in the *Journal of Genetics*, and it laid down what came to be known as Haldane's rule. He had gathered a sheaf of studies from experiments with hybrid animals, in which scientists had crossed one species with another to observe their offspring. The literature drew from all over the animal kingdom: guinea pigs mated with shiny guinea pigs; chickens with pheasants; wood grouse with black grouse; the puss moth with the lesser puss moth; cows with yaks. The tables of data spoke of a veritable orgy of cross-breeding.

Frequently in such crosses, one sex in the first generation turned out to be rarely born or was born sterile or was absent altogether. Whenever this happened, Haldane noticed, it was the heterogametic sex—with two different sex chromosomes—that was affected. In mammals, the male sex is heterogametic, inheriting an X chromosome from the mother and a Y chromosome from the father. (The female mammal, inheriting an X chromosome from each parent, is homogametic.) In birds, moths, and butterflies, conversely, the female sex is heterogametic. Haldane's rule, as he phrased it, was a keen observation rather than a link between cause and effect. In his paper, he submitted the broad possibility that as a species passed through an evolutionary fork, its two tines grew genetically different—at first gradually, when they could cross to produce healthy but sterile children; then significantly, when heterogametic offspring were rare or died before breeding age; and then vastly, when they failed wholly to produce any offspring. Into a single holding principle, Haldane scooped nearly 2,000 years of biological reasoning, dating back to Aristotle's nagging puzzlement over the sterility of the mule. Haldane's generalization has become one of the most reliable patterns in the formation of new species, and for a rule propounded a century ago, it has run into a remarkably small number of exceptions. It has also kept scientists busy; no single definite theory yet exists to describe why Haldane's rule is so persistently valid.

The paper lit Haldane's hot streak of publications on genetics: 10 papers over nine years, beginning in 1924, that helped invent the discipline of population genetics and cast the tools of its kit. The very first sentence of the very first of these papers sounded his ambition like a bugle: "A satisfactory theory of natural selection must be quantitative." It wasn't enough to just believe that Darwinian selection influenced a Mendelian change in a species. The truest validation would come only when scientists modeled the rate of that change and found that it agreed with the transitions of plants and animals in the real world. Haldane prized his ounce of algebra, but he knew it was worthless if nature defied its predictions.

AT NIGHT, WHEN *BISTON BETULARIA* does most of its flying, it resembles a small, pale, anxious ghost. Its wings are a creamy gossamer, with spots of black shaken over them: hence its common name, the peppered moth. A male peppered moth takes wing late every evening in search of a mate, the impulse to reproduce defeating even the threat of being consumed by birds or bats. During the day, it rests, motionless, high up on birch trees, whose grey bark grows so mottled with lichen that the moth blends in perfectly. The peppered wings are lifesaving camouflage, deceiving the eye of the bird passing in a hurry. When R. S. Edleston of 5 Meal Street, Manchester, went out on entomology sorties in the 1840s, the peppered moths he caught all had the regular speckled wings. Then, one day in 1848, he saw one moth that was bat-black. Other lepidopterists in Manchester started to report black moths as well, the pepper on their wings thoroughly submerging the salt. "Last year, I placed some virgin females in my garden in order to attract the males, and was not a little surprised to find that most of the visitors were of the 'negro' aberration," Edleston wrote in 1864 in a letter to the newly launched *Entomologist's Monthly Magazine*. "If this goes on for a few years the original type of *B. betularia* will be extinct in this locality."

Sightings of the uncommon black moth—the *B. betularia dou-bledayaria*, as it was then named—erupted in the magazine's pages over the following decades. Like a touring theater company, the *doubledayaria* variant made a string of heralded appearances across England: Cannock Chase in 1878, Berkshire in 1885, Cambridge in 1892, Norfolk the next year, Suffolk in 1896. Along with these anecdotes—for there was no real census of these moths, no data to speak of—came trains of speculation. Was it the humidity that was turning moths black? Was it a change in temperature? It wasn't until 1896 that a lepidopterist named J. W. Tutt suggested natural selection. The factories of the Industrial Revolution—Blake's dark Satanic mills—had poured smoke into the green and pleasant land. The sulfur dioxide killed the lichen on the trees, and the soot darkened their trunks and branches. The peppered moth now suddenly shone off its tree; it was *B. b. doubledayaria*, the crazy variant, that blended into safety. "The paler ones the birds eat, the darker ones escape," Tutt wrote. "Year after year it has gone on, and selection has been carried to such an extent by nature that no real black and white peppered moths are found in these districts but only the black kind."

Here was a perfect test case for biologists, trussed and dressed and ready for inspection. Haldane pulled it into the first of his 10 papers, a dense consideration of 13 varying scenarios of natural selection. Using the variable k to connote the degree to which a trait benefited an organism under natural selection, Haldane proceeded through his hypothetical situations, teasing out the mathematics for each one. How many generations, for instance, would a dominant, beneficial allele require to spread through a population if its k was .001—a low, slow selective advantage, in which 1,000 of its kind survived versus 999 of a weaker allele? What happened with self-fertilizing plants? Or in cases where a trait appeared in one sex only? These conditions mattered. They affected the speed with which a population changed, the speed with which an allele's frequency of appearance in the population rose. In the simplest of these scenarios, for example—with a dominant gene, its appearance unconnected to sex—Haldane calculated that if its k was .001, the gene needed 6,920 generations to pass from a frequency of 0.001 percent to 1 percent; 4,819 more generations to pass from 1 percent to 50 percent; 11,664 further generations to pass from 50 percent to 99 percent; and then 309,780 generations to furl through the final 1 percent of the population.

The fate of the peppered moth was a live model of this scenario: a basic Mendelian setup in which the all-black tint was the result of one dominant allele. Haldane could work out how quickly the ratios of moths had changed; *B. b. doubledayaria* didn't form more than 1 percent of the population in Manchester in 1848, but within half a century, it was the only peppered moth variant to be found. The moths reproduced once every year. From these numbers and his equations, Haldane worked out his *k* here to be .332. It was as if the *doubledayaria* variant gained such an advantage over its paler variant that it was able to produce 50 percent more offspring. *B. betularia doubledayaria* swamped the ordinary *B. betularia*.

The moth population transformed in such haste that some scientists had believed the environment to be acting directly on its genes, mutating them. Not so, Haldane wrote; for *B. b. doubledayaria* to become so widespread in this way within 50 years, one recessive gene in every five moths would have had to mutate—a rate that wasn't supported at all by experiments that had been done in the 1890s. Besides, Haldane's paper showed, such macromutations weren't necessary. Natural selection itself wielded tremendous power, and although it often worked slowly, it was also capable of a scorching pace. The *doubledayaria* variant was living, fluttering proof of that.

But it didn't persuade everyone, or not immediately. For a long while, some scientists dismissed natural selection as the force behind the shifts in the moth population. The year after Haldane's paper was published, one experimenter tried to induce melanism in a lab batch of moths, pushing pollutants like hydrogen sulfide, ammonia, and pyridine down their breathing tubes. Another botanist, John William Heslop-Harrison, an ardent Lamarckian, argued that the manganese and lead salts in Britain's smoky air were deforming the genes of the moth. He had, he claimed, fed moth larvae on leaves dusted with these same salts and had produced coal-black *doubledayaria* out of them. But the experiment was flawed and perhaps even forged.

(Later, Heslop-Harrison was discovered to have deliberately planted certain species on a remote Scottish island to prop up another of his theories.) Finally, in 1953, an Oxford researcher named Bernard Kettlewell led a study that endorsed Haldane's proposal. Birds were the peppered moth's prime predators, Kettlewell confirmed, and the moths were picked off selectively, depending on the efficiency of their camouflage against their trees. Against white bark, *B. b. doubledayaria* ran scarce; the numbers of the more conspicuous moth were always in decline. It had taken three decades for experiments to catch up to Haldane's paper, although by then, the matter felt like a formality—a tug pulling an ocean liner the final few meters into its harbor. The mathematics had progressed far further, natural selection had found other validation, and population genetics had become a bulwark of biological thought.

THERE ARE FEW CONCRETE EXAMPLES of the peppered moth kind in the nine remaining papers in Haldane's series. Instead, the mathematics grows increasingly conceptual, addressing itself to scenarios that Haldane devised: inbreeding or self-fertilization, in which selection plays out with enhanced effect; cases in which multiple genetic factors act on a trait; instances from the plant world in which generations overlap, so that an offspring fertilizes its parent. With systematic patience, Haldane took apart the natural world, examined its various schemes of reproduction, and calibrated the workings of selection in those circumstances.

He then threaded mutations into his calculus. Modern biology recognizes a mutation as a change in the genetic code, which occurs during DNA replication or repair, as a result of an external agent like radiation, or even spontaneously, through a chance shuffle of molecules. Haldane knew it only as a phenomenon that produced a new genetic factor. A mutation could affect both dominant and recessive alleles. Haldane showed that mutated dominants are more frequently "fixed" into a population—passed on with greater fidelity to subsequent generations—than mutated recessives. This preference of natural selection came to be called "Haldane's sieve."

A mutated dominant that proves strongly advantageous can still be snuffed out by sheer probability. Haldane revealed that even if a mutated variant helps an organism have twice as many offspring as its peers—a muscular advantage—the probability that the variant gene flickers out in subsequent generations is .203. Mutation isn't always destiny. Haldane also showed how inbreeding between a brother and a sister gives recessive mutations better chances of survival, which contributes to the diseases afflicting the children of siblings. He considered several special cases in which populations moved around. One scenario envisioned an event of intense selective pressure, such as a famine or a plague. Another examined the curious case in which two mutated recessives were individually disadvantageous but beneficial in tandem.

All his life, Haldane insisted he was profoundly unmusical, but there was something jazzlike in the way he went about the mathematics in these papers. He laid down his tune—his fundamental equation—and then riffed off it, flattening a note here or bending a chord there, shading in syncopations, pulling the melody through all its alternate lives. He was methodical, even plodding, in his explorations, but his simplest premises grew fuller and richer over his 10 papers, approaching the complexity of the real world. By explaining the potency of Darwinian natural selection in so many scenarios, he showed how it could account for the dramatic diversity of nature.

In 1932, Haldane assembled much of his material into a book, *The Causes of Evolution*, which remains a classic of its genre. He included an epigraph,

Darwinism is dead.
　　—Any sermon

and then proceeded to show how Darwinism was, in fact, vital and alive. It was a prototypical Haldane tract, stocked with casual nods to the Book of Daniel and William Blake and Vladimir Lenin and with untranslated German verse, ranging easily over the history of the world but pausing just long enough on practical examples involving primroses or apples. The book was an ode to the power of Darwinian selection. "However small may be the selective advantage,"

Haldane wrote, "the new character will spread, provided it is present in enough individuals of the population to prevent its disappearance by mere random extinction." This single sentence could have been the creed of the modern synthesis.

The Causes of Evolution is slim but capacious, and, sensibly, Haldane allowed himself to be flexible on a number of fronts. For one, he admitted that natural selection is only the most important driver of evolution, not the sole force behind it—hence the "Causes" in his title. He pointed out, more than once, the failures of Lamarck's theory of the inheritance of acquired characters. But he yielded some space to mutationism, the sudden appearance of new forms—"hopeful monsters"—by hybridization or drastic genetic mutations. "When they have arisen, they must justify their existence before the tribunal of natural selection, but that is a very different matter."

By accepting the possibility of large, discontinuous changes within the framework of natural selection, Haldane anticipated the theory of punctuated equilibrium, which the paleontologists Stephen Jay Gould and Niles Eldredge proposed 40 years later. If the fossil record between one species and its markedly different successor showed no transitional forms, Gould and Eldredge argued, that didn't always mean the record was incomplete. They explained with a diagram of *Gryphaea*, a gnarled oyster from the Triassic period, whose fossil record showed an abrupt split. One new form sported baroque coils in its shell, while another grew flattened. In this way, a lineage in evolutionary stasis may diverge with suddenness; a new species may arise rapidly.

Haldane also thought that natural selection could, on occasion, deplete an organism's fitness for its environment. If a population grows large and dense, then the rules shift, and its members use their selective advantages to compete—with disastrous results. "The geological record is full of cases where the development of enormous horns and spines . . . has been the prelude to extinction," he wrote. "It seems probable that in some of these cases the species literally sank under the weight of its own armaments." Once, when a student tried to explain an evolutionary trait by its benefit to its species, Haldane scrawled in the margin: "Pangloss' theorem." It was a criticism. To believe, as Voltaire had Pangloss believe, that "all is for the best in the best of all possible worlds" was a folly of evolutionary thought.

Having set up this model—of changes that aided individuals but impaired their population—Haldane looked to the obverse: "socially valuable but individually disadvantageous characters." There were men and women—Christian saints, for example, or winners of the Victoria Cross—who behaved in altruistic ways that would lop short their own lives just to protect those of others. Darwin had figured that altruism was a losing strategy in the grander struggle for existence; evolution would not favor the soft of heart. But maybe, Haldane speculated, altruistic behavior was a kind of Darwinian fitness after all. He never followed up on this surmise. Had he been drawn from the robust lineage of British naturalists, his student John Maynard Smith thought, Haldane would have augmented his idea with field studies or at least with further curiosity about the biological nature of altruism. But Haldane's notions of altruism were distilled out of literature and mythology, Maynard Smith wrote, so he simply tossed these remarks into the ether. Later, they were captured and mulled over by other scientists, who turned them into contested theories of kin selection and group selection, by which altruism helps protect an organism's genetic relatives or its wider population.

"I can write of natural selection with authority because I am one of the three people who know most about its mathematical theory," Haldane wrote in *The Causes of Evolution*. The other two were Ronald Fisher, at the Rothamsted Experimental Station in Hertfordshire, and Sewall Wright in the United States. Their work was built on common bedrock: the deep, measurable capacity of natural selection to remake a population. Their mathematics varied in technique, but there were also philosophical differences in their theories. Fisher thought that evolution relies almost entirely on natural selection and that natural selection is mostly deterministic, affected very little by randomness. In Fisher's world, organisms are continually changing, their genes trying to improve their evolutionary fitness to adapt to their forever-shifting environment. (British evolutionists were always hung up on adaptation, Maynard Smith once remarked, perhaps because of their country's lengthy history of natural observation. "I think it may be that, if one watches an animal doing something, it is hard not to identify with it, and hence to ascribe a purpose to its behaviour," he said.) But Wright believed that in addition to the demands of the environment, chance also plays a crucial role in the fate of fit genes. Particularly in small, isolated groups, the probabilistic vagaries of reproduction and survival—of the business of life—mean that a population could experience, at least temporarily, a setback or stagnation in its quest for fitness. Wright called the phenomenon "genetic drift."

Haldane landed roughly between the positions of Fisher and Wright. In *The Causes of Evolution*, he briefly acknowledged the influence of chance, but only in the cases of large, freak events. The fruit of the tropical *Shorea* tree ordinarily falls 100 yards away from its parent, but a typhoon could carry it 100 miles. A great migration might leave behind the unlikeliest of survivors, which might not otherwise have proven fit enough in the normal course of events. Independent of Wright, Haldane also concluded that the genetic changes in small, isolated groups aren't always neatly predictable. "Such changes, while they must ultimately stand the test of natural selection, are not themselves due to natural selection." And most species did, after all, exist and evolve in pockets, even if the fossil record delivered the opposite, false impression by throwing up the remains of large populations. "It is a striking fact that none of the extinct species, which, from the abundance of their fossil remains, are well known to us, appear to have been in our own ancestral lines. Our ancestors were mostly rather rare creatures," Haldane wrote. And then, to drive the point home, a quick, sly, Haldanian reference from literature: "Blessed are the meek: for they shall inherit the earth."

The relationships between Haldane, Fisher, and Wright were uneven. Wright, the son of an economist, was a quiet, modest man, and he and Haldane mutually admired and liked each other. Wright's 1931 paper, his most important investigation of evolution until then, was published after the bulk of *The Causes of Evolution* had already been written, but Haldane made sure to discuss the paper in an appendix. He was always prompt to credit, and to publicly approve of, the work of others. He was generous with Fisher as well, calling his book, *The Genetical Theory of Natural Selection*, brilliant and professing himself indebted to it—this, despite Fisher's adamant refusal to cite any of Haldane's work at all in the book. (It rankled. In Fisher's complicated papers, "which I was never brought enough to fully digest," Haldane once wrote sarcastically to an acquaintance, "he completely ignored my elementary contribution." In his letter, Haldane was pointing out an error by Fisher, "who is such an expert in higher mathematics [but] evidently cannot do simple addition.") Fisher had grown up in a family that, in his teenage years, pitched from wealth into poverty. He was perennially insecure, a devoted Anglican, and a Conservative—a ready match for Haldane's intellect, but his opposite in nearly every other way. They squabbled frequently. Fisher was always armed with a quiver of sharpened insults, and although they bounced off Haldane's thick hide, they got under Wright's skin. Fisher and Wright detested each other so much that they ceased to speak altogether. When Fisher visited the University of Wisconsin, one student remembered, "we were always careful to bring him into the building when Wright did not happen to be in the hallway. Wright was the gentlest of men, except in one regard—Fisher."

Out of this triangle of complicated individuals and their complicated mathematics, a radical new approach to evolutionary biology was born.

DURING THESE YEARS, in his pomp, Haldane seemed blessed with a superabundance of time: more time to write and think, more hours in the day than were ordinarily allocated to anyone. In 1927, while still at Cambridge, he accepted further responsibility: a part-time appointment at the John Innes Horticultural Institution, in the borough of Merton, just south of the Thames. The director of the institution, Daniel Hall, recruited Haldane for a salary of £400 a year, promising that he would succeed Hall very soon. Every two weeks, during Cambridge's terms, Haldane visited John Innes for a day and a night and additionally worked there over Christmas, Easter, and part of the summer. He commandeered a desk next to a young biologist named Cyril Darlington, and they spent nights discussing science, philosophy, and politics. Haldane was already a giant figure; his reputation barged into a room long before he set foot in it. "For about seven years," Darlington later said, "I regarded him as my infallible mentor. He took the place of my father and of Newton in my still immature mind."

John Innes was, when Haldane arrived, an institute tossed by change. Its founding director, William Bateson, had died the previous year, and Hall was a puzzling substitute, so poorly schooled in genetics that everyone assumed him to be a stopgap. (Initially, Hall told his colleagues that he only planned to "have a look round each day.") Bateson had run John Innes with a light, informal hand, with no strict divisions of work or contracts or terms of employment. The books were unbalanced. The costs of maintaining the institute—with its dozen acres of farmland and nine greenhouses, its labs and insectarium, its library and lecture theater, its 15 scientific researchers and 50 other workers—were rising. Staffers kept leaving, and no new hires filled these vacancies.

Into this climate of mild chaos, Haldane brought his knowledge of genetics but also his stature. (The staff, Darlington remembered, felt an "intense relief" at Haldane's appointment, sure now that John Innes would be "sheltered by his powerful figure.") He also brought his irreverence. One year, under the pen name of "John Boredom Wanderson," he coauthored a mock paper "On the Permeability of Colleges," using reagents like "100 per cent American tourists [alcohol free]" and "Normal undergraduates." (The authors concluded that "no colleges were ever permeable to omnibuses or hippopotami, and they were always permeable to sparrows" and that "in the presence of even traces of alcohol, undergraduates were usually precipitated from the membrane.") Another year, he set a genetics exam that demanded of students:

1. Give an account of lethal genes, with special reference to Sir James Jeans and Miss Ursula Jeans.
2. Discuss the genetics of local races, illustrating your answer from the University boat race. Account for the dominance of light blue over dark blue . . .
3. Distinguish between hermaphroditism, intersexuality, and gynandromorphism. Which is most prevalent in Middlesex?

To Brenhilda Schafer, the John Innes secretary, whom he described as "a walking bibliography of cytology," Haldane left a note. He needed some things; could she procure them for him?

1. Insignia of Order of Holy Ghost
2. Mr. Ford's bank balance
3. Clara Bow
4. Information on all trains between 11 a.m. & 5 p.m. from London–Paris
5. £5 (cheque cashed)
6. 4 safety-razor blades (Star or ever ready) from Woolworth
7. If you can manage, a box of matches
8. Karl Marx's "Kapital" (Everyman's Library)

John Innes needed a geneticist, to be sure, but it wasn't clear to everyone that Haldane was the kind of geneticist it needed. His research didn't consist of dogged rounds of breeding experiments; the last such studies he had run was when Naomi and he had raised guinea pigs in the hutches in Cherwell. John Innes was a place for practical genetics: its stock work consisted of cultivating primroses and tulips and crocuses, crossing and backcrossing them, tracking their traits through generations, and staining chromosomes to be squinted at under microscopes. Haldane didn't bother to mask his indifference to the horticultural aspects of John Innes's business. A colleague would introduce Haldane to some plants of note in his plot, only for Haldane to growl out some critical remark or say nothing at all before stumping off in another direction. While Darlington drew chromosomes next to him, Haldane did sums, scribbling across thousands of sheets of paper. There wasn't a calculating machine, or a slide rule, or even a book of logarithms in sight; he just ran his long divisions by hand, the digits spilling slowly, diagonally, toward the edge of the page. "And then there would be a break for lunch," Darlington recalled. "Bread and cheese, and the fruits of the garden, and coffee and talk. He was a good talker."

Haldane did have ideas for how to join theory to practice and how to knit one discipline into another. For the better part of a decade,

The *Primula sinensis* glasshouse at John Innes, c. 1930.

he had been thinking not only about how genes were inherited but how they worked. The body was formed and sustained by acts of bio chemistry, and if genes were an organism's master blueprints, they must control these biochemical processes. A French scientist, Lucien Cuénot, had bred mice and speculated, in 1903, that three genes produced a pigment or an enzyme each, the products interacting to determine if a mouse was black or yellow or an albino. A few others had suggested similar functions of enzyme manufacture for genes, and Haldane liked this theory. "We have very strong evidence," he wrote in a paper in 1920, "that [genes] produce definite quantities of enzymes, and that the members of a series of multiple allelomorphs produce the same enzyme in different quantities."

In the latter hypothesis—that the alleles of a gene differed by how much or how little of an enzyme they produced—Haldane was wrong; he rejected it himself later. But he pushed forward the concept that genes expressed themselves by building and releasing enzymes.

"The enzyme is a product of the gene, not the gene itself," he wrote in an unpublished manuscript in 1931. Then, in *The Causes of Evolution*: "To my mind it is probable that every gene produces a definite chemical effect, but we are far from being able to prove this as yet." The "one gene, one enzyme" theory, constructed out of the work of numerous scientists, has now proved to be too simplistic. Genes express themselves, alone or in cahoots with others, in multiple ways, some of which have yet to be recognized. But the notion of the biochemical handiwork of genes had to be raised first, before it could be studied and expanded into the modern field of molecular genetics.

John Innes was the ideal place to appraise the practical results of genetics and biochemistry. It held plot after plot of plants, their ancestries and breeding histories all carefully recorded, and their phenotypes—their heights or the colors of their flowers or their other features—plainly visible. Further, the institute had a crew of researchers who were adept at crossing plants for experiments. Many of these were women—unusual for the time, but Bateson had always collaborated with women, and he drafted several into John Innes from horticultural and agricultural colleges. The so-called "Ladies Lab," a segregated space at the opposite end of the institute from where the men worked, wasn't popular with everyone at John Innes. The women were regarded by some as glorified gardeners. Darlington, in particular, grumbled about them to himself; in his diary, he suggested that Bateson merely wanted to be surrounded by "suitably submissive and even adoring females" and that they dragged down standards "to a more leisurely age when science was the work of amateurs." But Haldane found his closest colleagues among the women. From Cambridge, he brought down a biochemist named Rose Scott-Moncrieff, and he also worked with Dorothea de Winton, who cultivated Chinese primroses, and Alice Gairdner, a cytologist studying flax and wallflowers. The women ran their own experiments, although Haldane assisted in designing some of them and analyzing their results.

His various roles in these studies showed how Haldane grew deeply absorbed even by small, discrete units of scientific research and how he valued the incremental progress of knowledge. Scott-Moncrieff was trying to puzzle out how genetic differences adjusted the effects of anthocyanin pigments, which turned flowers of one plant red and another purple or blue. (Haldane, correctly, suggested that the acidity of the petal sap—a biochemical agent controlled by genes—helped calibrate the shade of the pigment.) De Winton cataloged as many as 40 mutant versions of the Chinese primrose, pasting samples of petals and leaves onto beige record cards. She and Haldane published papers on how certain alleles correlated with the presence of particular anthocyanins and on evidence of genetic linkage in the plant. With Haldane, Gairdner discovered that two linked genetic factors were, in certain combinations, lethal to the snapdragon plant, and they worked out the probabilities of such combinations occurring. Every discovery, however limited, was a fresh card overturned in the quest to reveal nature's vast hand.

At the same time, Haldane's sense of the monumental never faded. In 1929, in an article for *The Rationalist Annual,* he brought together his knowledge of physics, genetics, biochemistry, and physiology to advance a viable theory for how life was born on Earth. This wasn't his realm of study in any focused way. Yet within eight pages, written for the lay reader in a vein of such smooth informality that he even slipped in a silly poem, Haldane thought up a functional model for the origin of life.

In the 1800s, a number of ideas had tried to explain the genesis of the planet's organisms: spontaneous generation from inanimate material, for instance, in the way that maggots appeared to arise from dead flesh; or panspermia, which held that basic forms of life had drifted in from outer space. Darwin thought that human knowledge hadn't expanded enough to consider the question properly. "It is mere rubbish thinking, at present, of origin of life," he wrote in a letter. "One might as well think of origin of matter." But he wasn't above speculation—about "some warm little pond, with all sorts of ammonia and phosphoric salts, light, heat, electricity, etc., present, that a protein compound was chemically formed."

By the twentieth century, several of these hypotheses had rusted and dropped away, but nothing plausible had replaced them. So Haldane made some suggestions. The atmosphere of early Earth would have plenty of carbon dioxide but no oxygen—and, by extension, no ozone, so ultraviolet rays from the sun could reach the land and sea unhindered. A laboratory in Liverpool, Haldane knew, had shown how ultraviolet rays acted on a mixture of water, carbon dioxide, and ammonia to form organic compounds. Were that to happen anywhere today in Darwin's "warm little pond," these compounds would be consumed by microorganisms. "But before the origin of life they must have accumulated till the primitive oceans reached the consistency of hot dilute soup," he wrote—using, for the first time in science, the word "soup" to describe the bath of nutrients that might have yielded first life. Without oxygen, the first "half-living" molecules probably drew their energy from fermentation, and they reproduced by sheer chance, coaxed along only by the sun gently cooking them.

Perhaps life remained marooned in the virus stage for millions of years before the first unicellular organism formed; perhaps many preliminary drafts collapsed before one filled out into a cell with working, cooperating parts. That event, the cellular transition, probably only happened once, Haldane wrote, or perhaps the descendants of that cell swamped any other contenders. But that single, successful cell was sufficient to kindle the creation of all life, passing from anaerobic organisms through photosynthesising beings to aerobic, or oxygen-using, bacteria. Scientists think today that Haldane was wrong on this point, that biochemistry might have brought forth life several times. But Haldane was guessing, and he admitted it: "The above conclusions are speculative. They will remain so until living creatures have been synthesized in the biochemical laboratory. . . . But such speculation is not idle, because it is susceptible of experimental proof or disproof."

A similar thread of logic had run through a monograph, "The Origin of Life," published in Russian in 1924 by a biochemist named Alexander Oparin. Haldane hadn't known of this paper when he wrote his own article, and the theory came to be known, in joint posterity, as the Oparin-Haldane hypothesis. Much later, when he first met Oparin during a conference, Haldane said, with typical generosity: "I have very little doubt that Professor Oparin has the priority over me. . . . There was precious little in my small article which was not to be found in his books."

In 1953, a 23-year-old graduate student named Stanley Miller told a seminar audience at the University of Chicago that he had roughly recreated the conditions of Oparin-Haldane's early Earth. In a 5-liter flask, he had sealed water, methane, ammonia, and hydrogen. Then Miller had simulated lightning, as a source of energy, by passing sparks of electricity within the flask. After just a week, he was able to identify 5 amino acids—the constituents of proteins— that had formed out of the experiment. Modern methods, working off Miller's original samples, have discerned at least 22 amino acids, including some that were never even known in Miller's time. It felt like wizardry, like Miller had murmured an incantation and called into existence the building blocks of life.

Someone in the seminar—one story has it that it was Enrico Fermi, who built the world's first nuclear reactor—asked if Miller's processes might have occurred in the Earth's primitive years.

"If God did not do it this way," Harold Urey, Miller's research adviser, said, "then he missed a good bet."

It isn't the only bet going today. Some scientists suspect that the atmospheric conditions of early Earth were very different from Miller's model: less methane and hydrogen, more carbon dioxide and nitrogen, which makes it trickier to synthesize organic compounds. (This may still have happened, though, in the vicinity of volcanic eruptions, which hurl methane and hydrogen sulfide into the air.) Moreover, amino acids are not easily catalyzed to go on to form nucleic acids, the bricks of genetic structures. Deep-sea vents, with their supplies of superheated water and hydrogen compounds, are now regarded as more likely crucibles. Variations on panspermia have persisted, particularly after meteorites—such as the one that crashed down near the Australian town of Murchison in 1969—yielded amino acids, presumably forged out in space under the ultraviolet glare of stars. Extraterrestrial molecules might have seeded Earth's bodies of water or added to the compounds that were brewing in them. But the mechanics at the core of the Oparin-Haldane hypothesis continue to provide a conceivable way to create, if not life itself, at least the chemicals adjacent to life.

Throughout his career, Haldane kept returning to these ideas, refining their biochemical principles in the light of new science. He agreed that methane was a more important gas in the early atmosphere than carbon dioxide. He laid out his minimum conditions for a cell to be considered alive: It had to hold a spiral of nucleic acids that could copy itself, that could produce a set of essential enzymes, and that could synthesize molecules to transfer energy. "I believe that life demands not only self-reproducing molecules but a self-reproducing system of such molecules," he wrote. At a conference in 1963, a year before he died, Haldane broached the thought that the first key molecule of life to be constructed might have been RNA—ribonucleic acid—which creates proteins, quickens reactions, and performs multiple other functions within a cell. The RNA would produce enzymes, which would then build more sophisticated molecules. "I can't say more than that," he said during his presentation. "I think we are all groping very much." That life grew out of an RNA world is now widely accepted. But how the RNA world came to be—how the stage was set for the drama of life to begin—remains an unsolved, slippery mystery.

And the better part of the soul is likely to be that which trusts to measure and calculation?
Certainly.
—Plato, *The Republic*

Numbers were safe. They were impossible to argue with. Two was greater than one in every conceivable universe. Everything can be quantified, Francis Galton believed, and that included Darwin's model of evolution.

"Whenever you can, count," Galton liked to say. And so he did. When he was in Namibia as a young man—his head not yet bald, but his full sideburns already worn down to his jaw—he appraised the shapes of women. He couldn't ask to run tape measures around them, so he watched them from a shifty distance, his sextant in his hand.

"As the ladies turned themselves about, as women always do, to be admired, I surveyed them in every way and subsequently measured the distance of the spot where they stood—worked out and tabulated the results at my leisure," he wrote to his brother. Later, in Britain, he constructed a beauty map of the nation, grading its towns on how attractive he thought their women were. (Aberdeen ranked at the very bottom.) In 1884, in his biggest study, he convinced more than 9,000 people to pay threepence each for the privilege of being measured in his Anthropometric Laboratory, their heights and weights and arm spans and lung power and a dozen and a half other attributes recorded in his ledgers.

Galton was a statistician of tremendous influence. In kneading his numbers, he discovered the concepts of correlation and regression, and in pioneering the large-scale use of questionnaires to source his statistics, he was an early enthusiast of big data. But his devotion to the quantitative was extreme. It led him to the opinion that every aspect of humanity could be expressed as a number—that intelligence could be plotted on a graph just as accurately as height or that artistic faculty was a metric. And if these traits could be measured, Galton thought, they could be bettered and passed down from one generation to the next. Farmers did this all the time, with their crops or their livestock, breeding strains that produced more grain or more milk. "Could not the race of men be similarly improved?" Galton wondered. "Could not the undesirables be got rid of and the desirables multiplied?" Quoted by themselves, as they've often been ever since he set them down, these sentences glint with menace; they hold the premise of every forthcoming act of racist violence. In fact, Galton softened his thought immediately afterward: "Evidently the methods used in animal breeding were quite inappropriate to human society, but were there no gentler ways of obtaining the same end, it might be more slowly, but almost as surely? The answer to these questions was a decided 'Yes,' and in this way I lighted on what is now known as 'Eugenics.'"

The word was Galton's own: *eugenics*, from the Greek for "well-born." He coined it in 1883, convinced that talented, intelligent parents always produced talented, intelligent children. Two decades later, the discoveries of Mendelian genetics appeared to vindicate him, revealing deep, unshakable patterns of inheritance. He had always been attentive to the contest between nature and nurture—indeed, he had been the first to turn that phrase, too—and now nature seemed to be the stronger force. To encourage scientists to study how to improve the mental and physical qualities of future generations, Galton promised a fellowship of £500 a year to London's University College. When he died, in 1911, he left the university £45,000— nearly a third of his estate. A new chair in eugenics was filled by Karl Pearson, the biometrician who had been Galton's protégé.

When Galton began his career, eyeballing Namibian women from afar, Darwin's theory of evolution had not yet been published; by the time Galton died, the theory, in its various misdigested versions, was fanning a perfect fever of eugenics in Europe and America. Newly founded eugenics societies distributed literature and screened films, trying to educate the public about the importance of breeding. The cause attracted promoters of lofty fame: economists like John Maynard Keynes and Irving Fisher, scientists like Alexander Graham Bell and Luther Burbank, industrialists like George East- man and John D. Rockefeller, Jr. Interpreting Friedrich Nietzsche's idea of the *Übermensch*, eugenicists found support for the superla- tive humans—"men of strong and beautiful bodies, wills and intel- lects," as one described it. George Bernard Shaw turned his hero in *Man and Superman* into a pamphleteer for eugenics. "Nothing but a eugenic religion can save our civilisation from the fate that has over- taken all previous civilisations," Shaw wrote in 1904, in approving response to a Galton lecture.

The infection sped through the veins of society. "You would be amused to hear how general is now the use of your word Eugenics!" Pearson wrote to Galton in 1907. "I hear most respectable middle- class matrons saying, if children are weakly, 'Ah, that was not a eugenic marriage!'" In America, in state fairs that handed out prizes for the biggest pumpkin or the fattest pig or the bonniest baby, eugenicists held Fitter Families contests. Participants filled out a Record of Fam- ily Traits, allowed their IQ to be measured, and gave blood and urine samples. Then they waited for the results, in tents or sheds, amid exhibits that explained how marriages between "Pure" and "Tainted" people turned out. If they won, families received trophies or medals and a citation with a line from Psalms: "Yea, I have a goodly heritage." In England, newspapers closely covered one Mrs. Bolce, an expectant mother who had studied Pearson's books and attended concerts and vaudeville acts, sure that this preparation would help deliver a better child. When England's first "eugenic baby" was finally born, she was named, somewhat pharmaceutically, Eugenette Bolce. Meeting her at age 7 months, a reporter wrote that she displayed "remarkable intelligence and already has a pronounced sense of humour."

The fearful and the prejudiced alike clutched at the straw of eugenics. There were those who worried about the physical decay of the species—who worried, as Darwin himself had, that as people ate better and saw more skillful doctors and lived less dangerous lives, the unfit were defeating the stern scrutiny of natural selection altogether. (The term *unfit* was itself meaningless. Darwin had defined fitness only in relation to an organism's environment, but his vocabulary was pilfered and twisted out of shape.) During the Boer War, army recruiters had admitted that three out of every five volunteers in Manchester, wishing to sign up to fight in South Africa, had to be rejected on the grounds of physical fitness. In 1903, the year after the war ended, the British parliament grew anxious enough to set up a committee on physical deterioration. The committee's recommendations were sensible: better sanitary regulations, better nutrition for children, exercise sessions in school. But that a committee had to be set up at all was a drastic sign, to those who wished to see it, of how soft John Bull had grown. "Now, more than at any time in the history of the British people do we require stalwart sons to people the colonies and to uphold the prestige of the nation," the *British Medical Journal* wrote that year.

Not that there was ever a convenient time for this kind of decline to set in, but the early twentieth century felt, to champions of British power, like a particularly bad moment for it. In rapid sequence, the Boer War and then the Great War killed thousands of Britain's strongest young men. The fighting was feared to have been dysgenic, wiping out the fittest in the land. The wounded soldiers who made it home after 1918 were so biologically valuable, the government thought, that it issued them "eugenic stripes" to be sown onto their sleeves. The stripes were meant to signal to women that these men would be exemplary mates. British realms were troubled enough, as they were, by revolts and contestations. If the country failed to produce a spirited new generation, her imperium would falter further.

These alarms rang overwhelmingly in the ears of those who ruled. When the British fretted about being effaced, they were really envisioning the higher classes being swamped by the lower. The geneticist Ronald Fisher spent a third of *The Genetical Theory of Natural Selection*, his seminal 1930 book, agonizing that the wealthier, better-educated elites of Britain were having fewer children than the less worthy poor. He proposed that the government pay its wealthiest families to reproduce more and to maintain their numbers. (Fisher himself, in his personal bid to remedy the situation, had six daughters and two sons.) In America, the panic betrayed the divisions of race rather than class— the inequalities between white people and black, the dread that immigrants and their children would inundate the country. The American Eugenics Society ran essay contests on the theme of the slowing fertility of "Nordic" races. A prominent psychologist spoke frequently about the incoming "yellow and Oriental peril." Theodore Roosevelt warned that the white middle class would be perpetrating "race suicide" by raising smaller families. In thin disguise, a white supremacist tract called *The Rising Tide of Color*, by a Klansman named Lothrop Stoddard, even crashed the decadent world of *The Great Gatsby*:

> "Civilization's going to pieces," broke out Tom violently. "I've gotten to be a terrible pessimist about things. Have you read The Rise of the Coloured Empires by this man Goddard?"
>
> "Why, no," I answered, rather surprised by his tone.
>
> "Well, it's a fine book, and everybody ought to read it. The idea is if we don't look out the white race will be—will be utterly submerged. It's all scientific stuff; it's been proved."

In the 1930s and 1940s, the Nazis turned eugenics into a sharp, cold-eyed rationale for genocide, so it's difficult now to look behind the Third Reich and recognize what an amorphous muddle eugenics was in the earliest decades of that century and how it drew ideologues of every inclination. Conservatives used eugenics as an excuse to preserve the hierarchies of society. Socialist radicals boosted eugenics because they thought it would work only if the barricades of race and class fell away, so that men and women could choose their optimum partners from the widest possible pool of candidates. Eugenics attracted serious scientists who wanted to discover how diseases like Tay-Sachs were inherited. It also attracted kooks who thought "shiftlessness" and "thalassophilia"—a love for the sea— were heritable conditions. Christian clerics believed that eugenics enshrined the divine duty of women to bear children, while feminists believed that eugenics required the liberation of women. Into the gene, this indiscernible particle that no one could see or fully understand, the world invested its grandest aspirations, its basest biases, and its plainest fears.

Only in relative terms were some ideas of eugenics more benign than others. The measures of Galton's so-called "positive eugenics" movement—essay competitions, sermons on inheritance, and Fitter Families contests—were designed to encourage the meritorious to breed. The criterion was implicit, but it was there, hidden in the sundering of society into the worthy and the unworthy. But then it emerged into the light as a bolder creature, in the shape of "negative eugenics," which tried simply to halt reproduction among those who had been deemed unfit—or maybe to remove the unfit entirely. "Far too many live, and far too long hang they on their branches," Nietzsche had written. "Would that a storm came and shook all this rottenness and worm-eatenness from the tree!" Eugenicists took it upon themselves to be that storm.

AMERICA KNEW THE JUKES as one family, but they were really several: the Sloughters, the Bushes, the Keysers, the Millers, the Ploughs, some of whom perhaps shared some ancestors in the tangled bloodlines of Ulster County. Not all of them were poor, but many were; not all of them lived in Ulster, in upstate New York, but many did, along the forested shores of the five sky lakes, and their lives were as hard and unforgiving as the brows of rock around them. They lived in shanties made of logs or stone, which grew so crowded that, in winter, they slept on rushes strewn on the floor, their feet to the hearth.

In the lore of Ulster County, the families all descended from one man, Max Keyser, who was born early in the eighteenth century and who was described more than a hundred years later, by people who'd never met him, as a hunter and fisherman, a hard drinker, jolly but averse to steady work. When, in the 1870s, a sociologist named Richard Dugdale first learned about the clan Keyser founded, he counted 709 people as belonging to it, spread over seven generations and straggled across the county. Dugdale referred to them, collectively, as the Jukes, and in an 1875 report, he announced his discovery of how streaked with crime and immorality the family had been. The Jukes had repeatedly been convicted of misdemeanors and larcenies petty and grand. They had been diseased or frail. In addition, Dugdale accused the women of being harlots, the men of being drunkards, the children of being illegitimate, and the family overall of simply being poor. He called one woman, Ada Juke, "the mother of criminals," because 92 percent of her descendants had supposedly turned out to be criminals, paupers, sex perverts, mentally ill, or illegitimate. To sterilize her in 1740 or thereabouts would have cost $150; by the 1870s, Dugdale reckoned, the Jukes' various misdeeds and flaws had cost the state of New York around $1.3 million. And this didn't even take into account "the entailment of pauperism and crime of the survivors in succeeding generations, and the incurable disease, idiocy and insanity growing out of this debauchery, and reaching further than we can calculate," he wrote. "It is getting to be time to ask, do our courts, our laws, our alms-houses and our jails deal with the question presented?"

Dugdale made some concessions to the Jukes' environment—their history of desperate poverty, their crippling lack of education. But his report was read as an indictment of the Jukes' heredity, as if wickedness swirled in their blood. Forty years later, near the height of the eugenics craze, a researcher from the Eugenics Record Office conducted a follow-up study and identified an additional 2,111

Jukes. The living Jukes were "unredeemed," the researcher, Arthur Estabrook, concluded, and taxpayers had coughed up $2 million to support, prosecute, or imprison them. Estabrook took photographs during his fieldwork. One shows three young children, their parents, and a grandmother standing outside their house, a box of mud and wood with a chimney sticking out. It looks like a sunny, cool day, and everyone is smiling and squinting in the light, unaware, surely, of the reproach they will soon encounter.

Other studies in Europe and America came to similar flawed conclusions. One family, the Kallikaks of New Jersey, were thought just as degenerate as the Jukes; they, too, formed a part of the growing literature of eugenics. Cesare Lombroso, an Italian physician, held that criminality was inherited, like blue eyes or the ability to curl one's tongue. For social Darwinists, who thought the principles of selection ought to be applied as rigorously to human communities as they were to natural species, the Jukes and the Kallikaks were unfit for propagation. In an exhibition in 1926, the American Eugenics Society set up signboards with flashing lights, which periodically reminded visitors of the threat facing them:

Every quarter of a minute, $100 in taxpayer money is spent on the care of someone with undesirable genes!
Every 48 seconds, a mentally deficient person is born!
A "high-grade person," though, is born only every 7½ minutes!
Be vigilant, citizens, for the West is scheduled for a dramatic Darwinian collapse!

It has never been an overlong journey from wild fear to misbegotten plan. Even in those years, when no one knew much about the gene, the temptation to manipulate heredity proved too strong. Beginning in 1907, more than 30 US states passed compulsory sterilization laws, trying to pull the "unfit"—the blind, the epileptic, the felonious, the mentally deficient, the physically deformed—out of the gene pool altogether. The laws were challenged, and then they were reinforced, by a Supreme Court that considered these sterilizations "lesser sacrifices . . . to prevent our being swamped with incompetence." At least 65,000 men, women, and children were sterilized against their will. In Britain, fewer eugenicists supported sterilization; they pushed, instead, for the government to deal in other ways with the "feebleminded"—a term in great vogue, deployed as if these minds were wilted stalks of vegetation. (Winston Churchill, for one, thought that criminals, tramps, wastrels, and others deemed to be feebleminded ought to be sent to labor camps.) In 1913, with an overwhelming majority, Parliament passed the Mental Deficiency Act, which proposed to set apart and institutionalize people with a spurious range of limitations: serious mental illness, epilepsy, alcoholism, criminal tendencies, or simply an inability to cope as well as could be desired with the hardships of life. The law didn't make such segregation mandatory. Even so, roughly 65,000 people were housed in "colonies" for the feebleminded at the height of the act's influence. Genetics had hardly started out, and already society's guiles and predilections were deciding how the science should be employed, how it should progress, what it should be and do.

WHILE DRAFTING HIS FIRST THEORIES in genetics, Haldane was also marshaling his ideas on the science's relationship with society. He had been alive to the implications of eugenics even earlier, in fact. When the measures of the Mental Deficiency Act were being drafted and a Royal Commission on the Care and Control of the Feeble minded was urging the state not to allow the mentally deficient to procreate, Haldane was still at Eton. But he was aware of these deliberations and their import. He had a friend whose father had committed suicide and whose mother was mentally unstable—a potential double whammy of heredity, Haldane thought at the time. "So poor devil hardly dare marry," he wrote in his Eton diary. "Eugenics will be pretty awful for the unfit, though really no worse than the way we treat them now."

He didn't often let drop such morsels of sentimentality. As a geneticist, Haldane designed a clever stance for himself on eugenics. He would admit, in passive, impersonal tones, the concerns of his particular class of Britons—as he did, for example, in a speech in 1928: "It is true that in England the rich breed more slowly than the average and the skilled than the unskilled labourer." He cited this as a fact, a speckle of demographic data; he made no judgments on the ethics of considering these differential birth rates a cause for worry. Then he revealed his own convictions, making it seem as if they happened to be progressive only because of the cool logic of science. You could, he said, make the rich richer and the poor poorer, but then the slums would harbor diseases that might eventually attack the kind of men sitting in his audience, listening to his speech. You could send armored cars through the slums from time to time, to fire upon women and children—and if Haldane flung this modest proposal out with a straight face, he doubtless enjoyed the shiver of shock the idea provoked. But really, scientific thought had already suggested the best approach to this vexing problem, so why not just follow that? "The correct remedy for the differential birth-rate," Haldane explained, "would seem to be such a raising of the economic standards of the poor as would give them the same economic incentives to family limitation as exist among the rich, and such an equalisation of educational and other opportunities as would lessen these latter incentives." If science someday proved that banishing every second child in the slums to Jamaica was an even better solution, Haldane appeared to imply, he would back that instead.

This was a spry mode of debate. It allowed Haldane to both affirm and reject his ties to the upper bourgeoisie, first validating and then upturning the anxieties of his class. And by shearing morality and sentiment from his assertions, Haldane became an objective authority, a brilliant man who could be trusted to convey hard truths in a forthright manner. So, thinking perhaps of another scientist's rough computation—that sterilization would require 700 generations for the incidence of feeblemindedness to drop to one in 1 million—Haldane dismissed the measure as untenable. "Many of the deeds done in America in the name of eugenics," he wrote, "are about as much justified by science as were the proceedings of the inquisition by the gospels."

In the 1920s, Haldane was still infected by some of the prejudices of his class and of his time. As a Darwinian biologist, he ought to have believed in the rewards of variation. A wealth of genetic types in a population renders it agile, prepares it better for adaptation and change. But he didn't question the essence of the eugenic belief: that there existed a single genetic type that was desirable above all others. So he commended the good work of the Eugenics Education Society "in persuading a certain number of intelligent people that it is their duty to have more children." And if the feebleminded are to be segregated, he said in his speech in 1928, "it should be in their own interests, and because they are unfit to bring up a family, quite as much as on eugenical grounds." To his credit, he would go on to revise this notion, arguing decades later that human diversity was not only desirable but was a signal of social liberty. He would recognize also that scientists had found no way to know which human characteristics would best serve its future evolution and no way to selectively breed them.

For eugenicists, race was an indicator of ability. The question of race would long bedevil Haldane—and indeed many scientists, who struggled throughout the century to understand how such visible physical differences could matter so little. Haldane knew that people's capacities were shaped by their environment as well as their genes and that ideas of "superiority" and "inferiority" were meaningless in comparing humans. He knew also that the word *race*, as it was used, was imprecise, that it never defined any sort of homogeneity in its members. And he scoffed at any nostalgia for bygone racial perfection: "The races of the past were no purer than those of to-day." But the science of the postwar years could not convince him to discard the biological validity of race altogether. It did not reveal, as it does now, that while one human population group may differ in some genetic aspects from another, these groups are not congruent with the old racial categories, or that genetic differences between so-called "races" are outnumbered by those within any one "race."

Haldane insisted that he was agnostic about racial differences and that no one had proved black people to be less intellectually able than white. But for a time, he didn't rule out any connection between race and intelligence either. And for all his caveats about the futility of race as a concept and about the importance of nurture in addition to nature, he could still lapse into shoddy generalization. He doubted, he once wrote, that "the black races of Africa" could ever match up to the level of white Europe. As for Australian blacks, he added, "they huddle around fires in cold weather. But they had never thought of making clothes from the skin of the animals they killed. I find it hard to believe that their descendants will produce a Watt or Edison." He was not a bigot, but he was also not fully exempt from the preconceptions held by those around him.

WHEN HALDANE FIRST SPOKE about eugenics, he used the word to mean not the coercive, limited measures of his day but the sophisticated genetic tinkering he thought humanity would inevitably adopt. It was February 1923, and Haldane was asked to talk to the Heretics, a Cambridge society. His address became so popular that he was invited to turn it into a small book. This was still nearly a decade before Haldane published *The Causes of Evolution*, and he hadn't given much thought to writing for the public. In fact, when his friend Julian Huxley's experiments once found their way into the popular press, Haldane warned that he would lose his standing as a reputable scientist and be taken for a quack. But then Haldane's book, published in 1924 as *Daedalus; or, Science and the Future*, sold more than 15,000 copies in its first year, and he was soaked by a sudden spray of fame. It changed his life. With *Daedalus*, Haldane became a man who did his thinking in public.

At the heart of *Daedalus* is a stunt of backward prophecy: extracts of an essay, supposedly written 150 years hence by a "rather stupid undergraduate," on the influence of biology on history during the twentieth century. The eugenic techniques of Haldane's day were crude and divisive, the student writes, but they prepared the world for what was to come. By 1958, every infectious disease was eradicated, and the yields of crops had multiplied. (Haldane could never resist jabs of whimsy, and he unleashed one here: The algal strain used to enrich agriculture once escaped into the sea, and "for two months the surface of the tropical Atlantic set to a jelly.") Everyone grew prosperous—the English in particular, so much so that the coal miners' union was able to run a horse in the Derby. But the real revolution was ectogenesis—the ability to fertilize a woman's egg outside her body and to grow it into a healthy infant. This was true eugenics:

> The small proportion of men and women who are selected as ancestors for the next generation are so undoubtedly superior to the average that the advance in each generation in any single respect, from the increased out-put of first-class music to the decreased convictions for theft, is very startling. Had it not been for ectogenesis there can be little doubt that civilization would have collapsed within a measurable time owing to the greater fertility of the less desirable members of the population in almost all countries.

Once again, Haldane held his ideas at arm's length. By looking back on these developments from the future, he framed them as inescapable; by looking at them through the eyes of a foolish student, he avoided having to judge their ethics. But Haldane didn't remain noncommittal about the value of science itself. *Daedalus* is known for many things—including its smart predictions about peak oil, humankind's switch to wind and solar, the hydrogen fuel cell, and semi-ectogenetic test-tube babies. But it is, foremost, a manifesto for society on how to deal with its scientists, how to allow them to light the way to human happiness.

In *Daedalus*, speaking of the 1920s and in his own voice, Haldane admitted that science had some mud on its face. It had enabled the infernal destruction of the war, and its pursuit of induced radioactivity could reduce the planet to a glowing cinder. Some people wanted to freeze research in its tracks—he mentioned G. K. Chesterton, who thought the hansom cab the acme of invention—but capitalism and nationalism wouldn't allow that. Then Haldane maundered a little, as he sometimes did, skipping from vacuum-jacketed hydrogen reservoirs to Milton to chemical stimulants to the Venus of Brassempouy, sketching in thick lines the past and future of his species. He tacked his way to his theme: that science, and biology in particular, was poised to deliver still more radical changes. "I believe that the biologist is the most romantic figure on earth at the present day,"

Haldane said. Nearly every member of his profession was grubbing around experimenting with lower species or staring fish-eyed into microscopes, hoping to uncover the fundamental secrets of existence: "There is real tragedy in his life, but he knows that he has a responsibility which he dare not disclaim, and he is urged on . . . by something or someone which he feels to be higher than himself."

Since science would march on regardless, society had to realign itself, retool its ethics. This had happened often in the past: "Our increased knowledge of hygiene has transformed resignation and inaction in the face of epidemic disease from a religious virtue into a justly punishable offence." To treat traditional morality—the morality of religion—as dispensable, or at least malleable, was the only way to utilize the powers of science. Haldane meant this not as a wholesale endorsement of every eugenic strategy, but as counsel that even to distinguish a wise strategy from a foul one would call for a new, secular morality. What this morality was, Haldane would not say or could not say. He admitted that the moral code might be altered for the worse, "giant flowers of evil blossoming at last to their own destruction." But these disruptions were to be expected, and humankind would either vault them or stumble in the attempt,

For the scientific world, Haldane wrote, the first modern man was not the sentimental favorite Prometheus but the less remembered Daedalus: architect of the labyrinth at Minos and humankind's original aeronaut. Daedalus had constructed the wooden machine within which Minos's wife mated with a white bull to produce the Minotaur; it was safe to say that no one had yet equaled this success of experimental genetics. Daedalus was the first to demonstrate that the scientist should not be concerned with the gods—and rightly, his monstrous and unnatural deeds earned him no punishment in this world or the next: "The scientific worker of the future will more and more resemble the lonely figure of Daedalus as he becomes conscious of his ghastly mission, and proud of it."

In the 1920s, when *Daedalus* was published, it screamed with significance; it still does today for many of the same reasons. Haldane delivered his speech at a time when Europe was still steaming and hissing from the war, and the readers of his book, not being scientists themselves, wondered if science would destroy civilization instead of elevating it. It was the first time in history that such a concern was even realistic, let alone urgent and terrifying, and since Haldane's time, that concern has only amplified. Haldane spoke directly to these fears, and he invested his authority in his brusque but essentially calming predictions of the future. Science was sure to spring nasty surprises and engender more death, but utter destruction was not a foregone conclusion. Instead of pure pessimism, Haldane offered the possibility that with the aid of science, our species could evolve by its own, deliberate hand into a higher plane—the first flicker of the modern movement known as transhumanism.

Parts of *Daedalus* carried the value of shock as well. Haldane's friend, Julian Huxley, had just been scolded by the BBC for daring to mention birth control on air, so discussions of sexless marriages and ectogenesis, not to mention the rejection of religious morality, agitated the solemn reader. Haldane loved the sensation he had caused, but his own father was deeply pained by the book's reception. "Oxford was talking of nothing else," Huxley later wrote, "and the family became the butt of donnish jokes and quips." Haldane's mother sent Huxley a letter about J. S.—the Senior Partner, as she called him—and his consternation over his son's mechanistic model of life, in such direct conflict with his own philosophy of idealism:

> Dear Julian,
> I find the S. P. is frightfully upset about Daedalus. Will you abstain altogether from poking fun at him on account of it? And if you can do so, keep people off the subject altogether when he is about?
>
> I knew he'd object, but had no idea till to-day how really unhappy he is—odd people these Liberals and no accounting for them!

Quickly, *Daedalus* became essential literature. After H. G. Wells, no writer—and certainly no scientist—had advanced any bold vision of the far future. Einstein owned a copy of *Daedalus*; in pencil, he marked up Haldane's emphasis of ethics. Aldous Huxley borrowed

Haldane's idea of ectogenesis and put it at the heart of his dystopian society in *Brave New World*, the intent to engineer higher humans corrupted instead to turn out serf-like Deltas and Epsilons. The philosopher Bertrand Russell felt compelled to respond with *Icarus*, in which he glumly predicted that science would only further empower rulers, give them greater control over their subjects, and allow them to indulge grander evils. The average human being was an Icarus type. Taught to fly, Icarus had destroyed himself with his rashness, and a similar fate awaited humanity. "Science has not given men more self-control, more kindliness, or more power of discounting their passions in deciding upon a course of action," Russell wrote. "That is why science threatens to cause the destruction of our civilization."

From a certain angle, *Daedalus* seemed like Haldane's way of thinking his way out of a dilemma. He had fought in Europe, and even if he claimed to enjoy the war, he also saw how it led, in his own words, "to darkness and confusion and the grave." At the same time, he was a scientist at a time when science was being assailed for precipitating the extremity of the war's carnage. He released himself from this bind by distancing the pure work of research from the emotional, political, and moral faults of its society. The scientific method was a thoroughbred, tested and true. But its speed and strength were reliant solely on the quality of its rider; on the rider's shoulders lay the responsibility and the blame for any adverse social effects. It was a defensive position, built on the back of his father's experiences, and his own, with political authority—the exasperations they had felt when their advice was heeded too late or was implemented imperfectly or was never sought at all. In *Callinicus*, a slim book published a year after *Daedalus* and a more vexing and unsound work, Haldane's defensiveness lapsed into scorn.

Like *Daedalus*, *Callinicus* was fleshed out of a speech. Haldane named it for a Byzantine chemist credited with inventing Greek fire, an incendiary weapon that, in the seventh century, set aflame the ships of an invading Umayyad navy. Callinicus was the progenitor of chemical warfare, which reached its horrible peak during the First World War. The use of chlorine and mustard gas so upturned the codes of the battlefield that during a conference in Washington,

DC, in the early 1920s, the great powers proposed a full ban on chemical arms.

But this was a mistake, Haldane argued, based either on convenient official propaganda or on ignorance of the science and statistics of war. Chemical weapons were, in fact, more humane than the pieces of metal soldiers hurled toward each other. Outfit a soldier with a respirator, and you protected him admirably; with most gases, in any case, men recovered with rest and fresh air, and no one was invalided for life. Mustard gas was different, a more potent poison even in small doses. "Someone placed a drop of the liquid on the chair of the director of the British chemical warfare department," Haldane wrote. "He ate his meals off the mantelpiece for a month." But even mustard gas killed only one man for every 40 it put out of action, he insisted. Shells killed one man for every three. Using gas, even on the largest possible scale, would cost less by way of life and property, abbreviate wars, and make their outcomes dependent on brains, not numbers. If only politicians and generals learned a little elementary science, got over their distaste for scientific thought and scientific method, they would see he was right about gas warfare—and, by extension, about other new, daunting applications of science to human life.

Haldane's analysis carried a certain icy logic, but he was also guilty of errors, presumptions, and elisions. He got the year of the discovery of mustard gas wrong, and he didn't mention that it could, in fact, deprive its victims of their eyesight or leave them with permanent respiratory illnesses. (He couldn't have known that mustard gas is also carcinogenic and damages DNA.) The prospect of disintegrating the atom and weaponizing it was so remote, he thought, that "when some successor of mine is lecturing to a party spending a holiday on the moon, it will still be . . . unsolved." He assumed that soldiers and even civilians would always have respirators on hand, failing to see the unlikeliness of this scenario in the real world. He accused the military of being so set against chemical weapons that it had suspended anti-gas training drills, but these had ceased for financial, rather than ideological, reasons.

Without evidence of the kind he was otherwise so particular about, Haldane also repeated the popular belief that black soldiers were impervious to mustard gas. "This is intelligible," he wrote, by way of thin reason, "as the symptoms of mustard gas, blistering, and sun-burn are very similar, and negroes are pretty well immune to sunburn." Perhaps Indians, only a few tones paler, would be nearly as resistant. The number of white men unaffected by gas was probably small—but enough, he wrote, to lead all these units of colored troops. Once again, Haldane was channeling the absurd racial notions of his age. This belief would persist. During the Second World War, the US government tested mustard gas and other chemical agents on 60,000 of its troops. Black and Puerto Rican soldiers were measured against control groups of their white colleagues in a mad attempt to find the secret of their supposed immunity.

But he was unlikely to convince anyone, Haldane concluded, in near-sullenness. In both *Daedalus* and *Callinicus*, he rendered scientists as martyred, maligned figures. Their work, and only their work, held the potential for true human utopia, but their inventions were repurposed, their difficulties ignored, their theories misunderstood. They could sing of salvation until their lungs gave in, but the world would plug its ears and refuse to listen.

AMONG THOSE WHO WROTE TO HALDANE after reading *Daedalus* was Charlotte Burghes, a journalist for the *Daily Express*. She had been commissioned to interview him for her newspaper, she explained. Could she call on him at Cambridge?

The letter was deliberately diffident; like asbestos, it concealed the heat of her exhilaration. She had been floundering over the plot of her first novel, about a futuristic society in which parents were able to choose the sex of their children. The science had to feel realistic, but it eluded her understanding. She borrowed *On the Origin of Species* from a library, but the book only bewildered her further. What she needed was a tutor. Then a colleague sent her an issue of a magazine called *Century*, and she read—devoured— an abridged version of Haldane's speech to the Heretics, with its prediction of artificially grown babies and its generally audacious visions of biology. This was her man! she thought. She looked him up, marveled at his accounts of swallowing sodium bicarbonate, and convinced her editor that he merited an interview. Already, she was one-eighth in love.

So it didn't matter that a couple of weeks passed and he never wrote back. Perhaps he had been traveling, or perhaps he was forgetful with his correspondence, or perhaps her letter had lost its way. She didn't wait further. On a hot Saturday afternoon, she caught the train to Cambridge.

In the biochemistry lab, on Tennis Court Road, they told her that he might be in Trinity College. In Trinity College, they told her that he was out but might be back by tea. She killed time, walking around the campus, growing progressively warmer. Her feet ached. Once an hour, she stopped by Trinity to check if he had returned. At last, she was shown up a flight of oak stairs, through the white door of his study, and into his enormous presence.

She shrank. She felt overawed. She tripped over her apologies. She was midway through a novel, she explained, and she wanted reference books in the biological sciences. She had sent him a letter and had no reply, which was why—

He made his own apologies. For all his brusqueness, he could be charming and solicitous if the mood seized him; it seized him now, when this raven-haired woman with flashing, dark eyes sat before him. He had been away from Cambridge, he said, and after his travels, he had started in immediately on new experiments, so he hadn't been able to attend to his letters. Resources on biology? "Of course you have read . . . ," and then he fired off title after title, the names singing past her ears like bullets. "I have read nothing," she admitted. "I don't know where to begin." Upon which he pulled half a dozen books off his shelf, dumped them in her arms, and said: "Well, you might do worse than start with these."

He walked her to the bus stop so that she could reach the station in time for the last train back to London. His strides outpaced hers, and she skittered to keep up. Nearly too late, she remembered that she had never interviewed him. He told her about an experiment he was running, for which he had drawn his own blood just that day. That was good. She could use that for an article. Then, as she was boarding her bus, he told her: "I have to go to London on Monday.

The last reporter who interviewed me made a complete hash of the story. If I find you haven't made an equal mess of yours, I shall call for you at your office at one o'clock and take you out to lunch."

He liked the article. He called for her. They went to lunch—lobsters and white wine—and discussed the poetry of Racine. She realized how desperately she had missed such conversation and how content she was when she was learning—just as content as he was, in fact, when he was expounding. They were two synchronous souls.

In Haldane, Charlotte saw an entry into the life she truly wanted. "What manner of man then is the modern woman's ideal husband?" she had asked once, in a column, before providing her answer: The finest husband ought to be "a guide, philosopher and friend." She cast Haldane in these roles instantly. In her letters, she told him about the articles she was writing or inquired about his experiments or suggested adventures: "We must have a shot at cocaine one day." She described the fugs of melancholy that set in as winter approached; she dosed them with whisky or with long, boiling baths, she said. "Send me something that will either kill or cure." She asked to borrow books, which he brought up with him when he came to London to see her: books on science, Plato's *Republic*, Samuel Butler's *Erewhon*, William Morris's *News from Nowhere*. They would discuss them, or he would deliver impromptu lectures in astronomy or geology, and she grew more convinced than ever that he was her predestined teacher. Every time they saw each other, their intimacy swelled. And then, when they parted, he returned to Cambridge, and she went home to Purley, where she lived with her husband and son.

GROWING UP, SHE HAD BEEN Charlotte Franken, the daughter of a German Jewish fur trader. The family had lived in London and for a while in Antwerp, always in luxury that they could not afford. The Frankens were forced to feel acutely conscious of their status as a double minority. As a girl, she heard so much discussion of the Dreyfus Affair that, she later wrote, "I knew about anti-Semitism long before I learned the facts of life." Then, during the First World War, her father was declared an enemy alien; his assets were seized, and he escaped the internment camps only by moving with his American-born wife to the United States in 1915. Charlotte stayed behind, miserable and alone. She worked as a secretary for a few years and in 1918 married a friend's cousin—a charming, penniless man named Jack Burghes, who had served in Arras and Ypres. Their son Ronnie was born soon after.

The Burgheses' finances were forever a stumble away from ruin. Jack Burghes, addicted to both alcohol and gambling, never settled into steady work, and their country house was too large for them and Charlotte's pair of Belgian maids too expensive. She thought she had inherited from her father a rosy, Micawberish trust in fortune and the future, but really, she laid her own road into an independent career. She sold articles and short fiction freelance and then took a position at the *Express*. For a time, when she was tasked with finding four feature articles a day to fill the leader page, she wrote many of these herself under an assortment of pen names. As a correspondent, she covered politics, becoming the first woman to report from the press gallery at the House of Lords. She volunteered to work every day of the week. In her spare time, she plotted her novel. Her marriage had turned so sour that there was nothing else to do.

Charlotte was bound to prove attractive to Haldane. She was well-read, she spoke French and German, and she had a blazing spirit and an inquisitive mind. She was a feminist and a progressive liberal. With all these qualities, she must have reminded Haldane of Naomi. Charlotte riled the establishment just as Haldane himself liked to do; the *Daily Express* was once sued, in vain, by a politician named Vera Terrington, who claimed that an article by Charlotte had made her look "vain, frivolous, and an extravagant woman." (The article's headline ran: "Aim If Elected—Furs and Pearls.") She displayed an avid interest in the social dimensions of science. That she was married when they began their relationship provided, for Haldane, a bonus needle with which to prick the world into consternation.

They wrote to each other constantly, and their letters were lively and funny and sensual. At the *Daily Express*, Charlotte would spot Haldane's comically unformed hand on an envelope shuffled in with her other mail, start to walk up to her office, and open it halfway up the stairs, unable to wait. When she typed out her own letters, she worried about her colleagues peering over her shoulder: "I feel positively indecent and then begin to reflect on the regrettable fact of the universality of this pastime, and to wondering whether it would remain attractive if it were practiced under conditions of greater publicity." He called her "Poppet"; she called him "Heracles" or "Doggie" or sometimes "Ponto," a nickname that rang with the echo of a private dirty joke. She recalled the nights they spent together—at Cambridge, where she would step out of the train and look for his great, globular head shining out above the others like a rock in a tide pool, or at the Adelphi Hotel in London. "I have a lovely memory of last night, or rather this morning, just before waking up, of your nakedness rubbing against mine," she wrote once; a different letter opened with the risqué endearment "My pink pillar." He responded with equal mischief, and she had to remind him, "When you write to me, always mark the envelope 'Personal'—your letters are, rather, you know . . ." Another time, she composed a doggerel:

> Little Miss Poppet
> Was sitting atop it
> Enjoying her big fat Pye.
> She put in her thumb
> And found a fine plum
> And said: "What a Ponto have I."

Their transitions were swift: from the pedagogical to the friendly to the amorous, all within months. But beyond that, Charlotte's marriage complicated everything. "My dear lover," she wrote in the spring of 1924, "I want to ask you to eliminate from our conversation for the next few months any discussion of a possible future marriage between us." Her emotions had been drowning her reason, but she was having a moment of clarity. "I am content with all myself to be, ad infinity, your dear mistress." Their careers had to come first, she wrote, soothing Haldane's concerns; he tended to fret that since he accorded his science so much more importance than everything else, she would soon feel resentful or neglected.

"Although I am repeating myself," she once told him, with clenched impatience, "I want to put on record the fact that I shall not be jealous of your work."

But they would certainly have children, they agreed. There were early alarms—"Hey love—there will be no baby this time, everything was quite normal. So it is 'as we were'"—but not very long after, they were planning their family. Haldane had always wanted children, with something approaching desperation, Naomi thought. When she had been pregnant in 1917, he told her, "Your approaching achievement fills me with envy." Charlotte, in an unexpected twist to her feminism, thought that motherhood was a woman's noblest role, and she told Haldane they would make "wonderful children" together. She didn't even want to wait for a divorce from Burghes; that marriage, in her head, was dead in all but law. Maybe she could move to Cambridge, and they could live in a house with a large garden for the children, or maybe he could take a professorship in London. This gumption may have faintly astonished Haldane, because she teased: "I warn you that if you do not consent, I may deceive you one night with regard to the precautions I may or may not have taken as usual." On several occasions in their first two years together, she felt prepared and hopeful. She had read in the *British Medical Journal*, she told Haldane once, "that they now give adrenaline to pregnant women for morning sickness, so tell Dixon or whoever makes the stuff to get some ready." Later, she signed a letter not with her name but with the Venusian symbol for womanhood; within the circle, she drew a small baby, curled like a bean inside its pod.

Haldane introduced Charlotte to his family. She seemed fond of Naomi, whom she called "dear little Motherpot," but it wasn't entirely evident that Naomi liked her. "I tried very hard to be pleasant," Naomi wrote in 1979, recalling Charlotte. She disagreed strongly with Charlotte's contempt for women who chose to never become mothers, and she wondered what effect Charlotte was having on her brother. Charlotte spoke so proudly of her ability to "vamp" men— to manipulate them with her femininity. Naomi worried: Was she now vamping Haldane? Naomi's conversations with Charlotte often turned into excuses for velvety dissuasion. Was Charlotte aware of how domineering Haldane could be? Did she know he sank so fully into his work sometimes that he would evince no interest in her or in anyone else? Did she recognize that what Haldane really wanted was children and not a wife and that he would likely be more devoted to them than to her? But Charlotte assured Haldane that none of this had escaped her notice.

Haldane's father worried about the morality of their relationship, Naomi told Charlotte. J. S. was in his mid-60s, still going underground and overseas to study human physiology, but he was slowing down and his health was flickering. He had already been saddened by the publication of *Daedalus*, and now his son was trampling over his sensibilities once again. His opinion dismayed Charlotte. "I shall really be unhappy if you quarrel with your father," she told Haldane, "as I think he is quite unhappy enough and for his sake and your own you must stay friends with him."

Louisa, more conservative still than her husband, frequently failed to be polite. At a social gathering, Charlotte noticed, Louisa shook hands but never once spoke to her. Then, as Haldane grew determined to be with Charlotte, Louisa took to giving her advice. Charlotte wrote to Haldane: "When I was talking to your mother, she said, 'You see, you are now entering an entirely new social world.' I said with emphasis, 'I am quite aware of it.' Then she rose and came to me and said, 'Please don't talk like that.' I realized to my astonishment that she thought I had made my remark in order to 'get my own back,' or for one of those sort of obscure motives that seem to actuate women." Charlotte resented having to be on probation in this manner, having to prove to Haldane's family she was good enough for them. "I still have to be convinced that, given my handicaps, anyone of you would have done a great deal better than I have."

WHEN THEY RESOLVED TO BE MARRIED, there was still Charlotte's previous union to dissolve. Jack Burghes begged her to stay with him—rather pathetically, she thought, and she was moved even as she refused. Although he knew of her relationship with Haldane, he refused at first to grant her a divorce or to let her take Ronnie away with her. She worked on persuading him, sending Haldane updates: "If he remains in his present frame of mind, there may even be a chance of his allowing to divorce me," or later, encountering fresh stubbornness, "He is by no means agreeable to as much as I had hoped." For Charlotte to simply desert Burghes would, in any case, not have been sufficient. At the time, the law recognized only adultery as grounds for divorce, and judges demanded detailed proof of the transgression, clawing through the meat of the affair with their fingers, holding up the bones for everyone to see. Having already committed adultery and taken pains to keep it secret, they now had to plan to be caught at it. If this was an invitation to notoriety, Haldane didn't mind. Hadn't he, after all, argued once in a debate in Eton, "For love to be worth having, it must be dangerous and potentially unsuccessful"? Charlotte saw, in fact, that he even found juvenile glee in the thought of being a corespondent in an adultery case, seeing himself "in a quixotic light as the chivalrous rescuer of a little woman in need."

When Burghes agreed, at last, to sue for divorce, Haldane and Charlotte drafted their scheme. They hired a detective agency to send a man to follow them during a tryst staged for his benefit, then booked a hotel for an overnight stay, so that its staff could later be summoned as further witnesses to their affair. ("Apparently it is impossible to get you identified by the chambermaid either on Saturday or Sunday morning as she never gets out till the afternoon," Haldane's lawyer, a fellow conspirator, wrote to him. "Could you manage 4 o'clock on Monday?") Even this plot of self-implication nearly went wrong. Arriving at the hotel, Charlotte decided she didn't like it and insisted on moving to the Adelphi, but they worried that they might lose their tail. Looking around the hotel's lounge, Charlotte noticed one particular young man and guessed correctly that he was the sleuth. Haldane walked over, told him about the change of plans, and even handed him a suitcase to carry as they walked to the Adelphi. The next morning, just to ensure that they were seen together in bed, the detective knocked on the door of room no. 4 and delivered them the day's newspapers.

Late in February 1925, Burghes filed suit against Haldane and Charlotte, accusing them of adultery. In an additional filigree of farce, they now denied the charge; to concede readily would have given their charade away. How could the judge be absolutely sure, they argued, of the sexual infidelities that may or may not have occurred in the privacy of their hotel room? But the judge paid no heed. In October, he ordered Haldane to pay £1,000 to Jack Burghes in damages, and he granted a divorce. Charlotte was at liberty now, finally, to marry Haldane—and she did, the following year—but this only inaugurated a new round of scandal and acrimony.

HALDANE COULDN'T SAY he hadn't been warned. His lawyer had reminded him, before the trial, of the tenor of the times. "The old-fashioned notion of a divorce suit being a compulsorily disgraceful dogfight dies hard," he wrote. "No doubt some will think that the sanctity of the home would have been better preserved by Mrs. Burghes being your mistress and not your wife and by any children of yours being born illegitimate." And surely Haldane had heard the hisses of reproach as his affair with Charlotte emerged into view; if anything, he encouraged them, announcing into the air as he left the common dinner table at Trinity: "I am going to sleep with my mistress tonight." His colleague in Trinity College, the philosopher C. D. Broad, waxed caustic: Haldane, he said, "was never a man not to blow his own strumpet." William Bateson, the geneticist, told another scientist: "I am not a prude, but I don't approve a man running about the streets like a dog."

Later, Haldane would insist he had taken precautions. Two months before the pantomime adultery, he told the head of his department, Frederick Gowland Hopkins, and A. C. Seward, the vice-chancellor of Cambridge, about his plans. Hopkins raised no objections. Seward merely said: "Oh."

Days after Charlotte's divorce was decreed, however, a Cambridge tribunal called the *Sex Viri*—the "six men," in Latin—wrote to Haldane. The *Sex Viri* assumed responsibility for guarding virtue and discipline at the university, and professors, when confronted with the court's judgments of their immorality, inevitably quit right away. Seward, a member of the *Sex Viri*, urged Haldane to do likewise. They had met—without summoning Haldane—and Seward wrote:

It is only fair to you that I should tell you that the clearly expressed opinion of the members present leads me to advise you, in your own interests to resign the reader-ship. . . . If you resign, the resignation would take effect as an ordinary resignation. If, on the other hand, you prefer to leave the matter where it is, and the Sex Viri should decide on deprivation, their decision would be published in the [Cambridge] Reporter.

Haldane did not resign. He might have if Seward had suggested it the previous year, before the trial, but as he had been given no sign then of the university's opinion, he felt under no compulsion to leave. Instead, at the next hearing of the *Sex Viri*—or the Sex Weary, as Haldane would forever, and with delight, call them—he argued that the divorce benefited both Charlotte and Ronnie. It wasn't persuasive enough. Once again the six old men met, and now they ejected Haldane from his readership altogether.

His hackles flared, his spine set in obstinacy, Haldane exercised his right to appeal, and he was, by this point, enough of a public personality that his battle with Cambridge made it into the London papers. The press covered the ensuing events faithfully. On February 20, the *Daily Telegraph* reported that Cambridge had formed a five-judge panel to decide Haldane's case: two scientists, an MP, the provost of Eton, and a high court justice named Horace Avory, who was known for being unemotional and merciless. Once again, Haldane noted, he was to be judged by old men who could hardly be called a body of revolutionaries. He hired a lawyer; Cambridge appointed its own counsel. "Nothing, it is stated, is known as to the procedure," the *Telegraph* wrote, "for there is no precedent."

On March 17, 1926, the special tribunal convened in London. "The utmost efforts were made to preserve secrecy as to what transpired in court," the *Daily Mail* recounted. "A number of persons were stopped by ushers stationed at each door, who were instructed to keep them locked. The public gallery was also locked and guarded." Within, Haldane called character witnesses, including his father and Frederick Hopkins, who sought to counter Haldane's image as a rake.

Why, he considered his own daughter safe in Haldane's company, Hopkins said. Other scientists testified to Haldane's professional abilities. Haldane himself let drop a cloaked threat. The National Union of Scientific Workers, to which he belonged, had decided to advise its members against applying for Haldane's vacant position. If Cambridge wanted a reader in biochemistry, and if it didn't keep Haldane, it might not get anyone at all.

Cambridge's lawyer, who had to prove specifically that Haldane was guilty not of immorality but of *gross* immorality, exerted himself. Standards had to be maintained, and professors had to set an example. He described a hypothetical undergraduate "caught in the toils of a woman, and brought up by the Proctor, the Proctor reprimanding him, and the young man saying: 'I am not as bad as "so and so." After all, I was only going to commit the offence of fornication. The professors and readers of the university can commit adultery without rebuke.'" How would that ever do?

The judges sequestered themselves for three-quarters of an hour before returning with the verdict. They upheld Haldane's appeal, "but this decision should not be taken as any expression of opinion that adultery may not be gross immorality." Haldane automatically regained his biochemistry readership. This happy outcome wasn't a foregone one, Haldane knew, and he wondered how the judges had arrived at it. Later, he wrote. "I am not sure whether Mr Justice Avory was more horrified by my conduct or by the fact that the Sex Viri, like the jury in *Alice in Wonderland*, had delivered their verdict before hearing the evidence."

The fireworks of Haldane's trial and his skirmish with his university burst plumb in the midst of his decade of prodigious work on natural selection and very soon after *Daedalus* had pushed him into the clinch of fame. These events made him an incandescent persona: the man who lifted the arras that hid the work of nature; the man who stepped down, into the everyday world, from his tower of ivory; the man who shrugged away convention and defied authority. They ratified Haldane's reputation as a brilliant, eccentric mind. You never knew what he could not do; you never knew what he would do next.

4.

Red
Haldane

OUT IN THE HOT, planed waters of the western Pacific, in January 1858, Alfred Russel Wallace left the coast of Ternate and made for the larger, squid-shaped island of Halmahera: 3 hours of rowing and sailing in a boat owned by a Chinese man and staffed by Papuan slaves. On Halmahera he rented a hut for 5 guilders a month in a village girded by low hills. The knots of limestone reminded him of Waloo, of home. Wallace, a naturalist, was there to work, but he only made a few preliminary observations of the local tribes and collected some insects before falling ill with fits of malarial fever, which sometimes left him feeling so cold that he rolled his lanky frame into blankets, even though it was a balmy 90 degrees outdoors. His thoughts slipped and tumbled around his overheated mind—his notions about species, tribes, and populations, about resources and poverty, all his obsessions eddying madly around each other.

Then, perhaps inspired by the malaria, Wallace thought of Thomas Malthus and of that clergyman's book, *An Essay on the Principle of Population*, first published in 1798. Wallace had read it more than a decade ago, but he remembered its arguments: that human populations grow in geometric haste to outstrip their supplies of food and that their numbers are kept in check only by the ravages of war, disease, and famine. Surely, Wallace figured, this law applied to other animals as well, "and while pondering vaguely on this fact," he wrote later, "there suddenly flashed upon me the idea of the survival of the fittest—that the individuals removed by these checks must be on the whole inferior to those that survived. In the two hours that elapsed before my ague fit was over I had thought out almost the whole of the theory, and in the same evening I sketched the draft of my paper, and in the two succeeding evenings wrote it out in full, and sent it by the next post to Mr. Darwin." Independently of Darwin—and fated forever to be mentioned second, if at all—Wallace had fashioned a model of evolution through natural selection.

Darwin, who had been working glutinously through his own concepts for two decades and who published *On the Origin of Species* only the following year, had absorbed Malthus as well. In the autumn of 1838, a couple of years after returning from his expedition aboard the *Beagle*, Darwin read Malthus's book—"for amusement," he wrote later, although in his Whig London social circles, it would have been impossible to avoid familiarity with Malthusian opinion. A recent law, passed by a Whig government, planned to cut expenditure on relief to the poor and to permit Dickensian conditions of squalor in workhouses. The law, a subject of scalding debate, was infused with Malthusian spirit. If the poor receive too much benevolence, Malthus had written, their population only increases, which creates still more poverty; he also recommended that workhouses "not be considered as comfortable asylums," but as a step up only for people in the most extreme distress. Struggle animates economic life, Malthusians believed, and Darwin saw that it might also animate life in the wild. Here at last, he realized, he had lit upon a theory by which to work. From 1838 onward, Darwin built his premises on these broad rails of thought.

At the very birth of evolutionary biology, then, the principles of political economy were in attendance. Through the next century, they appeared to knit tighter and tighter into each other, so that the impulse to discover how society ought to be organized was inseparable from the impulse to discover how nature organized itself. Galton's proposals for racial eugenics formed only a part of this reliance of politics on biology. There was the problem of rising poverty; until the end of his days, Darwin was gloomy about the future of humanity, Wallace found, because "our population is more largely renewed in each generation from the lower than from the middle and upper classes." There was the conundrum of the optimum economy, and evolutionary biology managed to fill the sails of different doctrines. The philosopher Herbert Spencer, among many others, concluded that the ruthless individualism of nature recommended a similar laissez-faire model for society; the phrase he coined, "survival of the fittest," was adhered as much to capitalism as to evolution. In contrast, Karl Marx read *On the Origin of Species* more than once and heard, in the struggle of species to adapt and survive, an echo of the class struggle. (He also grumbled about the book. In Darwin's natural history, he told Engels, "progress is merely accidental," whereas the course of human history was a result of coherent, material causes.) After Marx wrote *Das Kapital*, his disciples—socialist Darwinists—liked to claim that he had captured a force analogous to evolution. "Just as Darwin discovered the law of development of organic nature," Frederick Engels said at Highgate Cemetery in 1883, as Marx was being lowered into the earth, "so Marx discovered the law of development of human history."

It was not surprising that, in the nineteenth century, the social sciences looked to the physical sciences for affirmation. Science was acquiring a hard-boned rigor, promising wealth and health, revolution and salvation; its empirical precision was, John Stuart Mill wrote, "the general property of the age." But evolutionary biology, in particular, presented a neat metaphor for political philosophy. It was profound and radical, but it was also still amorphous enough to be interpreted as convenient. The themes of the two disciplines matched: They were both concerned with the coordinated destinies of millions of individuals. And by attaching themselves to the machinery of evolution, ideologues could paint its inexorable logic onto their own prophecies; they could claim to have wrung their wisdom out of the fundamental truths of the living world.

WHENEVER HALDANE SPOKE OF his political awakening, he told the story of the tram strike in the summer of 1913. At the time, horse-drawn trams trundled around Oxford, moving faster than a walking man but easily overtaken by a runner. The drivers and conductors earned less than £1 a week, and when some of them went on strike for higher wages, others continued to ply their trams. Over three days, fights broke out between the strikers and the strike breakers. The police had to descend upon the melees, armed with batons. Haldane, who was training for a boat race, missed all the excitement, so on the fourth evening, he marched up and down Cornmarket Street, hollering into the quiet the lines of the Athanasian Creed:

> *Quicumque vult salvus esse, ante omnia opus est, ut teneat*
> *catholicam fidem: Quam nisi quisque integram inviolatamque*
> *servaverit, absque dubio in aeternum peribit. . . .*

> Whosoever will be saved, before all things it is necessary
> that he hold the Catholic faith: which faith except everyone
> do keep whole and undefiled, without doubt he shall perish
> everlastingly. . . .

A hollering of Latin, recited from memory: It was a prime piece of Haldane braggadocio. A crowd gathered—large enough to block the street and prevent the strikebreaking trams from entering. Policemen tried to dispel everyone; Haldane insisted that they pushed some pious old ladies into the gutter. Nevertheless, the trams were stymied. Haldane had thrown himself successfully on the side of the agitators. Later, the university's proctor fined him 2 guineas for his act of public expression.

As a student, Haldane had been a member of Oxford's Liberal Club, but he wasn't otherwise politically active. After the war, he inclined toward a vague socialism—nothing too zealous, just sufficient to know, for instance, what had transpired at the international socialist conferences in Switzerland and Sweden. He gave up his allegiance to the Liberal Party and journeyed leftward to Labour. Once, in 1919, he even volunteered to act as a bouncer during a meeting addressed by two passionate critics of the government. When he spotted hecklers reddening the air with tomatoes, he knew what to do. Picking one of the smaller men, Haldane crept up behind him, stuck a finger in each of his nostrils, and pulled him backward, the man thrashing and helpless like a fish on a line. The others rushed to rescue their colleague. Haldane threw his bulk against them and fell down once or maybe twice; he also recollected "some rather half-hearted fighting with chairs" before the meeting room was cleared.

By 1928, Haldane had begun to give speeches to the Fabian Society, which advocated gradual rather than revolutionary progress toward socialism. He had taken his month-long trip to the Soviet Union and returned enraptured. His support for Labour had cooled, partly because, as he explained to the readers of the *Daily Express*, "I consider the present distribution of our wealth unjust . . . and I would rather see a unified industry controlled by the State than by financiers." He still denied he was a materialist—a subscriber to Marx's contention that history and even human thought were the pure products of material causes, of matter interacting with matter. But prompted by an emulsion of factors, Haldane was swivelling slowly toward socialism.

Like Darwin and Wallace, Haldane spotted a correspondence between political and scientific thought. Like Spencer and Galton, Haldane—a self-confessed utopian—felt that science, and his science of genetics in particular, ought to order society. Not in the way that eugenicists wanted, however; in fact, he was growing more and more critical of the blind ideas for genetic improvement that he heard around him. No board of doctors or magistrates could be qualified to direct the evolution of humanity, Haldane believed. He quoted William Bateson on the subject: "I would trust Shakespeare, but I would not trust a committee of Shakespeares." But he also saw that genetics was intrinsically political. This conclusion was inevitable, the historian Gary Werskey later wrote: "How, in a society based on hierarchies of class, race and sex, could a science of human differences either be kept out of the political arena or, more radically, be anything other than a political subject?"

First because of his father's work, and then because of his stints in the trenches, Haldane had known people of the proletariat to a degree unusual for anyone from his class. His social peers could say, in glib ignorance, that working men had no ethics or intelligence—that they were, as the writer R. Austin Freeman declared, such bad citizens that they were tantamount to an uncivilized, invading army.

Julian Huxley, the biologist and vocal eugenicist whose social background overlapped neatly with Haldane's, cautioned in 1931 of the propensity "for the stupid to inherit the earth, and the shiftless, and the imprudent, and the dull," and he suggested, like Ronald Fisher before him, that unemployment relief be granted only to working men who agreed to father no more children. Haldane, though, had spent enough time with such men to know that they deserved no such contempt and that they were products of their circumstances. If anything, he would reflect later, he had "learned to appreciate sides of human character with which the ordinary intellectual is not brought into contact." Even to have resided outside his country, in the mud of Europe or in India, was to have new angles added to his perspective. Rajani Palme Dutt, one of Britain's chief Communist theoreticians, observed that "of all British intellectuals showing some sympathy with Communist ideas and aims, an astonishingly small number had lived wholly in England," the writer Patricia Cockburn recalled. "For such sympathies to be aroused, it seemed to be almost *de rigeur* for a person to have had a chance to look at England from the outside." These experiences formed Haldane's views of politics and of the social worth of his science.

Human beings, Haldane held, will always be born unequal—dissimilar in appearance, mismatched in capacity, alike only in their membership to the species. Genetics demonstrated this; in fact, genetics insisted that this diversity was essential to the health of the species. So eugenicists could do no more than try to discourage the weakest 1 percent of the population from propagating their genes and to convince those of exceptional ability to have more children. But the real duty of biologists lay in determining the talents of young individuals, so that they could be steered into professions of optimum utility. And this was most easily achieved, Haldane thought, under watchful socialism. It needed the strong guiding hand of central planners. Socialists already believed in examining a person's utility to the state—a utility based on a capacity for work, not on frippery like riches or social status. "Any political movements which diminish the importance of inherited wealth are eugenically desirable," he wrote.

Of course, he didn't think that scientific possibility was limited to these clerical tasks of sorting and arranging people. If the human race managed to take its evolution into its own hands, progress knew no bounds. Haldane grew positively lyrical at the prospect:

> Less than a million years hence the average man or woman
> will realize all the possibilities that human life has so far
> shown. He or she will never know a minute's illness. He
> will be able to think like Newton, to write like Racine, to
> paint like the van Eycks, to compose like Bach. He will be as
> incapable of hatred as St. Francis, and when death comes at
> the end of a life probably measured in thousands of years he
> will meet it with as little fear as Captain Oates or Arnold von
> Winkelried. And every minute of his life will be lived with
> all the passion of a lover or a discoverer. We can form no idea
> whatever of the exceptional men of such a future.

For socialism to flourish—for the state to interfere intelligently in industry and society—it needed to possess a scientific way of thought. The socialists Haldane had heard and read for the most part recognized this. They respected scientists, and they sought a state that would be governed by technicians. Critics of the Soviet Union— particularly Bertrand Russell and John Maynard Keynes, Haldane's colleagues at Cambridge—drew from their brief tours of the country to make their points. "It is hard," Keynes wrote, "for an educated, decent, intelligent son of Western Europe to find his ideals here, unless he has suffered some strange and horrid process of conversion which has changed all his values." So Haldane did the same, harking back frequently to his own trip and to what he had seen as the Soviet Union's strident reverence of its scientists. The Russians were applying science to life in ways that were embryonic and crude, but they were the only ones even making the attempt, he said. In England and America, the public mind still obsessed over trivial matters like sport and religion; in the USSR, science was sunk deep into society. Soviet children studied more science than English children—and studied it not by memorizing their textbooks, the way they might have for the rules of French grammar, but by learning how science related to ordinary life. At Sverdlov, the Moscow university that trained new party workers, Haldane learned that the students spent many months learning cosmology, evolution, chemistry, and physics. He walked about central Moscow, finding more bookshops per hundred yards than in London, and in nearly every display window, he spotted books on science or technology. *Pravda* covered science better than *The Times*; the puzzle pages in the newspapers ran competitions to identify Newton, Einstein, and William Harvey from their pictures. Even the most slender opportunity for education was not lost.

Sometimes, though, Haldane observed but did not understand. Or he saw only what he cared to see.

Months before he arrived in Moscow, in 1928, the state's secret police arrested 53 engineers in Shakhty, a town in the North Caucasus so indelibly associated with collieries that its name translated literally to "mines." The engineers were accused of conspiring with the former owners of these mines to sabotage the extraction of coal from the rich seams of the Donetz Basin. Production had been falling; quotas were going unfilled. This could only be, Stalin decided, because these engineers, "the old specialists," weren't pledging themselves enough to Communism. So that summer, the very summer Haldane was in the Soviet Union, the engineers were put on trial. It would be—although he couldn't have known this—the first of a battery of show trials that continued far into the next decade.

But Haldane couldn't have missed the magnetic spectacle of the trial itself. The government advertised it as if it were a prestigious new ballet—in newspapers, on the radio, in newsreels, in announcements at schools and universities, on billboards. The trailers portrayed the engineers as dark-hearted villains, so that the whole enterprise acquired, as one American journalist described it, "a spirit of festival touched with hysteria—a crowd come to see a righteous hanging." When the trial—at which Haldane's acquaintance, A. N. Bach, was one of the prosecutors—finally began on May 18, audiences were issued tickets. Over six weeks, nearly 100,000 people witnessed some part of the proceedings—students, factory workers, Young Communists, delegations of farmers; everyone clattered through the polished halls of the once-haughty Club of Nobles in Moscow to watch the drama of treachery and indictment. Movie cameras recorded every disclosure for broadcast. The chief prosecutor, Nikolai Krylenko, dressed for effect, wearing puttees, riding breeches, and a hunting jacket, as if riding in hot pursuit of prey. The beam of the klieg lights rebounded off his shaved head. Each of the engineers would fold and confess, the state thought; in fact, 37 of the 53 rejected every charge that was loosed at them. The thundering Krylenko, the "glorious accuser," Aleksandr Solzhenitsyn later wrote, "stepping into what was for him a new field—engineering—not only knew nothing about the resistance of materials but could not even conceive of the potential resistance of souls." Only four men were let off. Five were executed in July, another four earned suspended sentences, and the remaining men were placed in prison for between 1 and 10 years.

Haldane recalled his time in the Soviet Union often, but he never spoke or wrote about this corruption of justice that ran concurrent to his visit. Only once, in a lecture to the Fabians in October 1928, did he even refer to the Shakhty affair. When he had been in Moscow, "a shop window display of books bearing on the trial of the Donetz colliery engineers was 'starring' a volume on the relation of geology to economics," he said. "They are trying to link up every branch of science with every branch of political life." That single, flat note of admiration was all.

In contrast to the Soviet Union, Haldane groused, the British establishment—the politicians, the bureaucrats, the army generals, everyone in temporal power whom he was fond of rebelling against—cared too little for science. Funds were sparse; facilities were poor; even the most basic scientific knowledge appeared to escape the government. A classic example: During the war, Britain allowed Germany to import fats, contending that the manufacture of explosive glycerine from fats was only a recent advance in chemistry. In fact, the discoverer of glycerine had, in 1779, obtained it from olive oil and called it "the sweet principle of fat." "Up till the war," Haldane remarked acidly in his lecture to the Fabians, "I do not suppose that a pint of glycerine had ever been made from anything but fat." But in his eagerness to prove his point, he resorted to the occasional dubious claim. "In India to-day far more first-rate research in pure physics is being done than in the majority of European countries," he said in 1928—10 months after Paul Dirac proposed an equation for the wave function of the electron, a year and a half after Werner Heisenberg broached his uncertainty principle, and barely a year before Louis de Broglie won the Nobel Prize for discovering the wave-particle duality of quantum objects.

In his advocacy for the scientific nation, Haldane was baffled by those who forecast a colorless, industrial quality to life in such a state. "It is only in so far as we renounce the world as its lovers that we can conquer it as its technicians," Bertrand Russell wrote in 1931. "But this division in the soul is fatal to what is best in man." Haldane, though, had experienced just the opposite. "Until I took to scientific plant-breeding I did not appreciate the beauty of flowers," he wrote, responding to Russell. "If I find out how to produce a certain change in the composition of my blood, I want to know what it feels like, to appreciate it as a fact of life as well as a fact of chemistry." Science provided truth, but comprehension and clarity also roused an aesthetic sense.

Haldane was confident that a socialist society didn't have to be sterile and uniform; if he suspected that even a little, he wouldn't have given socialism a second look. He was aware of his insatiable curiosity, and he fed it diligently—through his reading, but also through his personal relationships. The people who interested him were not "standardized people," he once wrote; they were people who deviated, who did unexpected things. Some of them joined jazz bands or chose to carry messages from Belgium to Holland during the war; others, it happened, were confirmed leftists. In Cambridge, the deviants were the socialists, burrowing their way into the dense, packed soil of the university's conservatism. The Cambridge Union gained its first socialist president in 1925, and then another in 1927; one teacher wrote that "the Russian experiment," felt to be "bold and constructive," was invoking keen interest among undergraduates. After a debate in 1925, 61 students voted class warfare to be a desirable prospect—far less than the 431 on the other side of the motion, because this was still Cambridge, but still a number with enough meat to stick in the gullet.

After the decade turned, the heat spread. By 1933, nearly a fifth of Britain's working population were unemployed, a crisis easily blamed on the capitalistic excesses that set off the Great Depression. Supporters of Labour thought that their prime minister, Ramsay MacDonald, had first hollowed their party and then abandoned it, in 1931, to snuggle into bed with the Conservatives. The nations of Europe began to rearm themselves, even though the memory of the First World War was still green and fresh. The Cambridge Union voted again and again on motions supporting peace and disarmament. In 1934, the university hosted an anti-war exhibition. That year, the *Girton Review*, published by the Cambridge college of that name, weighed in: "It is possible to conceive of a government, frightened by discontent and unemployment at home, welcoming war as a last resort to silence these murmurs by a stirring call to fight for the fatherland." Between 1933 and 1936, the Socialist Society on campus tripled its membership from 200 to 600, and its politics acquired a Marxist accent. The post of the society's treasurer was filled for a year by Kim Philby, who graduated with a degree in economics and went on to become the most notorious of the five Cambridge double agents spying for the Soviet Union. Even without the skulduggery, though, the radicalism of the left attained a gloss of romance and flamboyance. One student, who had briefly spent time in jail in Göttingen for participating in anti-Nazi protests, returned to Cambridge and enlisted in the Communist Party, then strode into Trinity College with a hammer and sickle pinned to his lapel.

Even if Haldane had resisted the appeal of these unstandardized people and their unexpected ideas, he couldn't have avoided the enthusiast under his roof. Already during their courtship, Charlotte had talked about her keenness for the Soviet Union. "I wish to heaven or hell, as you please, we could emigrate there for a few years," she wrote. She had a friend, she said, who was a big noise in the passport office; he could arrange papers for her to visit. There were times when the state of the world drove her to feel utterly Bolshevistic, and on one of those days, she visited a Soviet exhibition at Chesham House.

She brought back gifts for Haldane: a box for his pipes and tobacco, with a wooden rabbit crouched upon its lid, and a small, glazed statuette of a Russian peasant. Much later, in her memoir, Charlotte would claim that she returned from their trip to the Soviet Union with muddled feelings, but it didn't stop her from a slow migration toward the Communist Party of Great Britain (CPGB), which she finally joined in 1937.

Together, the Haldanes enlarged their well of empathy for the beaten-down people of the world. In 1933, once in the spring and again in the summer, they traveled to Spain, rattling by train through Andalusia, struggling to understand the deprivation that greeted them. Harvests had been poor; unemployment was high. This was a land without bread, *tierra sin pan*, as the title of a Luis Buñuel documentary put it that year. The poverty, Charlotte thought, was bad in Cordoba, worse in Granada, and almost universal in Seville. "The peasants were starving everywhere; they squatted in miserable huts of wattle and straw, in rags, barefoot," she wrote. "The town workers were even worse off, for they had neither work nor food. Everywhere was economic, mental and physical depression. . . . It was impossible to enjoy the beauty and the glamour as a tourist. As a human being it was impossible to remain insensitive to the misery of so many of one's fellow human beings."

Whether socialism could successfully lift these people up off their knees, Haldane couldn't say at the time. History provided no guide for this. At an intellectual level, socialism was, to Haldane, a monumental experiment. Either it would increase human happiness or it would not—but there was no way to tell until it had been tried and its results had been dispassionately studied. Haldane liked this ability to view the world as if it were a collection of test plots at John Innes; he liked this imposition of the scientific upon the social, the rational upon the arbitrary.

But there must have been more visceral reasons for his gravitation to the left. Among the radicals, Haldane, still shy and awkward beneath his bluster, found a sense of camaraderie—the kind he had never encountered at Eton or Oxford but had discovered in the trenches. Since the war, Haldane's adventures had been, as he put it once, intellectual and emotional only. By embroiling himself in socialist politics, he saw a chance to return to the front, to fight for a cause alongside his comrades. Haldane's sister diagnosed another factor. Naomi saw how he had begun to cut luxuries out of his life ("except," she noted, "the spiritual luxury of open quarreling with our mother"), how he had chosen to marry a woman from a background so dissimilar to his own, and how he had abandoned his once-beloved project of writing a biography of their father. She saw his past rear into conflict with his present, and she understood that his birth into Britain's privileged classes now discomfited him. Declaring himself a socialist, Naomi thought, was Haldane's way of breaking with that past, a way of hushing the guilt that pricked him.

In its endeavour, science is Communism.
    —J. D. Bernal

By deciding to tread the road of scientific socialism, Haldane wasn't alone. Within the heart of the British academy, a small, tight knot of scientists grew attached to the radical left, publicly espousing socialism and Communism. The historian Gary Werskey, writing in 1978, called them the "Visible College"—a reference to the Invisible College of the 1640s, a progressive group of natural philosophers who wished, with their work, to "take the whole body of mankind for their care." The Visible Collegians all had varying degrees of conviction in the redemptive power of science, and they reached socialism through different routes, but their collective arrival still gave the British left a valuable store of intellectual ballast. Their influence, in organizing and radicalizing their colleagues, was disproportionate to their numbers, and their message grew increasingly urgent as war crept up on Europe. By the end of the 1930s, the historian Neal Wood wrote, the movement to marry science to social planning "seemed almost to dominate the British scientific world."

The Visible Collegians were all men. In addition to Haldane, there was Hyman Levy, a Scottish mathematician at Imperial College in London, who joined the CPGB in 1931. Lancelot Hogben, an English zoologist, had worked in South Africa until, repelled by the country's racial policies, he returned to a chair for social biology at the London School of Economics. Joseph Needham, a biochemist, worked in Haldane's department at Cambridge. J. D. Bernal, an Irish crystallographer who was also at Cambridge, had joined the CPGB as early as 1923, after a friend converted him to socialism over the course of a long, squabbly night. ("My old life was broken to bits and the new lay in front of me," Bernal wrote in his diary.) These five aside, a wider, penumbral circle of scientists across Britain shared some of the left's concerns and agreed with some of its principles. The Cambridge Scientists Anti-War Group, for example, began by meeting in a basement room during the lunch hour and issued letters protesting the use of their research by the military. By the middle of the 1930s, the group counted 80 members, women as well as men. But none of them laid out their reasoning for socialism, or agitated for it, quite as strenuously as the five Collegians.

Like Haldane, the other Collegians had come to think that science was best nourished by socialism and that their own country was interested in research only for the aims of profit and war. We have, Haldane said in a speech, "in the entire British Empire only two professors of genetics," whereas the Soviet Union "is crawling with them." He didn't cite a source for this statistic, but Bernal explained the dismal state of British science at greater length in his book *The Social Function of Science*. Only 0.1 percent of Britain's gross national product was dedicated to research and development—an amount that Bernal called "ludicrously small," before pointing out that "the greater proportion of what is spent is wasted on account of internal inefficiency and lack of co-ordination." On the other hand, between 1933 and 1937, the government's expenditure on defense research increased by 50 percent. "Between one-third and one-half of the money spent on scientific research in Britain is spent directly or indirectly on war research," Bernal wrote—and there wasn't even a war on! (That changed soon, of course.) The Soviet Union, he calculated, was pouring nine times as much money into scientific research as Britain. And in its centrally coordinated economy, he believed, these funds were being managed efficiently, every ruble squeezed for its contribution to long-range plans.

The scientists also sensed a philosophical communion between science and Marxism, even if their views on the doctrinal details diverged. (They never came together, for instance, on the smudgy jargon of diamat—the abstract dogma that explained change in the world through the interactions of physical conditions. Diamat rebuked the earlier idealistic perception that reality was only an immaterial construct of the mind. In the official versions of diamat that Lenin and Stalin promoted, development arose out of the struggle of opposing material forces—a tenet they insisted on applying not only to history but to science and society. Hogben called diamat "obscurantist rubbish" and chastised Haldane for having "swallowed it hook, line and sinker." Levy scoffed at diamat's "almost medieval language" and woolly theory, then changed his mind and urged the application of diamat even to ordinary life. Bernal called it "the most powerful factor in the thought and action of the present day.") Like science, Marxism sought to organize knowledge and experience and to discern connections between events; like scientists, Marxists set as their objective the material improvement of the human race; like the laws of science, the laws of Marxism transcended national boundaries. More than one Collegian offered the thought that Marxism was really derived from the scientific method—a notion that perplexed some of their fellows in the party.

Trawling through Marx and Engels, Haldane found some particular, personal resonances as well. Marxism was utterly areligious. It privileged practice over theory—something that Haldane, as a boy, had seen his father do in his own field. Diamat concerned itself with understanding and controlling change—similar, in that overbroad sense, to the quest of genetics. It stood against war and against the capture of science for war, and these were both worries that had echoed through *Daedalus*. In that book, Haldane had shoved religion aside and called for another system of morality to guide science. In Marxism, he recognized that system.

Like Levy and Bernal, Haldane regarded diamat as a guide to everyday living. Its influence slipped easily into even his most ordinary thoughts. Once, sent a handsome new calendar as a gift, he wrote back: "I have only one possible serious criticism of the calendar.

It seems a pity to have to tear out the pages of such a beautiful and instructive production, particularly as there is no margin for binding it. However, I expect this criticism is undialectical." It was impossible to tell if he was jesting.

He persuaded himself that diamat, despite having been conceived more than half a century earlier, explained the known universe to near-perfection. That was proof of its fundamental veracity. He wrote and spoke about this subject plentifully, and perhaps it was evident even then—it certainly is now—how hard he was trying to stuff science into the vague rubric of diamat's language. He published an essay in the summer of 1937, for instance, titled "A Dialectical Account of Evolution," in which he performed torsions of interpretation to conclude that natural selection's processes fell into the dialectical triads of thesis-antithesis-synthesis. He produced a table:

THESIS	ANTITHESIS	SYNTHESIS
Heredity	Mutation	Variation
Variation	Selection	Evolution
Selection of the fittest	Consequent loss of fitness	Survival of noncompetitive species

These categories felt artificially neat even to Haldane. "I am perfectly aware," he wrote, "that these represent a certain abstraction from reality. Thus selection probably affects the rate of mutation. An animal species alters its environment to some extent, and the human species does so to a great extent. Thus the evolutionary process itself affects the environment, which in turn determines its direction." Nevertheless, he insisted, diamat was a clear, well-ground, powerful lens with which to scrutinize biology.

But one critic, the economist A. P. Lerner, scolded Haldane for turning diamat into a convenient pigeonhole for cherry-picked facts. Almost all mutations lower the fitness of an organism, Haldane had written, trying to portray mutation as a dialectical process called "negation." Lerner pointed out that mutations might be harmful in one environment but beneficial in another and that Haldane had skipped past this basic truth to coast into his argument. "Is there any process that cannot be shown to be dialectical in such a way?" Lerner asked. The items of Haldane's dialectical triads, similarly, carried no historical or logical relationship to each other. Sometimes the triad didn't even function as intended; "selection" was a force that intersected with "variation," not a contradiction of it. Applying the technique of the triad honestly would also have required, at the very core of genetics, that the development of an organism influence its heredity—the old Lamarckian idea, which found such favor under Stalin. But the genetics of Haldane's time had recognized that an organism's development did not materially affect its genes at all. This was the devilish problem, as John Maynard Smith, one of Haldane's students, admitted decades later: "Classical Mendelian genetics was damned undialectical."

Sufficiently convinced, though, Haldane evangelized on. In 1938, he delivered two lectures—the Haldane Memorial Lecture at Birkbeck College, named for his uncle, and the Muirhead Lectures at the University of Birmingham—and he used both opportunities as pulpits. (The latter speech was expanded into a book titled *The Marxist Philosophy and the Sciences*.) He explained the staples of Marxism: its view of history, the injustices of capitalism, the assured advent of revolution. ("In the week before this lecture was given," he added as a footnote in the published version of his Birkbeck speech, "over 600 Londoners joined the Communist Party.") But he also spoke of the value that Marx and Engels brought to science—of how, for instance, thinking like them might make it easier to understand and accept the mind-bending theories of relativity or quantum theory. A favorite trope involved picking up one of Engels's blurry laws of diamat and finding a matching example from the scientific world. To illustrate how quantity changed into quality—a postulate of Marxist thought—Haldane fetched one of his father's experiments, which showed that ordinary blood always held a volume of dissolved carbon dioxide: "If this is either doubled or halved serious symptoms arise. In fact, too much of it is a poison, but a certain amount is a necessity." The quantity of the gas determined the qualitative passage into life or lifelessness.

Most of his correlations between Marxism and science went like this, formulated cleverly and lit by a slight flame of insight, but made plausible only because the language of diamat was so baggy, so easy to wrap around nearly anything. With Marxism as a tool of analysis, Haldane felt able to integrate all his ideas, all his interests. Andrew Rothstein, one of the oldest members of the CPGB, wrote in a review of *The Marxist Philosophy and the Sciences*: "Like anyone on his first acquaintance with the Marxist method, thunderstruck at the new worlds which that touchstone opens in seemingly familiar things, Haldane hastens to run over the whole range of his knowledge, calling 'Open Sesame'—with great profit to himself and his audience."

Did dialectical materialism direct Haldane's scientific work itself? Werskey argued that Haldane, Hogben, Needham, Levy, and Bernal were mature scientists by the time they settled into Marxism. "Their investigations were already enmeshed in a web of fully tested methods, techniques and assumptions," so diamat stayed outside their laboratories. But in 1968, a biologist named C. H. Waddington ventured in a book review that the revolutionary new theory of the origin of life was first developed by Haldane and Oparin, Communists both. Was it just a coincidence that the theory was a materialistic one?

The timing, however, sat awkwardly with Waddington's supposition. By 1929, when Haldane published his essay on how life began, he had traveled to the Soviet Union, where he was introduced to Marxist thinking on science. He had been impressed by it, but he didn't embrace it instantly. "The process took me some six years, so it was hardly love at first sight," he wrote later. In 1938, even as Stalin's show trials reached their height of totalitarian horror, Haldane delivered the Muirhead Lectures and announced, "I have only been a Marxist for about a year," dating his conversion precisely, as if someone had asked him when he had moved into a new house. Besides, it wasn't as if the Haldane-Oparin hypothesis was any more materialistic than theories like panspermia and spontaneous generation that had come before it.

In fact, it's difficult to spot any overt materialist pressure guiding Haldane's research, even after his immersion into Marxism, or to detect any differences in the tenor of his work before and after his conversion. "Dialectics will not tell you what you are going to discover," he wrote once, in a letter. "It would be fatal to science if it were thought that there was any substitute for experiment and observation." The true substance of his theories, calculations, and experiments lived in a pristine space, untouched by anything except the scientific method of inquiry. Rather, it was that scientific method that Haldane spotted within Marxism, which was why he grew so pleased with its philosophy.

The most momentous effect of Haldane's Marxism was not how it pushed into his work but how it pulled him out. Once he had believed that a scientist was of more use to his fellows in the laboratory than outside it. The success of *Daedalus* and his subsequent eminence altered that view; Marxism renovated it further still. He felt activated by its spirit, its mission to direct science for public benefit, and its promise of might against the howling beasts of fascism. "I began to realize," he said later, "that even if the professors leave politics alone, politics won't leave the professors alone."

POLITICS STOPPED LEAVING Haldane alone when MI5 opened its file on him. The very first document that went into it was a brief typewritten note, with the spelling of "Pharmacologists" corrected in ink:

On March 22nd, 1928, Professor J. B. S. Haldane of Cambridge accepted an invitation extended to him through Madame Polovtseva of the S. C. R. to attend a conference of Russian Physiologists, Biochemists and Pharmacologists, which was to take place in Moscow from the 28th May to the 22nd June, 1928.

The SCR, or Society for Cultural Relations, had been founded four years earlier, by artists and intellectuals like E. M. Forster, Julian Huxley, Virginia Woolf, and Alexei Tolstoy, to improve ties between the people of the Soviet Union and the people of Great Britain. Even that intention of amity must have smelled faintly Communist to British intelligence, so MI5 kept a close watch on the SCR and its activities. After Haldane traveled to the Soviet Union, he moved into MI5's field of scrutiny as well, and he remained there for the next quarter of a century.

Surveillance report of Haldane, filed with MI5 upon his return to
England from a trip in January 1938.

Haldane's file is a thick one: page after page of reports and observations that rarely mention his work—for MI5 had no use for him as
a scientist—and follow his dalliances with the left in detail. It was as
if Haldane had an unknown audience scampering at his heels, trying to puzzle him out. Why did he go here? Whom did he meet
there? How Marxist was he? What was he up to? The answers were
never instantly obvious, but MI5 must have hoped that by doggedly
unpicking the stitches in Haldane's life, the garment would fall
entirely away and some shocking truth would show itself.

The file contains all manner of reports. There are clippings of pieces that Haldane wrote for newspapers. There is a list of addresses for the houses he lived in. There are snapshots of letters to Haldane and from him, intercepted and photographed in the negative; with their bone-white text on a scratchy black background, they resemble X-rays. There are accounts of his departures from, and arrivals into, England—short notes that are always variations on the same theme:

> Professor J. B. S. HALDANE and his wife . . . returned
> to this country this morning aboard m. v. "Parkeston." A
> discreet search of their baggage by H. M. Customs revealed
> nothing of interest to Special Branch.

Not once was anything of interest to Special Branch ever found in Haldane's baggage.

More kinds of items: gossip, assessments of his mood, responses to foreign intelligence agencies asking about Haldane. Whenever he gave a public speech anywhere in the country, an agent loitered in the audience and then composed a summary. Haldane's name was typed in upper case, and someone always penciled in, around the name, a pair of red brackets:

> A meeting was held on 21.6.37. at the Ambassador Cinema
> Ballrooms, Hendon Central Circus, Hendon under the
> auspices of the Hendon Peace Council, at which Professor
> HALDANE spoke on the subject – "The next War – must
> it come"?

The surveillance was a tiny part of MI5's quest in the 1930s to track subversive political activity at home. The British Home Office was reluctant to permit the agency to monitor fascist groups, but it readily dispensed warrants to peer into the lives of Communists. Thousands of workers in the bureaucracy, the postal service, the dockyards, and ordnance factories came under investigation. Scientists were scrutinized; so were poets and novelists like George Orwell, Arthur Koestler, W. H. Auden, and Cecil Day-Lewis. These plots of domestic espionage upon intellectuals were frequently shortsighted or ham-handed; Orwell remarked that "the policeman who arrests the 'red' does not understand the theories the 'red' is preaching." MI5 was confident that Britain's Communists had sinister ideas and that their loyalties rested in the east, in Moscow. All the same, it failed absolutely to detect the betrayals of Philby and the four other double agents who came to be known as Stalin's Englishmen.

As Haldane grew closer to the CPGB, and after he formally joined it in 1942, another avenue of espionage opened up. The party's head office, on King Street, had been bugged, and a wiretap snooped on the telephone calls made out of the office or patched into it. An MI5 agent even became a personal assistant to Harry Pollitt, the secretary general of the CPGB. Whenever Haldane dropped by King Street, a transcript of the proceedings made its way into his file. When he called or when others talked on the phone about him, those conversations were condensed for his file as well. MI5 became Haldane's first, if incomplete, biographer.

IN CAMBRIDGE, THE HALDANES lived in a house that had once been the village inn and that had hosted so many wild undergraduates in its time that it had acquired the name "Big Hell." They called it Roebuck House now, after it had been refashioned into a country home, its lawns tilting gently toward the river. But it still held a windowless room within, which was once the venue for loud, crowded cockfights. There was a greenhouse on the grounds, and when Haldane was working with primroses at John Innes, he suggested to Charlotte that she take in a thousand crossed plants, to note down their varying physical characters. He was harkening back to a time when he and Naomi had worked side by side at Cherwell. Charlotte agreed readily; then, after she discovered that she didn't have the mathematics to move the project beyond its first few stages, her interest expired.

She wasn't happy in Cambridge, she realized by and by. She thought it snobbish and provincial and felt that its society held her at arm's length, as if she was faintly tainted. "In those days," her son Ronnie later said, "you were either somebody or nobody if you were a woman. Charlotte wanted to be somebody. She had to be aggressive." So she built her own society—an assorted set of young writers, artists, scientists, and leftists who gravitated toward her hospitality at Roebuck House: an English major named William Empson, who went on to become a well-known literary critic; an aspiring novelist named Malcolm Lowry, who would write *Under the Volcano*; a sociologist named Charles Madge; a scientist named Martin Case, who had studied with Haldane and lived in Roebuck House and played jazz on the Bechstein piano in the drawing room. Not everyone liked Charlotte as much as she hoped; Madge told Empson's biographer, many years later, that he thought her "a dangerous and unpleasant person: spiteful, always gossiping about people in a destructive kind of way." But everyone appreciated her generosity. Empson wrote a poem for her birthday. Lowry, smitten, told a friend, "I don't think I have ever seen anybody so pretty."

Haldane called the gathering "Chatty's addled salon." He detested most of them, Empson had heard, but he was still courteous, responding to questions with "a brotherly gruffness" and occasionally sauntering down to the river for a swim with the gang. He swam slowly, gravely, his steaming pipe sticking up above the water like a tiny smokestack. His famed rudeness, Empson wrote later, "was mainly a refusal to be bored, hurrying the discussion on to where it would get interesting . . . he was curious to know what we did think, perhaps as part of a social inquiry, but at that age one will put up with a lot for being taken seriously at all."

Maybe it was Haldane's own affinity for experimentation or maybe it was his proximity to the daring young men and women in Charlotte's court that led to his light dalliances with drugs. For a fortnight, he gave himself regular fixes of heroin, but he hated how it interfered with his mental faculties, so he stopped. Much more regularly, almost until the end of his life, he swallowed Benzedrine to stay awake—once as much as 40 milligrams, so that he could drive 500 miles through the night—and he took hits of monosodium phosphate as a purgative and to lift himself from fatigue. But if any substance had to unbalance his mind at all, he preferred it to be whisky or beer. The Haldanes regularly closed down pubs late into the night with the members of Charlotte's court or with the members of the Cambridge University Benskin Society, a drinking club.

Once Martin Case, leaving the Red Cow and getting behind the wheel of his car, got a ticket for drunk driving. The police had one witness: the pub's night watchman. But on the afternoon of the hearing, Case gaped astonished as the night watchman, destroyed by drink, poured himself with difficulty into the witness box, loosed off a string of unintelligible sentences, and collapsed.

The charges were dismissed, and Case returned home to tell the Haldanes of his reprieve. Oh dear, Haldane said, he managed to get to court, did he? Then Case learned that Haldane had spent the morning with the witness in a pub, plying him with booze, liquefying his testimony. But don't be too grateful yet, Haldane said, pulling from his pocket a ragged piece of paper on which were scribbled the itemized expenses of his bender with the night watchman. He liked Case well enough, but he still expected to be fully reimbursed.

THROUGH THE 1930S, Haldane's writing surged, abetted by Charlotte, who started a small syndication agency for scientific articles, acted as Haldane's agent, and placed him in the leading popular journals. From a man who had once warned a friend that publishing in the popular press was a sure road to quackery, he became a scientist with a side line in journalism. "I think," he now wrote, "the public has a right to know what is going on inside the laboratories, for some of which it pays." His articles, scores and scores of them, were everywhere: in Britain, in the *Spectator*, the *Daily Express*, the *Daily Mail*, the *Manchester Guardian*, the *Rationalist Annual*, and *Discovery*; in America, in *Harper's Magazine*, *Forum*, the *Atlantic Monthly*, and the *New Republic*. (The geneticist James F. Crow once cracked wise: "Question: Who is the most widely read biologist? Answer: Haldane—he only had to read what he wrote to have read more than anyone else.") In December 1937, the *Daily Worker* announced that it had bagged Haldane as its new columnist, with a mission "to make science plain for you." It hadn't been Haldane's plan to place these pieces in the house newspaper of the CPGB; in fact, he had first approached the *Daily Herald*, only to be told that the idea wasn't quite fascinating enough. But the overlap of science and leftist politics suited him well. He would treat his subject, the *Worker* promised, "from the point of view of the active fighter for socialism. In an article on blood transfusion, for example, you will learn not only the scientific facts, clearly and simply stated, but what they mean to a Socialist."

That might have proved an ill-fitting combination of subjects had it not been for the fact that Haldane, as much as he loved socialism, loved science even more. He wrote out his *Daily Worker* columns—more than 300 of them—longhand, in his lined notebooks: as always, the text on the right-hand page, notes and corrections on the left. He struck through perhaps one word per page; it was as if he was pouring his thoughts onto the page fully formed. The columns wore their science lightly, and the political sermon usually came toward the end, like a small vestigial organ flapping weakly in the breeze. He grew so skilled at this craft that he once published a piece titled "How to Write a Popular Scientific Article," a distillation of his methods that was filled with good sense. His first piece of stern advice: know a great deal more about your subject than you put on paper. Then look for a familiar analogy; pull it out of the facts of everyday experience:

> Compare the production of hot gas in the bomb to that of steam in a kettle, the changes which occur in the bird each year to those which take place in men once in a lifetime at puberty, the precipitation of casein by calcium salts to the formation of soap suds. If you know enough, you will be able to proceed to your goal in a series of hops rather than a single long jump.

Some of his suggestions might well have come from a scientifically minded Hemingway. Go slow. Keep your sentences short. Use active verbs. Enunciate your theorem only after you provide its proof. Consistently, Haldane looked to the daily headlines for inspiration. Equally, he quoted Dante or Heraclitus when he felt like it—as a way, he said, of showing the continuity of human thought.

A fine way to appreciate Haldane's output is by snacking one's way through his articles, a taste of this and a taste of that, just enough to marvel at how great and varied the feast is. But staying for a while with one of his classics, "On Being the Right Size," is a wise idea. Its very premise is a stroke of brilliance. The first difference that we observe between animals of dissimilar species is their size relative to each other, yet no textbook makes plain *why* these sizes come in such a wide assortment and what the consequences of smallness or bigness are. "For every type of animal there is a most convenient size, and a large change in size inevitably carries with it a change of form," Haldane wrote. A human giant 60 feet high, of the kind Christian encounters in *The Pilgrim's Progress*, would weigh a thousand times as much as an ordinary man but have to bear that weight on bones only a hundred times as strong. With every step, a giant would break his femur. He would need sturdier thighs to conduct his rampages; his form would change.

Haldane discussed other cases organisms whose size is best adapted to their mode of functioning: gazelles, rats, algae, microscopic worms, crabs. As size increases, an animal's organs have to become more complicated: hearts grow more powerful, to pump blood around the body; intestines are coiled, so that they can be both long and compact; a hundred square yards of lung fold themselves snugly into a human being's chest. He called on simple arithmetic to show that if a large bird weighs 64 times as much as a small one, it requires 128 times the power to keep itself aloft. The life of a big bird is an expensive one, so natural selection limits avian size. "Were this not the case eagles might be as large as tigers and as formidable to man as hostile aeroplanes."

But there are also advantages to generosity of size. It's easier to keep warm, for one thing; a mouse has to eat a quarter of its weight in food daily just to maintain its temperature. (This is why the smallest mammal native to Spitzbergen, in northern Norway, is the fox.) The eye and brain become more efficient, too. "Such are a very few of the considerations," Haldane wrote, "which show that for every type of animal there is an optimum size."

"On Being the Right Size" is a model piece of writing by Haldane.

Having found a scientific riddle hidden in plain sight, the essay proceeds, slowly and transparently, to unpack it. Its tone is persuasive and didactic but—given how prickly Haldane was in person—surprisingly gentle. It is filled with examples that reflect the riotous variety of life. Its concessions to mathematics are smooth and sufficient: "Divide an animal's length, breadth, and height by ten; its weight is reduced to a thousandth, but its surface only to a hundredth." There are sparkles of wit, reminiscent of the playfulness of his Eton chemistry teacher, Thomas Porter. After recalling the illustrated *Pilgrim's Progress* of his boyhood and deducing the inadequate thighs of its giants, Haldane writes, "This was doubtless why they were sitting down in the picture I remember. But it lessens one's respect for Christian and Jack the Giant Killer." More than anything, the essay owes its very existence to one of Haldane's cherished beliefs: that no scientific question is too trivial to be explored and explained.

Some of Haldane's *Daily Worker* columns read better than others; some have aged gracefully, while others became outdated in Haldane's own lifetime. But they all share the essential attributes of "On Being the Right Size." Haldane's curiosity led him by the nose into every discipline—cosmology, chemistry, physics, theology, politics—and he pulled his audiences along. He sketched out the beginning of life as well as the end of the world. He introduced to his readers the lives and work of Newton, Archimedes, Copernicus, and Bateson. He chased down the most primal questions:

What does it mean when something is hot?
What is instinct?
What is it like inside the sun?
What is life, biologically speaking?
What, for that matter, is death?

And then he managed to supply both material answer and profound reflection. Instinct, for example—instinct is not the same as reflex. Our digestive process is a series of reflex actions. A female cat, taken from her mother at birth and fed from an ink dropper, will have no memory of motherhood, but when she bears her own kittens, she licks and suckles them by instinct. "We reserve the word instinct for actions of a kind which in ourselves are conscious and willed, and may be reasoned." In a more complex organism like a human being, instinctive behavior is less fixed. "We have to learn most of our behavior. And therefore we have greater possibilities for good or evil than any other animal."

Anything could set Haldane off into a spell of inquiry. "My wife has just started a raging cold. By the time this article is printed I shall probably have one too," he wrote, and began to wonder why the human nose is such a vulnerable point. "A few days ago winged ants were swarming in the outskirts of London," he wrote, and that led him into ruminations about insect societies. "In several daily newspapers of July 1, 1938, there was a story of an Egyptian called Rahman Bey, who threw himself into a trance, and stayed for an hour at the bottom of a swimming bath in a metal tank," he wrote, before telling of his father's experiments, and his own, on breathing. He kept pace with research in astronomy—he seemed to keep pace with research in everything, really; so after reading a report about two American astronomers who were looking for faint blue stars, he wrote a column about the significance of stellar color spectra. He thought about honorific titles like His Holiness the Pope or His Majesty the King, remarked that "I should like to be able to speak of his Ferocity the Major-General, his Velocity the Air Marshal, and his Impiety the President of the Rationalist Press Association," and then pondered over our propensity to lift humans into saint-hood or godliness.

Haldane's ability to stalk confidently over the full map of human scholarship gave him his renown as a polymath and made it seem indeed as if he knew all that there was to be known. Anec-dotes attached themselves to this reputation, enlarging it fur-ther, like iron filings fattening a magnet. His colleagues liked to describe Haldane's overstuffed attaché case, in which he always had a stray piece of paper on which, during moments of quiet in the train or in the pub, he knitted away at his algebra. Martin Case told an interviewer:

On one occasion, J. B. S. was playing Patience in a room where two Indians were arguing on the variations in the local language in one part of India. One gave a number which went into three figures; his companion differed—he thought it was one more.

Haldane looked up from his Patience and said: "Oh no, it's . . ." giving a figure different from either of the Indians. The following day, an interested onlooker checked. Haldane had been right.

For Haldane, his all-encompassing reach wasn't just a matter of spongelike retention. It was a view of the world. He saw a unity to all knowledge; he saw the same fundamental processes meshing into each other in every cross section of nature. Here was another reason he appreciated Marxism. Marx had believed exactly this, and in *Private Property and Communism*, he had made a prediction that sang to Haldane: "Natural science will in time incorporate into itself the science of man, just as the science of man will incorporate into itself natural science: there will be *one* science."

HALDANE DIDN'T FORGET the *Daily Worker's* mandate, so whenever he wasn't carried away entirely by the science, he scored some political instruction into the coda of his column.

Sometimes these lessons made resounding sense. "Human behaviour depends much more on environment than ancestry," he wrote at the end of his essay on flying ants. "That is why it is possible to bring a people from capitalism, or even feudalism or barbarism, to socialism and democracy in one generation. The ants are stuck in their state of society, and we are not." Sometimes, when he tackled the scientific aspects of society squarely, the political segue was inevitable. In a series of pieces, he lambasted the government for permitting workers to labor in unhygienic conditions and for allowing disease to flourish by failing to provide adequate housing for the poor. Sometimes it appeared as if he was being provocative just to madden critics who complained that he was always dragging Marx or socialism into his articles. He suffered from gastritis, he once wrote, "for fifteen years until I read Lenin and other writers, who showed me what was wrong with our society and how to cure it. Since then I have needed no magnesia."

More often, Haldane's transitions from science to ideology felt merely gratuitous or abrupt. In one article, having discussed insects and parasites, he declared: "This development of parental care and social life from parasitism is one of the most amazing in the whole history of evolution. It may of course be compared with the evolution of human society through slavery and class oppression to Communism." Just that one thought before a concluding paragraph on skeletons, as if he was satisfying a quota. Describing the work of C. V. Boys, the British experimental physicist, Haldane added: "There can be little doubt that he would have been far better appreciated in the Soviet Union, where the skilled workers constitute the nearest equivalent to an aristocracy that exists, than he was in Britain." When he wrote a piece on blood, he stated flatly, without support, that a Russian doctor's successful transfusion of blood from violently deceased corpses into living recipients "was clearly influenced by dialectical materialism."

Haldane sweetened his columns with kindness for the Soviet Union. Now there was a nation that was democratic, he said, far more so than Britain or France. Russian professors mended their own cars; Russian mechanics investigated the principles of physics underpinning their jobs. The Soviet Union's laboratories were better organized, its farms more scientific, and its speech more free. Readers also found, in these pieces, his defense of the state of genetics in the Soviet Union and of the opaque work of Trofim Lysenko. In an article titled "Spring," he wrote without doubt of Lysenko as the inventor of vernalization, saying that he had "made it possible to grow wheat in northern regions of the Union where summer is very short." He was breaking his own rule here, confirming a scientific result without studying the original data or a successful duplicate experiment. His columns were never intolerant of Lysenko. On the one hand, he could write: "There is no indubitable evidence that [mutations] ever arise in children in sympathy with bodily changes in their parents (the alleged transmission of acquired characters), and plenty of well-established cases where they do not." On the other, he refrained from scorching Lysenko's contention that acquired characters can be transmitted to the next generation. Lysenko dismissed the role of chromosomes in heredity, the keel that kept genetics afloat; Haldane, softly, ascribed Lysenko's stance to "a misunderstanding." Lysenko's contrarian theories were valuable because they forced other scientists to reexamine their beliefs, argued Haldane, a perennial contrarian himself.

Well after news of Stalin's brutality leaked into the West, Haldane danced around the matter. In a column called "Genetics in the Soviet Union," he named two Russian geneticists who lost their posts as Lysenko climbed to power. One, Haldane admitted, was rumored to have been shot. But they had, in any case, not done work of great originality, he wrote a few paragraphs later. This angled justification, carefully kept apart from the mention of the execution, bears an undertow of shock for the reader even today. Haldane went on to argue that British scientists were fired all the time for ideological reasons. He referred to Sydney Cross Harland, who was ejected from his job "for marrying a Chinese wife." In fact, Harland's departure from the Empire Cotton Growing Corporation's research station in Trinidad was a gnarled affair, involving his marriage to one of his assistants, personal disputes, and a libel suit. Haldane must have known all this. Harland was an old friend, an accomplice from the late 1920s, when they had together plundered the pubs of London's West End. Haldane knew his case well enough to resign from the corporation's research committee, protesting its treatment of Harland. Nevertheless, he chose to boil Harland's story down to his mixed marriage to suit the point he was making. To a close reader, "Genetics in the Soviet Union" sent a vivid signal of the rift developing within Haldane—the conflict between his scientific integrity and his political fealty—and his liability to muddle one in favor of the other.

HE WROTE OTHER THINGS AS WELL; as ever, the day seemed to open wider just for him, keeping him stocked with extra hours. In 1932, he released a science fiction story, *The Gold-Makers*, an approximation of a Wellsian yarn in which a new technique to distill gold out of seawater threatens to disrupt the world's mining concerns. (The inventor plans to use his profits "to start endowing science as it ought to be endowed"—a pet cause for Haldane, although in the tale, he doesn't recognize how those profits will plummet if the markets are suddenly flooded with fresh gold.) The story coughs and dies, incomplete, as if Haldane got called away to more interesting business. He released it, he later said, because he thought it was "rather unlikely that I shall ever write enough fiction to fill a volume." He was very nearly correct. Later, he toiled for a while on a scientific fantasy, and although it, too, remained unfinished, Naomi had it published posthumously in 1976 under the title *The Man with Two Memories*.

She admitted it made for laborious reading. Through an artless plot device, an earthbound professor accesses the mind of Ngok Thleg, a biologist in an alien civilization that existed elsewhere in the cosmos millions of years ago. To describe Thleg's world, Haldane mustered every utopian notion that had ever occurred to him. He populated it with beings who were clones, bred according to the need for their abilities, and he gave them such evolved control over their own bodies that they could suspend the flow of blood to any part of their body or secrete milk at will. They altered themselves chemically as well; they were, Thleg tells the professor, "from your point of view, a race of drug addicts." Haldane fussed so much over the details of his futurism—titanium cities, education through hypnosis, adhesive zippers—that he overloaded his narrative and stalled it. Despite its fine-grained creation of a world, the novel isn't a feat of the imagination; it is a heavy-handed program for how science can remake every cranny of life into the most ideal version of itself. When *The Man with Two Memories* breaks off, Ngok Thleg's world has just gone through an earthquake, and he has been sent to tend to an underground garden on a nearby moon. If Haldane stopped writing because he was getting bored by his own tale, it would be difficult to blame him.

Haldane's only complete work of fiction reached the most unexpected audience. *My Friend Mr. Leakey*, a children's book published in 1937, tells three stories of a biologist, much like Haldane himself, who

meets a magician in London. His own job of creating new kinds of
primroses and cats, the biologist says, is nearly as odd as Mr. Leakey's,
but magic is magic, and nothing compares with a man who heats his
food on the breath of a pet dragon, flies around the world on a carpet,
and turns a ferocious dog's teeth into rubber. (Haldane being Haldane, he spirited in motes of scientific instruction—descriptions of
the nests of penguins, asbestos boots worn by the dragon to spare Mr.
Leakey's carpet, and even evolution. "Only a few million years ago our
ancestors were animals," Mr. Leakey says, "and I expect our descendants will be animals too, and rather nasty ones, if the human race
doesn't learn to behave itself a bit better.") Haldane treats his young
readers the way he was treated as a boy: with utmost seriousness. So
he takes pains to build stout rafters of logic that hold up all the high
nonsense of magic. "From now on and until further notice, all modern fairy-tales should be written by Professor Haldane," a reviewer
remarked. People contacted Haldane asking to excerpt the stories into
anthologies, to read them on the radio, to translate them into Polish, Spanish, Czech, and Swedish. Children wrote to him as well.
"Dear Professor Haldane," ran one letter from Julian Tunstall, who
had inscribed two fat-lettered words per line, "I hope you will write
another book about Mr. Leakey. I think your first one is jolly good."

In its totality, Haldane's writing earned him an oracular fame of a
kind that made him the successor to H. G. Wells. Like Wells, he was
a man who peered constantly into the future and relayed his visions
of what the world looked like. But unlike Wells, he did so as a professional scientist with a loud political voice. In the business of explaining science, Haldane had peers: the physicist James Jeans wrote
popular books on cosmology, and the astronomer Arthur Eddington
lectured on relativity and its bearing on religious faith. But no one
ventured to tie together the ongoing developments in both science
and politics the way Haldane did. No one laid out quite as clearly
or prolifically how science could solve the immediate problems that
people faced.

The letters poured in—hundreds a week, his secretary reckoned.
His readers regarded Haldane as a fount of all scientific wisdom,
as a man who knew and did it all. If you suffered from hemophilia

and read about an experimental treatment that had yet to be clinically tested, as Charles Balsiger of Upton did, you wrote to Haldane,
begging him to conduct the tests and bring the drug to market. If
you thought you had discovered a connection between solar cycles
and mutation rates, as Edward Jackson of Weymouth did, you sent
Haldane your calculations. (He had read through Jackson's arguments and even shown them to his colleagues, Haldane replied, ever
scrupulous about writing back and ever ready to be blunt: "Frankly
none of us can make head or tail of them.") If you were a truck
driver in Australia and had some questions about the speeds of cosmic rays, as H. Rawlinson did, you put them to Haldane, and he
would send back a short, limpid paragraph of explanation. If you
wanted to learn more about biometry, as Muriel Finn of Greenford
did, you asked Haldane what books you ought to be reading, and he
would engage in a correspondence about book titles. "If you want
some calculations to do, I will give you some to do which will serve
you as an exercise, and also act as a check on my own calculations,
as we should probably make different mistakes," Haldane offered to
Finn, an absolute stranger.

Sometimes he had to plead for less mail. On one occasion, he
received a desiccated larva for identification and had to announce to
his public: "Please send me no more caterpillars." In another piece,
he remarked that scientists still knew too little about how traits were
inherited in cats. (He was fascinated by the subject. Once, on a trip
to Europe, he started a fresh notebook to record every cat he saw,
just as he had filled journals with plant species as a boy. "Cats, F.
Domesticus, Gatti Italiani," it said on the label, and inside, through
his travel to Geneva and Rome, he listed them. "Black, yellow eyes,
V. dei Pastini," or "Chemin de Velours, Tortoiseshell Agouti Stripe
S2 Orange Eyes"—nearly a hundred such entries.) He said, in his
column, that he wanted two kinds of cats for breeding purposes—a
tortoiseshell male and an albino, and he emphasized: "Please write
before sending any cats!" But his lab and then his house grew over-
run with cats all the same—cats on every surface, perched on the
backs of chairs or draped across desks, so many cats that even the
notoriously messy Haldane thought the premises had begun to smell.

So he wrote another column, explaining that he had too many cats and that, if they weren't adopted, they would have to be drowned—a prospect he detested. Once again, a gout of correspondence ensued: "May I have one of your yellow cats please?" "I should very much like to have the silver tabby female kitten." "I would like a tom kitten, black, please."

If editors desired a piece about science, they wrote automatically to Haldane. If reporters wanted an opinion on any scientific subject, they telephoned Haldane. What did he think of this mysterious fog in Belgium that was thought to be killing people and cows? What did he think of expressing race in terms of blood? What did he think of patent medicines? He was instantly quotable. No one can determine if, to a question about what the living world told him of its Creator, Haldane really replied: "He has an inordinate fondness for beetles." But however apocryphal, the quip has never stopped sounding like him.

His calendar brimmed with lectures on both science and politics; in a two-month span in 1939–1940, an MI5 agent sent in reports of Haldane's speeches, one each in Manchester, Derby, Birmingham, Portsmouth, and Aberdeen, and several in London, Edinburgh, and Glasgow. Sometimes he arrived at a public lecture already voiceless from previous orations. If, for some reason, the event was poorly attended, he scooped the scattered audience members into an intimate circle, removed his tie, and gave them his speech nonetheless. This was rare, though, because Haldane invariably drew big crowds. Once, he gave a speech titled "A Dialectical Approach to Biology," and nearly 200 people turned up. The Edinburgh University Socialist Society solicited Haldane as an honorary patron. So did the Sheffield Left Book Club, the Aberdeen University Labour Club, the World Student Association for Peace, Freedom and Culture, the Grand Amateur Boxing Tournament, the Anglo-Austrian Youth Association, the Marx Memorial Library, and a dozen other organizations wishing to bank on his prestige. Few intellectuals of his time were more widely known than Haldane, up and down and across the strata of society. He had acquired a quality altogether rare for a scientist: celebrity.

AND AS BEFIT A CELEBRITY, a squall or two of public drama always seemed to hang about Haldane. He rose to every opportunity for a spat, using a barbed letter to the editor or a needling article to settle scores. The more esteemed the opponent, the greater Haldane's relish. In 1930, he entered into a quarrel with Lord Birkenhead, the former lord chancellor of the United Kingdom, charging him with plagiarism. "A strange feeling began to oppress me," Haldane wrote, of reading Lord Birkenhead's book *The World in 2030 A.D.* "Certain of the phrases seemed oddly familiar. Where had I seen them before? Finally I solved the mystery. They were my own." With evident delight, he reproduced some of the 44 similarities he noted between *The World in 2030 A.D.* and *Daedalus*. His accusations provoked a scandal, reported by publications as distant as the *New York Times*. Lord Birkenhead replied in the *Daily Express*, a newspaper with a far higher circulation than the *Weekend Review*, where Haldane's piece had run. Strategically, he rebutted none of Haldane's points; instead, he rode down Haldane's grasp of history. Haldane was instantly baited. In the *Weekend Review* again, in dudgeon, he flashed his degree in the classics—a first, no less—and said that he had no objection to anyone following in the footsteps of other thinkers, but added, "I object to them stealing my boots to do so, and I am amused when they do not know how to put the boots on."

HALDANE WAS GROWING fitful at Cambridge. Ten years had passed since he had joined as a reader, but the university seemed to be in no mind to dub him a fellow at Trinity. The elders were perhaps still peeved at his triumph over the *Sex Viri*. Charlotte's displeasure with Cambridge was compounded now by her wish that Ronnie, nearly 14, enroll in a London school. Haldane could quit Cambridge at any time to teach somewhere else, she thought. So he resigned in 1932, in the middle of a year of furious productivity: two books, more than a dozen scientific papers, and a confirmation as a fellow of the Royal Society, the last a salve for the snub by Cambridge.

In August, he went to America to lecture at Columbia and Cal Tech and to attend the Sixth International Congress of Genetics. He was feted everywhere. At a dinner in his honor, he was introduced to the 250 guests present as the man "to whom our country owes more than to any other foreign scientist for his work in the popularization of science." His own views about America were less warm. He did admire the opera houses of New York and the handsome facilities in the universities; at Columbia, he was in such a hurry to visit a lab on the 13th floor that he ran up the stairs instead of waiting for the elevator. As a young society, America was vigorous, he thought, but it was also intolerant, and its sterilization drives were just one example of that. Its insistence on the glories of capitalism was another. He was bored by the marvels of the consumer market—the newest make of car or the newest model of radio. The American ideal, he wrote in an essay, was too exclusively economic, so he could never accept it. (The same was true of Communism, and he rejected that as well, in the same sentence. But on this, he would change his mind.) He deplored the damage that Americans inflicted on the English language. And then there was Prohibition—as crude and illiberal a step in its own way, he thought, as sterilization.

The genetics conference had been preceded by a eugenics congress, which Haldane had pointedly refused to attend. At that event, Ronald Fisher read a gloomy message from Leonard Darwin: If eugenics reforms weren't adopted with eagerness, Western civilization was scheduled for doom. Everybody present tended to agree—this was, after all, a eugenics congress—and so Haldane spent some of his time in America in vocal disagreement. More than once, he laid into the American school system and its aspiration to turn out students identical to one another, as if they were coming off a production line. The ideal society, he said at Cornell University, is a genetically diverse one. Different organisms are suited to different environments. In the same way, it is best if people have distinct talents to contribute to their community. For this reason, there is no such thing as the perfect human being.

What about Leonardo da Vinci? someone asked him. He was a man of numerous skills. Why shouldn't genetics aim to produce a race of Leonardos?

Had Leonardo lived in 1932, Haldane said, he would have been sterilized in some parts of the United States. "It is only in a society of great diversity that the Leonardos can be born and developed."

At one point during this discussion, Haldane spotted F. E. A. Crew, a geneticist from Edinburgh.

"Crew," Haldane called out, "what is the perfect man?"

"There isn't any," Crew replied. "Define us a heaven and we will tell you what an angel is."

After Haldane returned to England, he joined University College London as a professor of genetics, a post created especially for him. The place, he grumbled, was "as full of bloody Communists as Cambridge." His soul hadn't yet been dyed in deepest red.

At John Innes, too, Haldane's patience was nearing tatters. He had expected, when he joined, to succeed Daniel Hall quickly as the director, but Hall showed no inclination to leave. Under him, the institution was descending into second-rate work, Haldane thought, and he refused to be polite about it. He complained that Hall killed pedigree plants by using them in soil tests, that Hall's wife wandered about picking the fruit off the trees in the experimental plots, and that the staff played ping-pong in the lab, imperiling valuable microscopes. Periodically, the council that governed John Innes asked Haldane for his plans to transform the institution, to which he responded with an outline of its future: a reorganized staff, new appointments, and a residential system like that of a Cambridge college. "It seems to me entirely futile to attempt any serious reconstruction so long as the present Director remains," Haldane wrote in one letter to the council.

But every time, the council absorbed Haldane's views yet kept Hall on. Worse, it dropped hints to its fervent wish that Haldane would leave on peaceable terms. Once, while he was still on the staff, a committee even asked him to suggest a replacement for himself. (No first-rate geneticist is available, Haldane snapped.) These slights fanned his fury, confirming his worst opinions about the men and women in charge everywhere. When he wrote to William Bateson's widow, asking for the use of her late husband's slides, his letters vibrated with resentment. Beatrice Bateson showed them to a friend. "Who can wonder at JBSH getting kicked out of things if this is typical of his behaviour?" the friend wrote back. "The letters give a strong sense of some mental kink—incipient insanity?" Required to submit his department's report for 1936, Haldane composed a terse, 12-line passage. His secretary was so struck by his aggression that she wrote a poem about it. The last verse ran:

> But I have you on toast
> For I shall not retire
> Till I'm offered a post
> At the same pay or higher:
> And a person of my reputation
> You haven't the courage to fire.

She was correct. John Innes didn't sack Haldane. Instead, in October 1937, he quit, so that he could begin his new appointment as the Weldon Professor of Biometry at University College.

When he was in sour humor, Haldane spared no one. He quarreled with his mother and grew distant from her. He quarreled with his father, although he insisted they were merely disagreeing on questions of science. J. S. died in 1936, after a succession of mishaps: first a bout of whooping cough, contracted from one of Naomi's sons; then a fall from his high bed; then pneumonia. After doctors placed him in an oxygen tent, Naomi told a newspaper about his illness. Haldane scolded her for it. In the dining room of Cherwell, right beneath J. S.'s sickbed, brother and sister fought in whispered rage. They toppled backward into their childhood. She bit his arm. He twisted her wrist. When J. S. passed away, later that night, Naomi saw that he wore "a look of intense interest on his face as though he were taking part in some crucial experiment in physiology which had to be carefully monitored." They took his ashes up to Cloan. On the train, another altercation: upon learning that Naomi had booked herself into a first-class sleeper carriage, Haldane upbraided her for being a capitalist and thundered away into the third-class car.

If you watched him, you could see his temper rise, like mercury in a thermometer. He had no control over it and no wish to control it. At breakfast, something in the newspaper would displease him, and he would chew his lips in annoyance. Then he would find an argument to have with Charlotte until his irritation had spent itself. At dinner, if he was upset, he snatched up his plate and left, so that he could sit on the staircase to eat and stew in solitude. Haldane didn't realize, Charlotte wrote later, "what a terrifying effect his size, his dome, his bushy eyebrows had on smaller and lesser mortals."

His gusts of anger intimidated Ronnie. Haldane was a difficult stepfather, and his own nature interfered with his desire to be paternal. He played games of Halma and badminton with Ronnie, and they rambled through the countryside together, Haldane calling out the names of mushrooms and wildflowers as they went, just as J. S. had done on their own hikes in Scotland. But it was impossible for Haldane not to dominate the boy entirely. He outpaced Ronnie, outshouted him, outthought him, and even tried to outdo him. Once, Charlotte recalled, after Ronnie presented her with a fairy tale he had written, Haldane decided he would match the gift; *My Friend Mr. Leakey* was written not for Ronnie but out of a perverse spirit of competition with him. Even a child's accomplishment, Charlotte wrote, was treated as a challenge. "Everything you could do he could do better and, indeed, except for dancing and playing music, he could." By accident or design, the license plate of Haldane's car read "EGO 848."

Haldane and Charlotte wanted children of their own, and when they were unable to produce them, the marriage wavered and dimmed. It's unclear why Haldane never became a father. His sister speculated that a childhood attack of mumps had something to do with it; Ronnie told his mother's biographer, years later, that Haldane may have suffered an injury in the war. When Charlotte realized that their prospects of parenthood were small, she wrote, "I began to look for intellectual and emotional compensation in other directions." She embarked on a string of affairs, including one with Haldane's student Martin Case. The plot of Aldous Huxley's *Antic Hay*, of a Haldane-like professor whose wife was bedded by the men around him, had come to pass.

Haldane may have known about Charlotte's lovers. In fact, their marriage may have turned open with mutual consent, and Haldane certainly sought out his own companionship as well. (It wasn't uncommon in Haldane's circles. Naomi and her husband lived in one such flexible arrangement.) Everybody around Haldane noticed how fond he was of children and how much it saddened him not to have any of his own. His disappointment also corroded his ties with Naomi. When meningitis claimed one of her sons, she drove to Cambridge, desperate for comfort from her brother. Instead, Haldane accused her of being negligent, blinding her with so much grief and guilt that she couldn't think straight. On her way home, she smashed her car into a wall. "I came near to killing myself," Naomi wrote. "I think he could hardly bear it sometimes when it was I who had the children. He had to take it out on me." The irony was both apparent and painful: Haldane, who spent his waking hours thinking about how genes pass from parent to offspring, could have no genetic heirs of his own.

HALDANE'S POPULARITY BROUGHT HIM LETTERS from the Labour Party—four in three months in 1939, all asking if he would stand as a candidate. An election was due the next year. The correspondents assumed he was a Labour member, but in this, he informed them, they were mistaken. Haldane turned them down, offering various reasons. For one, he was too closely linked with the Communist Party. And he would be a dreadful politician, he thought: "My personal character is such as not to endear me to a number of people, who state, probably correctly, that I am rude." But mainly, he would have to give up research, and "at the moment, I am turning out more than I ever did in my life." They pressed him repeatedly: "We can think of no one else who could make the running at all. Do say yes to us!" and "We really do want you, hard." He didn't change his mind. He had work to do.

After his move to University College, Haldane left biochemistry and physiology behind to focus on problems in genetics. His work relied more and more on numbers, Haldane wrote—"too mathematical to interest most biologists and not sufficiently mathematical to interest most mathematicians." In his earlier papers, Haldane had browsed through various scenarios of reproduction, calculating how natural selection responded in each case. How does selection treat a trait produced by multiple genetic factors? How does selection play out when populations move and change? In a 1935 paper, Haldane considered an adjacent question: At what rate did new mutations arise in the genes of human beings?

The Arab physician Al-Zahrawi had described hemophila in the tenth century, writing of an Andalusian village in which the fathers and sons bled so easily that if a boy had his gums rubbed harshly, he was likely to die. Since then, hemophilia had been recognized as an inherited, mostly male disorder, but any further knowledge of its inheritance always relied on anecdotes—about a family where the disorder skipped a generation or about a boy stricken with hemophilia even though no one in any previous generation had shown symptoms of it. Careful data were never available—not just for hemophilia, but for any heritable disease. Naturally, it was the biometricians, with their devotion to statistics, who set this right.

"Whenever you can, count," Francis Galton had said. Beginning in 1909, the Galton Laboratory at University College published a series of data banks for disorders that were suspected of being passed down from parent to child. The *Treasury of Human Inheritance* included chapters on diabetes, polydactylism, pulmonary tuberculosis, insanity, cleft lip, and "ability." The fattest chapter in the first volume, on hemophilia, was written by a physician named William Bulloch and a pathologist named Paul Fildes. Together they read through 949 known descriptions of hemophilia, dating back to 1519, drawn from the medical literature of the Western world. Discarding cases where the stamp of hemophilia was too weak, they mapped out the pedigrees of 44 families with clear histories of the disease. In an appendix, they provided fine-grained case studies for each member of these families. ("After the extraction of a tooth, he bled for a week"; or "Fell from a low stool and cut his tongue against his teeth. Bleeding set in and in spite of a variety of remedies he died on the fifth day.") Bulloch and Fildes proved, in their study, that hemophilia overwhelmingly afflicted men.

Geneticists discovered, soon after, that the disease followed a Mendelian pattern. Hemophilia is a recessive disorder, arising from a flaw in a gene that encodes blood-clotting proteins. The gene lies on the X chromosome, and since women have two copies of this chromosome, the effects of a faulty copy of the gene on one chromosome can be offset by the functional gene on the other. (Only rarely do women have two chromosomes with mutated genes on each.) Men, however, possess only one copy of the X chromosome; its counterpart, the Y, doesn't carry a clotting gene at all. If their single gene on the X is a mutant copy, their blood loses the capacity to clot.

But hemophilia came with a paradox. If its sufferers bled profusely enough to die from a single cut, only a small number of them would ever live long enough to have children and pass their genes on. The disorder ought to write itself out of existence. (There were roughly 35 to 175 male hemophiliac babies born per million in Haldane's London. If Haldane had to account for the proportion of hemophiliac men who died before they bred, he wrote, he would have to conclude that every man in England at the time of the Norman Conquest must have been troubled by the disorder.) Instead, the numbers of hemophiliacs in the population stayed close to constant—which meant that the gene must mutate suddenly, unaffected by any parental history of the ailment. The balance between this pressure of mutation and the countervailing force of selection against the faulty gene must account for the frequency of hemophilia in a population.

Another scientist, C. H. Danforth, had suggested years earlier that the mutation rate for a human gene could be quantified out of this mutation-selection balance, but he hadn't found the kind of data to help him do it. For Haldane, working along the same lines, hemophilia presented a useful example to explore. Scientists had already estimated mutation rates in maize and fruit flies. Humans were harder to study, because they were more complicated and because their life spans were longer. But classic hemophilia ran along clean Mendelian lines, and Bulloch and Fildes had gathered generations of detail. In 1927, moreover, Haldane had assembled a general-purpose equation to yield the mutation rate required to keep a gene's frequency in a population in equilibrium. Now he put that equation to work.

From Bulloch and Fildes's data, he obtained the average fertility of hemophiliac men and of women who carried the gene for the disorder, and he had an idea of how many male babies out of every million turned out to be hemophiliac. (Some of these numbers were rough but sensible guesswork.) The rate at which new gene mutations occur so that the hemophiliac population stays steady is roughly one in 50,000, he concluded. In other words, out of every 50,000 new X chromosomes in a generation, one X chromosome's clotting gene spontaneously mutates into a variant that causes hemophilia.

The concept of a mutation rate has grown more complex since Haldane's time. The rate at which genes mutate in humans varies by factors such as gender, age, and exposure to agents like radiation. (In fact, it was Haldane who, in 1947, showed that the rate of mutation in germ cells—the cells producing sperm or eggs—is higher in men than in women.) Some mutations appear to do nothing; these errors in the code have no effect, but they are still aberrant. Different mutation types may occur at different rates. The incidence of mutation even fluctuates across a single genome, the full set of genes carried in the cells of a single person. In its time, though, Haldane's 1935 paper provided the first clear estimate of a mutation rate in human beings, and both his method and his results became canonical. "No greater compliment can be paid to a scientist than to take his original ideas for granted as part of the accepted framework of science during his lifetime," Haldane wrote later, with evident satisfaction.

When tools became available to peer into the human genome, a study proved Haldane's estimate to be astonishingly resilient. In 2009, combing through a village in China, a team from the Wellcome Trust Sanger Institute in England found two men who were distantly related to each other. Their common ancestor had lived 200 years earlier, and his Y chromosome had passed largely unchanged from father to son, across 13 generations, to these men. Using sequencing technology, the researchers compared common stretches of their Y chromosomes, each stretch running to 10,149,085 nucleotides. Only four of these showed mutations—a number that translates into one mutated nucleotide per 30 million in every generation. Haldane's estimate of mutations in the hemophilia gene was equivalent to roughly one mutated nucleotide per 25 million. The full human genome contains roughly 6.4 *billion* nucleotides. As far as it was possible for a man using pen and paper in the decades before the structure of DNA was discovered, Haldane came within spitting distance of what we now think is the true figure.

Mutations within the genome are unavoidable, but they are also desirable. Even if a man and a woman have scrubbed their world free of radiation, chemical additives, and potent medication, the DNA in their sperm or eggs will still most likely replicate inexactly—a hydrogen atom positioned wrong, a nucleotide changed, a sequence deleted or inserted, a stitch missed. Every human baby is born with between 60 and 100 de novo mutations—60 to 100 ways in which its genome is different from its parents'. Some of these lead to disorders like hemophilia or cystic fibrosis or sickle cell anemia. But the same mutagenic processes create gene variants that might confer an advantage on an organism—an advantage that renders it fitter for its environment and that natural selection tries to lock into its species. Life must tolerate the harm of some mutations so that it can evolve by the benefits of others.

In a 1937 paper, "The Effect of Variation on Fitness," Haldane made the first investigations into this fundamental trade-off. What consequence, he wondered, did deleterious genes have on the fitness of the species as a whole? If a mutated gene was lethal or nearly so, natural selection would ensure that it remained rare; the individuals suffering its effects would fast die away. If the variant was only mildly harmful, its impact on the individual would be negligible, so instead of being erased by selection, it would hang around in the gene pool, occurring more and more frequently. As a result—and this was Haldane's brilliant, counterintuitive insight—the mild variant imposed just as much load on the population as a far more deleterious alternative. Only in the less common case when a mutant near-lethal gene appeared at a high frequency would a species suffer significantly under an extraordinary load. He proved this for a triad of cases and then offered a reasoned guess: "If we could achieve the aim of negative eugenics and abolish all genes . . . which seriously lower fitness in our present environments, we might expect a gain in fitness of the order of 10 per cent., though this might lower our capacity for evolution in a changed environment."

The concept that Haldane outlined came to be known, after the

American geneticist Hermann Muller rediscovered it in 1950, as "genetic load" or "mutation load." Its principles have been regarded by some as sensible and by others as misleading. Scientists have used genetic load to gauge the effects of radiation on humanity and to expand population genetics. They look to it again now as they worry that humans, having found ways to medicate and control the effects of their significantly harmful mutations, are accumulating a number of less potent ones, shredding their fitness for any harsher conditions to come—a possibility Haldane mentioned as early as 1941. A population of diverse kinds of fitness is valuable, he wrote in his book *New Paths in Genetics*. A gene of slight disadvantage in one environment may be of supernormal fitness in another, and the presence of these genes in the gene pool renders a species elastic in a forever-shifting world. (The same can be said of society, he added, veering for a moment from science to politics: "Differentiation of function is a prerequisite of civilization, though the Soviet Union has shown that class divisions are not.") Quite possibly, he wrote, science would find some method to prevent mutations altogether:

> If so, I hope that this road will not be taken. For mutations are the raw materials of evolution, and if our descendants forgo the possibility of further evolution in order to abolish a few congenital defects they will show themselves to be as short-sighted as we are to-day.

Critics of the genetic load model thought that experiments didn't bear out its calculations. It felt to them like too closed and static a system, whereas breeding and selection in a population are dynamic affairs, with many forces pushing and pulling at each other. One scientist argued that a genetic load was not a burden but an asset. By culling young members of the species, it eased the strain on resources and staved off overcrowding—a ready response to Malthusian conditions. As our estimates improved of the number of harmful de novo mutations in the genome of each new human—at least two, on average—scientists also wondered how our species continues to survive with such a heavy genetic load. "Why have we not died 100 times over?" was the question asked in the title of a paper in 1995. Haldane's description of genetic load has come to seem like a piece of a larger, incomplete puzzle.

In his 1937 paper, Haldane introduced one other thought, almost in passing, that now glints with prescience. Evolution is a slow process most of the time and hard to spot within the span of a human life. "Our only reason to hope for observable evolution," he wrote, "is that owing to glaciation, agriculture, fishing and industry, the balance of nature has recently been upset in a manner probably without precedent in our planet's history; and hence on the Darwinian theory we should expect that evolution was proceeding with extreme and abnormal speed."

He was right: Drastic environmental changes are now making evolutionary alterations visible to us. Songbirds in northern Europe, mice in southern Quebec, and vinegar flies in Australia have all evolved to adapt to the strong selective pressures of a warming climate. To expand on one example: From 40 years of observations of the great tit in the Netherlands, researchers found that as spring began earlier every year, caterpillar larvae matured earlier as well. This upset the birds' schedule of laying eggs, which was timed so that the chicks hatched just when larvae were plentiful to eat. Increasingly, the chicks now emerge to find fewer larvae, and as a result, some starve before getting any older. The hardier birds, though, grow up to face less competition for food, and natural selection will prize their robust qualities. The parents that lay eggs earlier find that their offspring survive better as well, so selection shifts the breeding schedule of the bird population.

With humans, the situation grows more complicated, because we change our environment just as our environment changes us. We cure many of our diseases, we build habitats in any climate and terrain, and we move ourselves with machines. Our diets are becoming homogenous, our families are growing smaller, and our children are increasingly the products of mixed populations. These cultural changes, scientists such as Stephen Jay Gould have argued, are pulling human beings out of the hands of natural selection. By this logic, evolution by natural selection should be slowing down.

In fact, though, there is evidence that evolution has accelerated in humans, as compared with chimpanzees, our closest relatives. Over the last 10,000 years, pegged roughly to the spread of agriculture, parts of our genome have shown more variation; over the last 5,000 years, human evolution has become 100 times faster. Among the new traits gained by humanity in the past 10 millennia, according to a 2007 study, are blue eyes, a limited protection against malarial parasites, and the adult ability to digest milk. The reasons for this acceleration are still speculative. Perhaps by controlling nature and cultivating it, we improved our diet so much that our population swelled dramatically. In our larger numbers, more mutations flickered in and out—more gene variants, more traits for natural selection to sift, more chances for beneficial alleles to roll out through our species. (Darwin and Ronald Fisher had foreseen this. In breeding cattle, Darwin wrote, the chance of useful variations appearing and being retained climbed as the population increased, "hence, number is of the highest importance for success.") Had Haldane learned about this recent briskness of our evolution, it might have pleased his Communist spirit. By changing the material circumstances of its world, the human race changed the very marrow of its being. Haldane would have read that as a ringing vindication of diamat.

HEMOPHILIA FEATURED IN ANOTHER of Haldane's research questions, this one on genetic linkage. He had begun his career with a paper on linkage, watching for albinism and pink eyes in the hutches of mice in Cherwell. In 1936, with his colleague Julia Bell, Haldane made the first measurement of genetic linkage in humans. Other scientists had worked out that the genes for color blindness and hemophilia sat on the X chromosome. But how near to each other were they? How likely were they to be bequeathed as a linked pair?

Once again, Bell and Haldane collected pedigrees. Doctors sent them the names and addresses of their hemophiliac patients, and an optician was dispatched to their houses to test the color vision of these men and their brothers. They were from just six families—a small sample, Bell and Haldane admitted. Their method fixed on the event of genetic recombination, in which chromosomes dice and

swap stretches of genetic material between themselves during reproduction. The offspring's chromosomes, as a result, differ from those of
the father as well as the mother. Linked genes, nestling close to each
other on a chromosome, run a higher probability of being exchanged
jointly—of emerging from recombination still in tight proximity. So
Bell and Haldane developed a way to estimate how often the genes
for color blindness and hemophilia became separated during recombination, and they used probability models to analyze each of their
six pedigrees. For genes that bear no linkage with each other, the
recombination frequency is 50 percent. For the two genes they were
examining, the scientists saw, the frequency ran to 5 percent—an
indication of how brief the span of chromosome is between them.

By gaining a numerical grip on linkage, Bell and Haldane
became cartographers, penciling the very first squiggle of coast upon
humankind's gene map. For years thereafter, other geneticists followed their example, using pedigrees and recombination to assess
the relative distances between genes. (Haldane himself, in a separate paper, produced the first partial map of the positions of genes
on a human chromosome.) This was always going to be a limited
method. Most human traits are polygenic, their final effect a complex expression of many different genes, and not at all like the tidy
one-to-one match between a mutated gene and thin, runny blood.
Only after the birth of new technologies—to cut and splice DNA
with enzymes or to crunch through sequences of nucleotides—did
the ambition of charting the full genome take wing. In 2003, the
Human Genome Project declared it had a map that was 99 percent
finished. The small remainder is still unknown territory, although
scientists think they have most of the functional sections of the
genome plotted and placed.

Haldane was always looking for data, in journals or from
researchers—any opportunity, as his colleague Lionel Penrose noted,
"of examining the material collected by others and using his own
methods of analysis upon it." This reliance on the rigor of others let
him down in a 1936 paper on linkage titled "A Search for Incomplete
Sex-Linkage in Man." From a collation of medical records, Haldane
extracted cases of ailments that he thought were only partially linked

to sex: xeroderma pigmentosum, in which the skin grew roasted by the sun within minutes; or retinitis pigmentosa, wreaking damage to cells in the back of the eye; or Oguchi disease, which left its sufferers blind at night. He was cautious about some of the raw records—"Inaccurate data are considerably worse than useless," he wrote in his paper—but concluded that his cursory proof was at least "partially successful." It wasn't; later studies showed that he had trusted the data too much. A flawed paper was, for Haldane, so uncommon that it stuck out in his career of publications like an untuned trumpet in an orchestra. A quarter-century later, when Haldane was visiting Glasgow, the geneticist James Renwick poked the bear: "Is there anything new on partial sex-linkage in man?"

Haldane took it well and laughed. "Look here," he protested, "I said it was a *search* for partial sex-linkage in the title of my paper!"

Whenever he wrote his work up, he extended his conclusions into the future, strewing crumbs for others to follow. Having measured the linkage between the genes for hemophilia and color blindness, Haldane and Bell saw how much potential this held for diagnosis. Not of hemophilia, of course, because the symptoms of hemophilia made themselves known before color blindness became evident. But if a similar linkage existed, for example, between the genes determining blood groups and the adult-onset disease called Huntington's chorea, doctors could predict, by blood type alone, which of a patient's children would develop the disease. (And they could, Bell and Haldane added, "advise on the desirability or otherwise of their marriage"—a tablespoon of the mild eugenics that Haldane always supported.) For a long time, the use of such linked markers to detect the genes for disease remained, as a speaker at the Royal Society put it in 1988, "a pious hope included in the final paragraph of a grant application." Then in the 1980s, technology caught up enough to read DNA sequences in close linkage with disease-causing genes, identifying them before any symptoms appeared or even in the womb. Scientists found a swatch of DNA, dubbed *D4S10*—a linked marker that signaled the presence of the gene associated with Huntington's so strongly that tests using the marker to predict the disease could be accurate, at their best, 96 percent of the time. In 1993, again through linkage analysis, the trigger for Huntington's was located precisely—a defective, overlong version of a normal gene, at the very tip of our fourth chromosome.

There were plenty of these instances of Haldane's throwaway foresight. A paper in 1933, on the genetics of cancer, contained the proposal that our bodies reject transplants because of the antigens in the grafted tissue. If these antigen proteins feel foreign—if they're of a type different to the host's—then an immune response attacks the new tissue, Haldane suggested. The notion, occupying just a paragraph, is now the basis of antigen matching, a routine procedure before transplants to ensure that organs and tissues aren't rejected. In 1941, in *New Paths in Genetics*, he outlined how a gene could replicate by acting as a model for itself. He couldn't yet know, of course, that genes are made of twirling threads of DNA; he thought they were single lengths of protein. Still, he was able to foresee how genetic proteins, like crystals, could derive their physical structure and composition by copying a template. And how would an observer distinguish a parent gene from its copy? Perhaps by feeding a cell with heavy nitrogen, Haldane wrote, so that, during replication, the new gene's proteins would be built out of those heavy atoms of nitrogen rather than ordinary ones. The copy and its template could then be told apart by measuring how dense the genetic material was—how much heavy or regular nitrogen it held. In 1958, the scientists Matthew Meselson and Franklin Stahl set up an experiment that was a tweak away from Haldane's ingenious solution, labeling DNA strands with heavy nitrogen and tracking how they reprinted themselves. For its elegance and clarity, science historians call it "the most beautiful experiment in biology"—a glorious irony, given how Haldane, its progenitor, was among the clumsiest experimenters of his time.

For Haldane, the model scientist was always Louis Pasteur, who in the nineteenth century developed vaccines, discovered how to halt the contamination of milk, and tripped up the headlong spread of disease. His influence, Haldane thought, was supreme—greater even than Darwin's. Darwin changed the intellectual beliefs of his time, and his appeal was almost entirely to reason. Pasteur transformed the state of humanity and the structure of society. He published few experimental results, but every experiment he ran was final, decisive; no one conducted those same tests after him and came away with different results. It was rare, Haldane wrote in an unpublished book on Darwinism, that a scientist stimulated both reason and emotion. But Pasteur did. He ingrained in men and women a fear of germs, a revulsion of filth, and a desire for cleanliness. His work spoke to their deepest instinct for self-preservation.

Haldane himself wasn't a scientist in this mold. He published often and widely. In the 1930s alone, he wrote a paper on the link between quantum mechanics and philosophy, another on an economic theory of price fluctuations, several on statistical models, and a paper each on the cosmology of space-time and the future of warfare. (After the zoologist Karl von Frisch showed that bees communicate with each other through intricate dances, Haldane recalled Aristotle's description of bee waggles and, in the *Journal of Hellenic Studies*, inspected it in the light of modern science. He translated the Greek himself, of course.) He spread his energies so widely, in fact, that it appeared as if he might have done more monumental work—might have been a Pasteur—if only he had focused himself better. But this would be a misreading of Haldane's science. Most of his papers through the 1920s and 1930s advanced theories in genetics and natural selection; that was his focus, and he held steadily to it. If civilization did not pivot around his results, as it did around Pasteur's, that was only because such moments are uncommon in the history of science. They were particularly rare, during the time Haldane was active, in a field like genetics. In those formative years, what genetics required was the soundness of basic research; the big discoveries couldn't sprout until the fundamentals had been seeded and watered. This was Haldane's challenge, and he rose to it with gusto. Basic research, he once wrote, is "of all things most supremely worth doing for its own sake and that of its results." With every paper, he pulled genetics forward, often by long lengths, and he scattered ideas that sparked and sustained the work of others. Haldane's genius, the biologist Peter Medawar remarked, was not to bring new land to cultivation. It was to enrich the soil.

ON THE CONTINENT, rulers obsessed over genetics in other, darker ways. By the middle of the 1930s, the fingers of fascism, long and strong, were stealing around the neck of Europe. Books had been burned, opposition parties banned, and race laws passed. In 1936, Hitler renounced the Treaty of Versailles and occupied the Rhineland. The same year, Germany and Italy signed the Rome-Berlin Axis Pact, promising to stand together, jackboot by jackboot, in any war to come; then Germany and Japan agreed to oppose Communism. In England, Oswald Mosley's Blackshirts organized so many marches and rallies that the government had to pass a Public Order Act to curb the fascists at home. Through July and August, the police counted nearly 600 political meetings in London. A number of the Blackshirts' meetings—as many as 60 percent in August, the Special Branch calculated—were disrupted by Communist groups. The summer of 1936 was tense and inflamed, like a muscle threatening to bring the whole body down.

One Sunday, Charlotte and Ronnie walked through Hampstead Heath, curious to hear one of Mosley's followers address his public. It wasn't a big crowd, but the speakers might have been at Nuremberg, so uninhibited were they in their anti-Semitism. The audience looked indifferent to the poison billowing into the air. Ronnie didn't visibly react that afternoon, Charlotte noticed. The next day, though, he enrolled in the Young Communist League.

In July, Spain boiled over: the military rebelled against a Republican governing alliance that included the Communists. In quick time, Germany and Italy threw themselves behind General Francisco Franco, supplying the Nationalists with aircraft, weapons, and tanks. Suddenly the fascist-Communist divide crystallized into a real martial conflict. For any Briton who wasn't devoted to Mosley, but particularly for members of groups on the left, the Spanish Civil War sounded a bugle to action. Activists knitted sweaters, sold tokens to send milk to Spain, and sang Christmas carols to raise money for Spanish relief. At rallies, intellectuals and activists hectored their audiences to contribute to the cause. W. H. Auden wrote a poem, *Spain*, and the proceeds from its sale went to the Spanish effort; the sculptor Henry Moore sold some of his work and sent his earnings to Spain as well. The first British volunteers trickled into the country in August, prepared to fight the Nationalists, or at least to assist the Republicans and the people of Spain.

Ronnie decided he wanted to head to Spain as well.

"You silly little fool," Charlotte said. "What use would you be? Why, you can't even shoot."

Haldane refused to interfere. Ronnie was 17 and free to act as he pleased. For advice, Charlotte turned to Harry Pollitt, in the CPGB. If Ronnie was to go to Spain, they decided, he would need a gas mask.

"Gas masks. Gas masks," Pollitt said. "That's the one thing I need to know about. I want gas masks. Can you tell me how I can get them? Hundreds of them?"

"I don't know anything about it," Charlotte said. "You'd better meet J. B. S. He does."

As yet, no gas masks were available. The Haldanes dropped Ronnie off at the railway station with no protection for his lungs. Then they drank in the pubs till closing time. After that, requiring further solace, they went home and split a bottle of whisky.

Like other leftist intellectuals, Haldane had joined the chorus against Franco's Nationalists. He added his name to a letter to *The Times*, alongside Julian Huxley, Virginia Woolf, E. M. Forster, and two dozen others, expressing "grave concern [that] a persistent attempt is being made . . . to enlist the sympathies of Britain for the military rebels, on the ground that the [Spanish] Government is Bolshevist or Communist." He delivered pungent speeches at fundraisers, and he published articles attacking the forces of fascism. But after volunteers from Britain and other countries started traveling to Spain, Haldane saw his own opportunity to enter the fray. The trickle became a stream; the Communist International drafted the early volunteers into a newly raised International Brigade, which only attracted more sympathizers. Haldane thought he might be of use. He knew what a gassed field of battle felt like and how a human might survive within it. He recalled, no doubt, how much he had enjoyed being on the front during the previous war and the warm companionship of men in battle. For a man of Haldane's age, the

Spanish war promised a sense of Byronic adventure and romance and a theater for masculinity. Charlotte thought Haldane was seeking a frivolous, pseudomilitary excitement. "He got himself a most curious outfit," she wrote later, with a sniff of disdain, "a motor-cyclist's cap with a visor, and a black leather jacket and breeches."

Be careful, his colleague Julia Bell warned. Don't go getting killed out there.

Haldane exploded. Only if a few people like him did get killed, he told her, would the British really understand what a menace fascism was.

On December 9, 1936, the *Daily Worker* carried a slug: "Professor J. B. S. Haldane, the expert on fighting gas warfare, has agreed to go to Spain next Monday as adviser to the International Legion." An MI5 staffer solemnly typed up this note and added it to Haldane's file.

By boat to France, then a train to Paris, then another to the town of Perpignan, then a bus into Spain. (For a while, the man in charge of marshaling the volunteers in Perpignan was Josip Broz Tito.) From the border, Haldane took a truck to Figueras, where a train could be caught to the International Brigade's headquarters in Albacete, southeast of Madrid. In Albacete, housed with hundreds of other volunteers, eating with them on long benches in the Casa Salamanca, Haldane felt once again like an ordinary soldier. "I found myself acting as interpreter between the Italian who spoke French and a Hungarian who spoke German," he wrote later.

After a few days, he left Albacete to make his way via Alicante to Madrid, held by the Republicans but under Nationalist onslaught. On the train to Alicante, he was shaking with influenza, so he was given a first-class ticket. But the carriages were stuffed full of refugees. Children relieved themselves on the floor. The windows were sealed tight against the frigidity of December, so inside the train, the air grew cloudy with the smoke from the Spanish cigars called *dinamiteros* ("dynamite"). "To a man like myself, already suffering from laryngitis, they were . . . nearly homicidal." The only place Haldane could find to sleep was on the floor in a corridor, people stepping over him or on him throughout the 12-hour journey. From Alicante, Haldane hitched a ride on a truck and rode 10 hours to Madrid cold and uncomfortable, sandwiched between two cases of ammunition. He arrived in the capital two days before Christmas.

Madrid had been roughed up, but it was still on its feet. The signs of the war were everywhere—in the wounded facades of buildings, in the shelled pavements, in the wailing ambulances, in the chipped nose of a stone lion on the Castellana, in the browning hair of once-blonde women who could no longer buy peroxide. The front, locked in a stalemate between the Republicans and the Nationalists, lay a couple of miles west. But in the winter air, as cold and crisp as tin, the *tacka-tacka-tack* of rifle fire sounded as if it was just around the corner. The city was running out of food. Long queues formed daily—sometimes as early as midnight—to receive the beans and bread handed out at noon. Not long afterward, the bombing began and continued nearly every day: German or Italian aircraft bickering overhead, dropping bombs that looked, Norman Bethune thought, like great black pears. Bethune, a Canadian doctor and a staunch Communist, had come to Spain to provide medical assistance to the Republicans. When the bombs struck, there was little to be done, he said in a broadcast on January 2, 1937. "If the building you happen to be in is hit, you will be killed or wounded. If it is not hit, you will not be killed or wounded. One place is really as good as another." After the dust and smoke cleared, "from heaps of huddled clothes on the cobblestones, blood begins to flow. These were once live women and children."

In the Palace Hotel, the Republican army set up a hospital, with more than a thousand beds in its rooms and hallways. The grand dining room was transformed into an operating theater: eight tables, staffed by volunteer surgeons, anesthetists, and nurses who worked by the light of crystal chandeliers. In other hospitals, surgeries were moved to the basements, so they could proceed uninterrupted while the city was being bombed. Bethune organized supplies of blood for the sick and injured. He enlisted a thousand residents of Madrid to donate blood once a month, and these stocks, mixed with a sodium citrate preservative, were stored in ampules in refrigerators, to be rushed across the city when needed. Haldane attached himself for two weeks to Bethune's unit and promptly made improvements. The donated blood had to be screened for malaria and syphilis, he insisted, and if Bethune didn't have the necessary equipment, another facility ought to do it. A private lab was located, and hundreds of men, women, and children were saved from potential infection.

Later, when he wrote about his time with Bethune, Haldane recalled an incident so stirring it had the suspicious perfection of cinema:

> A Spanish comrade was brought in with his left arm shattered. He was as pale as a corpse. He could not move or speak. We looked for a vein in his arm, but his veins were empty. Bethune cut through the skin inside his right elbow, found a vein, and placed a hollow needle in it. He did not move. For some twenty minutes I held a reservoir of blood, connected to the needle by a rubber tube, at the right height to give a steady flow. As the new blood entered his vessels his colour gradually returned, and with it consciousness. When we sewed up the hole in his arm he winced. He was still too weak to speak but as we left him he bent his right arm and gave us the Red Front salute.

Between the bombing runs, Madrid breathed. The trams, painted a sunny yellow, sauntered through the roads. In the shops, a woman could buy fox furs or jewelry or handmade shoes or expensive perfume. The cinemas still ran movies: the Marx Brothers' *A Night at the Opera* and Greta Garbo's *Anna Karenina*. Occasionally, hotels rustled up enough fuel to heat their water, and once the news flew around, people queued for a wash. Haldane himself, he told a newspaper proudly, didn't bathe for three weeks after Christmas. Scientists continued to work, and Haldane visited a couple of them. One biologist was investigating a polymorphic beetle, and another was breeding *Ecballium elaterium*, or the exploding cucumber. (How appropriate, Haldane thought.) Sometimes trucks of food brought rare treats, courtesy of the Soviet Union. Their bully beef was better than any he had ever tasted, Haldane thought—good enough to turn a man into a socialist in 5 minutes. The big cafés like Molinero's and Chicote's had little food to offer, but they nevertheless filled up in the afternoons

with soldiers, volunteers, and Madrileños, all smoking and drinking
and singing. Virginia Cowles, an American journalist, thought the
people too nonchalant for war. Life would go on, it seemed, until
there was no one left to live it.

Haldane didn't have to work long to protect Madrid's citizens
from gas attacks. He advised the Republican army and Prime Minister Juan Negrin, and he held a few demonstrations of gas masks. In a
photograph, he stands with an assistant in front of a hillock of rubble.
The assistant, Hazen Sise, wears a mask with a long snout, looking
like a sad anteater; Haldane, having swapped his motorcyclist's hat
and breeches for a regular cap and baggy corduroy trousers, appears
to be explaining the use of the mask. But there simply weren't enough
masks to go around, so Haldane searched for other means. He sawed
the bottom off a wine bottle—Madrid never ran short of those—and
stuffed the glass with grass and charcoal. That would, he thought,
filter the fumes while he breathed through the bottle's mouth. In
a hospital, Haldane found a fume cabinet, shut himself into it with

Haldane with a volunteer, Hazen Sise, demonstrating a gas mask in Spain,
c. 1936–1937.

his bottle contraption, and asked for chlorine to be pumped in. He lasted just a couple of minutes. Having signaled wildly to be let out, he emerged retching and coughing. The incident was embarrassing; afterward, to British newspapers, Haldane would insist that he had suffered from laryngitis. "Don't forget to contradict the story about my being gassed," he told one journalist, who promptly printed his instruction verbatim.

But gas, it transpired, was not the greatest danger in Madrid. To suffocate a city, Haldane wrote later, Franco needed enormous quantities of chlorine shells. "At least ten times as much gas is needed to poison people in houses as to poison them out-of-doors," he calculated, and certainly on the upper floors of a building, everyone was safe. Besides, Franco recognized that he could do far more damage with incendiary and high-explosive bombs— dozens or hundreds dead within minutes and their city shredded to boot. "I will destroy Madrid rather than leave it to the Marxists," Franco had promised, and he certainly seemed to be trying. Every time the raids started, some people scampered into cellars marked *refugios*, but others peeked out of their windows to watch the bombers shriek past. Haldane had a difficult time convincing them to take shelter. He darted into basements himself, at first, but then he felt he had to assume the role of unperturbed spectator as well. He was, for that time, a citizen of Madrid, so he had to be just as brave as those around him.

Haldane was prone to romanticizing the Spaniards who remained in the capital. He was not the only one; Bethune, in an empurpled broadcast, described the looks of "fortitude, dignity and contempt" worn by the people of Madrid. They have "endured from the arrogance of wealth, the greed of the church, the poverty and oppression of centuries," he said. "This is just one more blow, one more lash of the whip. They have stood these blows, these lashes before, and they will stand them to the end." Bethune never considered the possibility that families simply had nowhere to go or that they were reluctant to reveal their emotions to the visiting internationals. And he said nothing of the Republicans' own, energetic repression—how they hungered for rumors and tips about fascist sympathizers, how they collected and jailed them, or even executed them, without the flimsiest of trials.

Bethune himself was later accused of espionage and expelled from Spain. Haldane, too, was part of a gang of foreign suspects rounded up by the police. He wrote later of these men and women, "I think, one was shot"—a fleeting clause, as if the killing deserved not one further drop of his attention. In his own broadcasts, Haldane, like Bethune, presumed to occupy the minds of Madrid's residents. He gilded their thoughts. He valorized them, but in a way that felt diminishing and patronizing:

> Here perhaps is what they feel. "I may be killed, but I'm going to die some time anyway. And now I have a chance, such as comes to very few people, of taking part in one of the great events in world history. I'm not going to leave Madrid. I'm not going to send my children away. I'm not even going to interrupt my afternoon walk for a few fascist bombs." You may think they exaggerate a bit. I am not sure. Honour is worth dying for, and every citizen of Madrid feels that he or she is willing to die for the honour of democracy.

HALDANE WENT TO SPAIN AGAIN in the spring of 1937. Charlotte left England as well, sent by the CPGB to Paris so that she could receive British volunteers and direct them on to Spain. Ronnie was still in Spain; in February, during the Battle of Jarama, he took a bullet to his left arm. In those weeks, the Republican cause consumed the family wholesale.

Among the International Brigadiers, Haldane provoked both admiration and exasperation—this aging, thickening gent on a wander in a field of war. Haldane recognized how out of place he was, but even when he joked about it, he let slip a grain of truth. "Just a spectator from England," he told people. "Enjoyed the last war so much I thought I'd come to Spain for a holiday." Fred Copeman, who commanded the British Battalion, wrote in his 1948 memoir: "It was impossible not to be affected by the sincerity of this man. He was one of the greatest living scientists, intensely shy, and yet capable of expressing the most intricate problem in simple language." Thirty years later, however, Copeman told an interviewer from the Imperial War Museum that Haldane could also be "more bloody nuisance than he was worth." He posed a problem for Copeman. Haldane was a valuable asset to the cause; don't let him get killed, don't allow him into the dangers of the front line, Copeman was instructed. And yet Haldane seemed to be there always, brandishing a small revolver that probably couldn't hit a cow at 10 paces. "I would go up," Copeman said, "and every time I would say: 'What bloody good do you think you are? First of all you're taking two blokes' room, two blokes can sit where your fat arse is, so get down out of it and get back to Brigade headquarters.'"

Helping to hold the front had been one of Haldane's purposes in coming to Spain. He couldn't keep away. Once, at the Gran Via restaurant, Virginia Cowles ran into Haldane having lunch. "Think I'll hop down to the battlefield and have a look round," he said to her. "Do you want to come?"

They walked west, toward the parklands of the Casa de Campo. Shells sailed over their heads—only little ones, Haldane said to Cowles, so come along. In the trench on the front, young men poked their rifles through a bund of sandbags; yards ahead, in no-man's-land, three Republicans lay slain. Haldane wanted a better view of the Nationalist positions, so he suggested that Cowles wait in the trench while he reconnoitered. So she did, until she was discovered by a French officer. "This is no place to stand, Mademoiselle," he admonished, and led her away. Eventually, they encountered Haldane in a dugout with three other soldiers. He was sitting on a low stool, drinking out of a bottle of wine.

"Hullo," he said to Cowles. "Where have you been?"

Haldane's bravado peeled away sometimes, baring not so much cowardice as good sense. On another sortie with Martha Gellhorn and Ernest Hemingway—"like college kids on an outing," Gellhorn wrote of these dashes to the front—Cowles met Haldane again. They were holed up in an abandoned building, watching the tanks in the distance, when Haldane came up the stairs and joined them. "From the debris he dragged a dilapidated red plush chair, placed it in the middle of the room, and sat down in full view of the battlefield," Cowles wrote. "He put his elbows on his knees and adjusted his field-glasses."

Haldane was too easy to spot, Hemingway warned. Then, once again, he told Haldane: "Your glasses shine in the sun; they will think we are military observers."

"My dear fellow," Haldane replied, "I can assure you there isn't any danger here in the house."

Within minutes, shells began to hail down. One burst into the apartment next door. Cowles, Hemingway, and Gellhorn threw themselves on the floor. Haldane pelted out and never returned; they came upon him, much later, in the bar of their hotel.

Twenty years earlier, he had felt excited and alive when shells bit into the earth around him. But in Spain, the bombing discomposed Haldane. He was older now, more unsettled by the prospect of his own death, and rendered more desolate by the deaths of others. He was also alive to the lopsided nature of power in aerial bombing: the aircraft remote and untouchable, miles overhead, and their victims bewildered below. He saw how people fled in confusion for shelter; he saw old women who should have been bedridden having to stumble to safety. He was, as ever, on the side of the weak. "Air raids are not only wrong. They are loathsome and disgusting," he wrote. "If you have ever seen a child smashed by a bomb into something like a mixture of dirty rags and cat's meat you would realise this fact as intensely as I do."

Once, Haldane told Julia Bell, he was sitting on a park bench when a rabble of airplanes passed overhead. A bomb dropped so close that an old woman occupying the same bench was hit by shrapnel and killed. The experience shook Haldane. It wasn't that he was frightened of getting killed, she recalled later; it was just that he still had so many things he wanted to do. When Haldane returned to England in April 1937, he couldn't drive along a road without scouting its sides for cover from bombardment.

In December 1937, Haldane went to Spain once more, traveling to Valencia to appraise the town's air raid shelters. With Harry Pollitt of the CPGB, he then visited Teruel, a town in the mountains of eastern Spain that had just been won by the Republicans. Teruel was in the teeth of a hostile winter, the coldest in two decades. At night, temperatures sank to 20 below zero Celsius. Oranges froze solid. The International Brigadiers, held in reserve, spent the nights sprinting up and down the fields until they were warm and exhausted, then huddling together and sleeping until they grew numb with cold, then waking up and running again. Pollitt and Haldane arrived just in time for Christmas, bringing cigarettes, chocolate, and letters for the soldiers from home. Pollitt wrote of a Christmas feast—wine and pork, utter luxury!—but this decadence was denied to the rank and file. "I was pretty browned off with the old Christmas business, having an officers' dinner and a squaddies' dinner," one soldier, Frank West, remembered later. "When the shit was being served up, we were all together, weren't we?"

"Within the last year the British Battalion has changed from a group of extremely brave, but untrained men, into a unit which would be a credit to any professional army," Haldane wrote after returning from Teruel in January 1938. Holding the town during those precious days was a rare, feeble glitter of hope in that phase of the war, but it was rapidly smothered. When the Nationalists recaptured Teruel in February 1938, 10,000 Republicans lay dead in the streets and in the houses. In April 1939, having pocketed Madrid at last, Franco announced that he commanded all Spain. One more country had fallen into the hands of a fascist ruler.

THROUGH 1937 AND 1938, the fate of Spain, the swelling fortunes of fascism, and the imminence of a continental war drove many Europeans and Americans more conclusively toward Communism. This was not a time for conceptual debates over politics or for "half-truths and hesitations," as the poet Laurie Lee put it. It was a time to double down, to be committed to any force that threatened fascism. In early 1939, in a poll, 83 percent of Americans said they would want Russia to win in any war against Germany. "The only choice," the historian Eric Hobsbawm wrote, "was between two sides, and liberal-democratic opinion overwhelmingly chose anti-fascism."

On paper, Haldane remained unaffiliated with the CPGB, but

his allegiance to it was clear, particularly to himself. He had earlier thought Communism a fine fit for science and society and deemed it desirable for those reasons. Now it felt imperative and urgent for the very sake of civilization. With Communism, he wasn't a man testing the water with his big toe; he had stripped off to his trunks and had waded in up to his neck. He was, in the phrase of the time, a "crypto-Communist"—a stout supporter of the party who just wasn't carrying a card yet. The CPGB wanted him, of course. "Professor HALDANE has been urged to declare himself a Party member . . . but he has not yet consented," an MI5 agent reported in 1939.

Haldane grew more and more scathing about fascism as Europe marched toward war; in those years, he was more activist than scientist and one of England's most thunderous voices raised against Germany and Italy. He scoffed at the weakness of Neville Chamberlain, who tried strenuously to appease Hitler. (In one essay, discussing primitive animals, Haldane knifed the prime minister swiftly and kept on moving: "In fact, compared with a limpet, Mr Chamberlain is quite progressive.") He was withering with anyone who supported Franco, Hitler, or Mussolini. In a May 1938 letter to an unspecified recipient, Haldane wrote:

> Dear Sir,
>
> I have to thank you for your invitation to dine . . . but must respectfully decline, as I should find social relations with you very difficult. During the last two years I have seen so many Spanish women and children murdered by Italian bombs . . . that intercourse with those who support it in Parliament is somewhat repugnant to me.
>
> It may be, Sir, that my objections to lies told to justify murder . . . are old-fashioned. It may be in the public interest that British ships such as the Endymion [torpedoed by a Nationalist submarine in January 1938] should be sunk, and friendly relations immediately opened with those who sunk them. But you will perhaps understand that those who do not take this view find a certain embarrassment in associating with those who share in the responsibility for such events.

When scientists and other refugees fled the Nazis and arrived in London, Haldane found positions for them. As early as 1933, he and Charlotte invited a Jewish woman named Maria Lessing to stay with them on the recommendation of a mutual German friend. When scientists were referred to him, he read their publications carefully. Some he hired himself, among them Hans Kalmus, a Jewish biologist from Prague, and Juan Negrin, the Spanish leader who had trained as a physiologist. Others, such as Hans Grüneberg and Ernst Chain, he sent on to universities and laboratories around the country. Chain had come to England with just £10 in cash. When Haldane narrated their meeting, he wrote that he told Chain: "I don't think I can help you much, but there is a man called Florey at Oxford who is certainly interested in this kind of stuff, and I would advise you to have an interview with him." Chain and Howard Florey subsequently shared a Nobel Prize with Alexander Fleming for their work on isolating and producing penicillin. Haldane's memory of their meeting must have garbled with time; Chain and Florey met later, through another scientist. But Chain would say, "The whole of my career in England is really due to Haldane." And Haldane himself wrote, with endearing pride, that assisting Chain was "what posterity may regard as the best and most important action of my life." Florey and Chain rescued millions of people from disease and death, and so, Haldane wrote: "Perhaps all my discoveries will be forgotten and I shall be remembered only in the words of the ancient Greek poet Pindar: 'He once nourished the contriver of painlessness, the gentle limb-guardian Asklepios . . . the heroic conqueror of manifold diseases.'"

Standing against fascists had to involve standing against their principles of racial purity; in fact, Haldane was the kind of man who might have changed his opinions about race just so that they did not match those held by the fascists he hated. Through the 1930s and 1940s, Haldane weighed and reweighed his thoughts on race. He didn't abandon his belief that race was a valid way to sort humanity; just look, he wrote, at the "absolute differences" of physiognomy between white men and black men or at the varying immunity to disease conferred by race upon an Englishman in West Africa or a West African in England. But his discussions of these qualities were very distant from the links that racists were claiming between skin color and innate aptitude. Rather, he seemed to be using the word *race* the way geneticists today refer to a particular kind of "population group"—a subset of the larger species, as defined by geography or place of origin.

In the interests of scientific objectivity, Haldane had once claimed to keep an open mind about disparities in the aptitude of races. Now he emphasized, more and more, that no proof of such disparities could be found. He saw, earlier than most, the importance of environment and education; poor Native American children did badly on tests, he would later write in an essay, but "the children of the Osagi [Osage] tribe, who used some of the money got from oil found on their territory to build schools, did equally well" when compared with their white classmates. He offered a speculative idea that modern genetics has since confirmed, even to the extent of using language very similar to Haldane's: "If we could pick a representative for musical capacity, or any other such character, we should find that the representatives of different races varied little compared with the variation found within a race." Toward the end of his life, Haldane committed himself more firmly still. "Most geneticists," he wrote in 1963, dismissed the notion of "large genetically determined differences between the mental and moral capacities of different human races." As the science progressed, he permitted his views to change.

So much about race and genetics was still occluded, and yet the Nazis invented racial principles, claimed them as stone-hard fact, and used them to create policy. It made Haldane furious. Any claims to Teutonic superiority, for instance, were empty of meaning. The Germans were not a distinct race, he wrote, and neither were European Jews; the populations had bred into each other and overlapped for centuries, and boundaries between them were meaningless. As president of the Genetics Society, he exhorted a colleague to rebut an article in *The Times* that argued that a Jew was "one who has more than ten per cent of Jewish blood." (The colleague had to propose the idea to the society first. "Let loose your motion," Haldane urged.) The Nazi doctrine of "*Blut und Boden*," or "Blood and Soil," was also laughable. The soil of Friesland was just like the soil of northern

Holland, he wrote in an article, and the blood of human beings differed only in its groupings of A, B, O, and AB, which occurred across races and geographies. To credit any of these unproved ideas of racial hierarchy was to succor the Nazis, Haldane thought. He picked up a British teaching manual once and was dismayed to find proclivities of custom and culture attributed to race. These beliefs "are meat and drink to anti-semites. They may be useful to British imperialists, too. But why call them 'truth'?"

In his speeches, Haldane amalgamated all these scientific lessons with his stories of the Spanish war to summon his audiences to rise against fascism. Often without a microphone, relying on his bullhorn of a voice, Haldane towered on the dais, shaking his fist or slapping his thigh for emphasis. If he carried notes, he didn't need them. His tone still tended to that of the pontificating professor, but to hear an unafraid scientist speak his version of the truth was rare enough to be impressive, and he was known so widely to be brilliant that his listeners drank his words in. One young academic at the University

Haldane and Charlotte at a rally of the Communist Party of Great Britain, 1939.

of Bristol told an interviewer, "We couldn't get a hall big enough for him to talk in."

Sometimes his rallies ended in clashes with the police; after one of these, in which Haldane saw that his fellow demonstrators had taken flight, he made his own rapid exit, huffing in complaint: "The trouble with these pacifists is that they won't fight." Other times, his lectures drew sympathizers of Franco or Hitler, who tried to howl him down. "These filthy traitors are supporting the murder of British seamen," he yelled back at an "Aid for Spain" meeting in Shoreditch. "Throw them out!" Once, he wrote in a letter to his publisher, a gang of fascists attacked him on the street. "I can still use my fists, and was not much hurt, though lame for a week or so from kicks."

In March 1938, during a speech at Trafalgar Square, it looked briefly as if Haldane would start some kind of mutiny all by himself. Even as Hitler's troops were invading Austria, he said to a gathering of 1,200 people, British ministers were lunching with their German counterparts. When he got this agitated, he spent all his breath in the first half of a sentence and had to speak the second half while breathing in. "We are being governed by criminals," he declared, urging everybody present to march to the German embassy, right then and there, to signal their protest.

"Are you prepared to come to the embassy with me?" he shouted, frothing now.

"Yes!" the crowd roared back.

"Come on then," Haldane called out, and tried to lead a rush through the cordon of policemen. But "this was frustrated," an MI5 agent later reported. The crowd dispersed. "Shortly afterwards HALDANE was seen in earnest conversation with a group of demonstrators, apparently of the extremist type, and he appeared to be very annoyed at the turn of events." In that moment, Haldane must have felt in his veins the hot, charging blood of a young Lenin— must have felt that anything, even an impromptu revolution, was possible. Then the moment fell apart, the world reasserted itself, and he became once again a scientist with a lab to watch over and equations to solve.

THE PRELUDE TO THE WAR, and then the war itself, set off a series of displacements in Haldane's life. When hostilities opened in September 1939, London was still quiet, the Blitz yet to come. University College, however, decided to evacuate its buildings, sending its staff and students to other campuses around the country. Haldane was ordered to relocate his lab to Wales; he did not obey. Officials ordered the slaughter of the animals kept by zoologists, so scientists dumped their tropical fish and left large, dead tortoises to rot on the roof. Haldane, again, refused to kill the mice and fruit flies that his researchers had bred carefully for years. The university cut off electricity, telephone, and heating services to its buildings, and an official told Haldane that he wasn't sure how long his lab's workers would continue to be paid. The libraries were shuttered. "In order to get into the Biological sections of this College Library," he wrote in a letter, "I fear I should need an Order from Gauleiter Evans, and this would not be easy to obtain." By October, the college felt gutted, a friend told Haldane with sadness.

Still, Haldane stayed. The evacuation was a waste of time and money, he told a colleague, because his building couldn't be repurposed for military use and held nothing of value to anyone except him. If the fighting went on, of course, several of the staff would be called away to farm or to manufacture munitions, and his experiments would have to stop then. But the war was still a fledgling one, while for him, "another three months' work will give really valuable results." He decorated the door of his office with a mischievous poster: "Freedom is in peril—Defend it with all your might." No one could decide if he was referring to the college's administration or to the Nazis.

The following autumn, however, London's horizon grew shaded with bombs. The docks burned, the railways snapped. The planes of the Royal Air Force and the Luftwaffe bit and snarled at each other in the sky. In a single raid on September 8, shrouded by darkness, the Luftwaffe killed 412 people, injured 747, and lit the city on fire. The German bombers visited every night after that, for eight weeks.

It was impossible to remain in London, and when Haldane

searched for alternatives, he received an offer. Ronald Fisher, his colleague at University College, had relocated to the Rothamsted Experimental Station in the town of Harpenden, in Hertfordshire, and he invited Haldane there as well. In October, the Haldane lab moved into pinched quarters in Rothamsted. The entire department had to be folded into a single room, and he and his colleagues had to stay together in a large, gloomy Victorian house. Haldane took with him a crate of books, including the collected works of Marx, and a single suitcase of clothes.

Naomi visited him in Harpenden and judged his house a piggery. Seven adults, two children, and a baby shared a single bathroom, and no one had any privacy. Haldane lived in an attic with his white cat Mitsi, and when Naomi was there, he ceded the room to her and slept on the sofa. She noted a gnawed block of cheese on the mantelpiece, a slate-stiff pillow on the bed, an old hairbrush that had once belonged to their father, and a pair of frayed braces. "I wonder if he has any aesthetic perceptions," she thought, "or if he crushes them: standing there saying that mankind should be judged by algebra as their highest achievement or reading a piece from Engels." He insisted on shining his own shoes and doing his own dishes. The household hired no help—partly because of a shortage of labor, she guessed, but partly, no doubt, because of ideology.

Charlotte didn't go with him. Their marriage, strained by the frustrations of childlessness and infidelity, had abraded so far by then that he had told her he didn't intend to live with her again. In 1939, she told Haldane she wanted a divorce, but the party stymied her. She was still a member, and the party insisted that a breakdown of such a public marriage would injure the Communist cause. So they remained for a while longer in their festering union. In any case, they were pursuing practically independent lives. Charlotte's career as a journalist had blossomed, and now that Ronnie had grown up, she traveled for long stretches: to France to cover a congress against Fascism, to China to write about Communism in that country. And Haldane, for his part, fell in love with another woman.

As an undergraduate at University College, Helen Spurway had attended one of Haldane's lectures in 1934. She was 22 years younger than her professor, grave and fierce—equally determined even then, the story went, in her academic ambitions and in pursuing Haldane. She received a doctorate in 1938, working on fruit fly genetics, and then became a research assistant in Haldane's department. She was a meticulous observer, her colleague John Maynard Smith remembered in an interview years later. Perhaps not a great theoretician, he added, but marvelous with animals. "We used to have great arguments," Maynard Smith said. "I used to say: 'Look, the way to do science is to think of an important problem and then find the right organism on which to solve this problem.' Helen would say: 'No you must allow the organism to lead you. Keep the organism, keep the plant, and it will tell you what is interesting.'"

Like Charlotte, Helen had grown up in hardship, and she resented the bourgeois classes the middle class lawyers, doctors, and clergymen who conspired to keep the poor poor. She wasn't easily approachable. She spoke in a curt, harsh manner, although she later told a student that her permanently raised voice was a result of her mild deafness. Naomi thought Helen to be "suspicious of all kindness, prickly, distrustful, in some ways terribly objective." But Haldane warmed to that kind of objectivity, the honesty that came at the cost of politeness and comfort; indeed, he prided himself on having it. When they first went to visit Louisa, in Oxford, Haldane warned Helen that his mother was awful. Don't pull your punches, he told her. But Louisa was very friendly, Helen learned, to her surprise. Naomi had made up her mind that she would be kinder to Helen than she had been to Charlotte; possibly Louisa had come to the same decision.

Helen was, to Haldane, a colleague as well as a partner; later, he would remark on what an ideal arrangement this was. One of Haldane's friends, Ivor Montagu, wrote of the kind of couple they made, unusual yet singularly well tailored to each other. They dropped in on him sometimes in his cottage in Hertfordshire, and Montagu realized that, like Haldane, Helen was a pedant, tyrannizing her speech into precision. When Haldane wasn't sitting in the garden, filling bits of paper with equations as the sun browned his skin, they conducted biological expeditions into the countryside:

One day they came in to the cottage unexpectedly after collecting *Planorbis* (the pond snail) from the woods a few miles to the north, and it was preposterous to hear them courteously discussing their specimens as they ate their tea in their tiny, low-ceilinged room, he so huge, she so angular and slim. "Do you not think, Doctor . . . ?" from the one, and, "Yes, but what is your opinion, Professor . . . ?" from the other.

The chronology of their romance is not fully clear, but by October 1940, when they moved to Harpenden with the rest of their colleagues, Haldane and Helen were certainly together. Soon after they settled into their large, shared house, one of the children revealed, over the breakfast table, the results of his mathematical inspections. He had counted the number of beds in all the rooms, the 5-year-old said, and the number of people living there. He was coming up a bed short. Haldane had to extract a promise of secrecy from everyone in the house. After the nastiness of the *Sex Viri* scandal at Cambridge, he didn't need more officials snooping into his newly adulterous life.

Quite aptly for Haldane, his final break with Charlotte came over a matter of ideology—the personal coiled thickly into the political. Throughout the tumult over the *Daily Worker* and its ban, Haldane never wandered from his conviction that Stalinist Communism was a beneficial force—not even when Charlotte returned from a trip to the Soviet Union late in 1941 and told him what she had seen. Scientists were being co-opted or dispensed with, she realized. Publishing houses were abruptly dissolved. Babies died from starvation. One day, she saw a procession of poor people dragging themselves along a road, and she was overcome by a sense of shame. In England, she had celebrated the Soviet Union as a beacon for the world's workers, but this tableau made her a liar and a fantasist.

When Charlotte was back in London, she told Haldane about her experiences. For 3 hours, she talked about how education was no longer free, how party sycophants were enriching themselves, how Soviet society was separating, like oil and water, into a thin class of the powerful and an uncountable mass of the powerless. The scien-

Haldane with Helen Spurway.

tist Nikolai Vavilov, Haldane's friend, had disappeared; she feared he had been arrested and possibly killed. Haldane reacted with surprise and refused to believe her. Her party comrades heard her reports and decided that she had been turned by British intelligence or that she was angry about being denied a chance to interview Stalin.

Disenchanted, Charlotte left the CPGB in 1942—a ramifying move, because it led not only to the end of her marriage with Haldane but also to his becoming an official member of the party. Now that she had renounced Communism, a divorce became a party priority, as if she was a malignancy to be snipped quickly from the hale body of Haldane. She filed suit that year, but the country was deep into wartime, and an official divorce was only granted toward the end of November 1945. Just 10 days later, he would write to his university. "Dear Mr. Provost," his letter ran. "This is to inform

you that I have just married Dr Helen Spurway, Honorary Research
Assistant in my department."

In May 1942, a month after Charlotte's departure from the fold,
the party officially inducted Haldane as a member to secure his image
and to demonstrate that he suffered no ideological infection from
her. "Famous Scientist Joins Communists," a newspaper headline
announced. The party distributed a seven-page booklet, titled "He
Speaks to Millions," filled with plumes of praise: how he had set into
science at a young age; how he had, like an epic hero, withstood the
pains of his experiments upon himself; how he had taken a hard look
at fascism and come alive to the necessity of practical politics. "A
scientist should think of what he is doing in relation to the whole of
life, not some minor portion of it, and realise his full responsibility to
the people for the changes that his work must inevitably bring about,"
Haldane said to the booklet's readers. Having ridden alongside for
years, the fellow traveler had clambered aboard the wagon at last.

5.

The
War
at
Home

ONLY TWICE DURING THE WAR did the Luftwaffe attack Harpenden, and only once at night—the very night Haldane moved there from London in the autumn of 1940. An incendiary bomb hit the roof of the entomology building, but the fire watchers on duty spotted it and smothered the flames with buckets of sand.

Haldane knew that the aerial bombing of England was inevitable. He had seen Germany and Italy profit from the strategy in Spain, and he feared similar destruction now: houses bursting apart, children and women remaindered to lumps, every morning, afternoon, and evening laced with dread. The British government worried about raids as well. Even before the Spanish war, in 1935, an Air Raid Precautions (ARP) Department was created to plan for the contingency of aerial attacks. Three years later, a new law instructed local councils to build ARP shelters. Volunteer wardens, with their ARP lapel badges and their service helmets each painted with a "W," were trained to guide people into their nearest public shelters, rescue the injured, and enforce blackouts.

In February 1939, the first Anderson Shelter was installed in a house's garden in Islington. Named after Sir John Anderson, Britain's home secretary, the shelter was formed from panels of corrugated steel, bolted together into a long box with a curved roof. It looked like the carapace of a metal tortoise. Half-sunk into the garden and covered by sandbags, these shelters were designed to protect six people apiece during a bombing raid. Between February and September, 1.5 million shelters had been distributed, free for any household earning less than £250 a year and £7.10 a unit for everyone else. Another 2.1 million were allocated before the end of the war. They were pushed into lawns, nuzzled up against homes, and even brought indoors by families all over the country.

But the shelters were riddled with problems. Not everybody had a garden, for a start, and if the steel structure was merely leaned against a wall, it became instantly less effective. When buried in the garden, it was apt to grow damp or flood with rain; on deep winter nights, it refrigerated its inhabitants. After dark, when air raid sirens began to wail, families found it difficult to abandon their warm homes and stumble across to their Andersons. When London was surveyed in 1940, the government discovered that just a quarter of the population used Anderson Shelters. Nine percent used the larger, brick-and-mortar community shelters. Most people slept on at home. If they were fated to die, they said, they preferred to die in comfort.

Haldane didn't think the Andersons were as safe as they could be. They staved off the shock waves of explosion and the dangers of shrapnel and collapsing masonry, but a direct hit on an Anderson crumpled it like a tin can. And six people sitting huddled within a metal hutch, with bombs going off all around them, were sure to feel frightened and isolated. The government had first tested the Anderson's sturdiness with goats inside them, but these reports, Haldane told the *Herald Tribune*, were "far too optimistic. . . . The best way to clear up the psychological questions would be to conduct experiments on human beings." Ever the willing guinea pig, Haldane offered to sit in an Anderson while it was bombed, but he couldn't perform this experiment himself. "If I started to buy explosives, I should probably be sent to Ireland," he said. "The government ought either to investigate the question thoroughly or else say: 'We don't know whether people in the shelters will be killed or not'—and then shut up."

Why couldn't Britain plan better shelters, of the kind he had seen in Spain? In Valencia, they had hollowed vaults 15 feet below the ground, and 750 people fit into each one; above them was a concrete roof, then packed earth, then a further 3 feet of cement. In Barcelona, the Republicans had scooped out 1,400 tunnels and lined them with brick or cement. Haldane had toured one, and he saw they had provisions for lighting and ventilation and that they kept dry even in bad weather. Britain needed shelters like these. To keep 5.5 million people safe, he calculated in his book *A.R.P.*, London would need 780 miles of tunnels sunk 60 feet deep; the network would cost the state's purse £11 or £12 a head. He also designed his own version of Valencia's subterranean vaults, walled and roofed with reinforced concrete at least 8 inches thick. The Haldane Shelters, as they came to be known, were best placed underground, but they could serve on

Haldane holding a model of a Haldane Shelter, July 1940.

the surface as well. They had seats, gas-proof entrance locks, ventilation, lavatories, and a first-aid post, and 900 people could crowd into the largest. A basic Haldane Shelter, he claimed, could be knocked together in a month for just £7 a head. In 1938, the year *A.R.P.* was published, Haldane cofounded and chaired the National ARP Coordinating Committee, a body of scientists, architects, and engineers who lobbied the government to bless these options.

At first, Anderson refused even to consider them. In December of that year, just two months before his eponymous shelter was launched, he told Parliament that tunnels were too difficult to construct and that he had no desire to compel his compatriots to a "troglodyte existence deep underground." The following April, when the borough of Finsbury applied to build communal, subterranean shelters, Anderson wouldn't hear of it. When he learned of the decision, Haldane took to the *Daily Worker* to breathe fire. "You Have Received Your Death Sentence," the headline announced. "In his career at the Home Office, in Ireland, and in Bengal, Sir John Anderson doubtless approved of a number of death sentences," Haldane wrote. "But until this month he had never condemned thousands of people to death by a single stroke of his pen."

The government might have been oblivious to the kind of ruin the Luftwaffe could distribute. Haldane wasn't. He remembered the people struck down in Spanish streets, the bleeding bodies pulled from rubble, and the haste with which death could supplant life; he remembered his terror when shrapnel killed a woman before his eyes. In September 1940, Haldane visited Naomi in Scotland, and she saw how the war frightened and tired him. Not a hair, of the few that were left, had saved itself from gray. He appeared suddenly to feel the years bunched into his bones. Even a child could—and did—interrupt him, never an easy feat otherwise. "His views about the possible future are extremely messianic; he thinks things are going to be much, much worse before they're better," Naomi wrote.

During this time, Haldane kept a patchy account of his days for the benefit of Mass Observation, a project that recorded everyday life in Britain. He professed himself overall rather gloomy, weighed down by the thought of mortality. Was he taking holidays? "I shall take all I can. It may be my last chance." What was he planning for the near future? "I have no further plans, as it seems futile to plan more than a few weeks ahead." What were his dreams about? He didn't usually recall them, but once, he dreamed about the end of the war—no details, just an absence of fighting. What made him happy about the state of the country? "The participation of women in the war effort, making for sex equality." And what made him sad? "The failure to get rid of the group in the ruling class responsible both for the outbreak of war and our unpreparedness for it." Once he graphed his happiness over the course of a week. Monday, he was in London, and although he wasn't bombed, his train was late; on a scale, he gave himself 20 out of 50. Tuesday, he ran an experiment, then had beer and a steak: 40 out of 50. Then a nasal cold: 10 out of 50; then ephedrine to dose the cold, inducing artificial euphoria: 50 out of 50. Up and down he swung, a slave to his mood.

Five days after Naomi saw her brother that September, the Blitz began, and Haldane, still in London at the time, scrambled into his basement every time the bombers came nattering overhead. One night, a part of University College's library went up in flames. Another night, with incendiaries showering down, Haldane decided to seek refuge in a community shelter, a paltry structure made of brick and wood. As soon as he arrived, a bomb demolished a section of the shelter, injuring several people within and killing an old woman with a weak heart. After that, he took to spending his nights in the shelter at the zoological gardens nearby, where his friend Julian Huxley served as director. (Charlotte would insist, in her divorce suit, that Haldane refused to take her along; he would claim she was lying.) A couple of weeks later, Haldane moved to Harpenden. The danger was too real and too close, and he wished in no way to tangle with it.

The ARP Coordinating Committee agreed, in time, that excavating miles of new tunnels was difficult, but it stuck to its prescription of Haldane Shelters made of reinforced concrete. The government's obstinate refusal to build these was entirely in line with Haldane's opinion of politicians. Rather than listen to scientists, they took their own, unlettered counsel. His ARP Coordinating Committee wrote up a memorandum for Anderson's department, full of practical suggestions. The farthest they went, he complained to an acquaintance, was to admit after six months that some of the document's principles were correct. "As however this has so far made no difference to their practice," he wrote, "it cannot be said that they are taking the advice of the Committee."

This wouldn't happen in a socialist economy, Haldane was sure. Only in a capitalist state would the government protest that cement prices were too high, not admitting that the big manufacturers had formed a cartel to gouge the exchequer in the middle of a war. (When Haldane described the Cement Makers' Federation as a ring of monopolists, they threatened to sue him for libel.) Only in a capitalist state would the rulers ensure their own security while skimping on the safety of the working classes. Once, when a deputation of labor activists and Communist Party members went around to the Home Office, to press their plea for Haldane Shelters, the air raid siren began to yowl. Find a shelter on the street outside, the visitors were told; the shelter on the premises couldn't hold everyone. But there was no time, so the activists were squeezed into the Home Office's shelter—which was precisely the kind of Haldane Shelter they had come there to demand.

For most Londoners, the nearest tunnels were those of the Underground, but at first, when people spent their nights on station platforms, the government balked. The crowds hindered the movement of troops, officials claimed, so the entrances to the Underground were shut every night. As a result, queues began to form at midday, when families bought tickets ostensibly to catch a train, remained on the platform all night, and emerged the next morning. Eventually, the government yielded to the pressure from the throngs as well as from Haldane's committee and permitted the Tube's stations to host hundreds of thousands of people overnight. If you didn't have time to sprint to the Underground, though, you risked being bombed in your bed or catching pneumonia in the Anderson Shelter. A *Daily Worker* cartoon compared the Andersons to dog kennels. "Now be British and be bombed while we go into the country to carry on the government," a politician tells a family of four, presenting them with a new Anderson and prancing away to the limousine that awaits him.

Haldane became the engine of the movement for stronger shelters. He wasn't a structural engineer or an architect, but he was Britain's most visible scientist, and he had a reputation as a man who never flinched from telling the truth. Invested with this authority, he consumed reams of newsprint to explain the virtues of deep shelters, and he devoted his public lectures to air raid safety. He would begin by listing the ways in which someone might die in an air raid—blown up, turned inside out by the shock wave, buried under falling debris, salted with shrapnel. The government's shelters were no good, he said. Once, in Bradford, he held up a lump of mortar. "I don't like to steal municipal property," he said, "but a lady who was with me poked out a piece of something—I won't call it mortar: I don't know what it is—from between the bricks of one of these shelters, and here it is." Then he rubbed the mortar between his fingers and crumbled it into dust.

Having thus won the audience's attention, he advised them on how to seek protection. The government's handbooks were plenty useful, he said sarcastically. A section on evacuating pets, for instance—no doubt that cheered the working men and women of Britain. He painted a picture: the poodles and greyhounds of Mayfair being hurried to safety in Rolls Royces through streets clotted with London's poor. If the people wanted their government to keep them whole, they must agitate for concrete shelters, he said. Haldane had pulled audiences even when he spoke about knotted ideas of science and philosophy. Now, when he was applying himself to matters of life and death, his congregations multiplied further. Hundreds, sometimes thousands, of people came to hear what he had to say. "Panic speeches," his critics called them—and that was just fine by him. If someone had listened to his panic speeches when he was raising the alarm about German meddling in Spain, he thought, Europe would not be imperiled by fascism.

His correspondence swelled further. A man wrote from Shanghai about the experience of being bombed from the air there. A man wrote from Charlottenlund, asking if *A.R.P.* could be translated into Danish. A man wrote from Manila, thanking him for his clear and logical book. A man wrote from the Australian Association of

Scientific Workers, asking how they might dig shelters into the rock under Sydney. Many letters contained suggestions, as if he were crafting British policy on the subject: Please find enclosed a blueprint of a perambulator that can protect babies in gas attacks. Please find below my scheme for aerial mines. Please find described a way to proof our houses against gas. Please find within a plan to redesign a building to shelter its occupants better. Somehow, Haldane made time to respond. In the wartime shortage of paper, he reused the sheets lying around his office, so a letter was very likely to have scrawled equations or a typed bibliography on its other side. I suspect a 40-foot wire would be snapped by a rapidly moving airplane, he might write back. I'm not sure your perambulator will be gas-tight enough. The least satisfactory part of the plan is, I think, the thinness of the walls. In his willingness to believe that scientific originality could come from anywhere, Haldane was the ultimate democrat.

The campaign for safer shelters eventually squeezed the government. Members of Parliament argued about the merits of Haldane Shelters through 1940 and 1941—the government defending its decision not to build any and legislators trying to tear these reasons apart. Other newspapers joined the *Daily Worker* in its crusade. Churchill grew worried; his citizens were writing to his wife Clementine with such regularity that she had taken up the habit of dropping in on some of their shelters to see them for herself. Corroding public morale during a war or pitting Britons against each other would lead "to serious trouble," Churchill told his war cabinet. Londoners came together in protests on the streets. During one air raid early in the Blitz, nearly 100 East Enders trooped into the Savoy Hotel's basement shelter and refused to budge. "If it's good enough for the rich it's good enough for the Stepney workers and their families," the group's leader announced. Flummoxed, the Savoy let them stay until the All Clear sounded. Waiters brought them tea and toast.

In October 1940, Churchill moved Anderson out of his post, but the new home secretary, Herbert Morrison, handed Haldane only a qualified success. A few new deep shelters would be built and linked to the Tube tunnels, he said, and he would implement a system to allot bunks, provide better water and light, and spruce up the community shelters. But deep shelters could play only a "limited part" in defending the capital, Morrison said on the BBC. "Political schemers sailing under all sorts of official-sounding names who seek to destroy our will to take risks in freedom's cause are playing Hitler's game," he said, referring plainly to Haldane and his ARP Coordinating Committee. (Morrison, the *Daily Worker* shot back, "promised his listeners more bunks and ladled out more bunkum.") Another mixed concession: Even before Anderson's departure, the government had agreed to approve funds for Haldane Shelters "in suitable cases," but under Morrison, few such grants were actually given. Some neighborhoods around England raised funds to build their own. With their many broadsides against the government, moreover, Haldane and the *Daily Worker* earned themselves a powerful antagonist.

IN AUGUST 1939, Hitler sent his foreign minister to Moscow to sign a nonaggression pact with Stalin. The radical left of Europe splintered. By the creed of Communism, the Soviet Union was supposed to resist fascist nations, not enter into treaties with them. "So now I'm cured of socialism, if I needed to be cured," Jean-Paul Sartre wrote in distress in his notebook. But even if other fellow travelers were falling away, Haldane stood firm. When the Second World War finally began, in fact, his warm and constant praise of the Soviet Union gave him a faintly treasonous air, which gratified him. Perhaps he would be arrested, he thought. What a commotion that would cause! He made plans to distribute his money among the members of his staff, so that the law wouldn't confiscate it all if he was taken into custody. But no one came to hustle him to prison.

The war trapped Haldane in a contradiction. Just months earlier, he had been raising the roof with speeches about stamping out fascism. Now Great Britain was pitted against Germany and Italy in that very effort, yet the Soviet Union had decided to stand by and watch. How could Haldane possibly defend Moscow's policy? In the *New Statesman*, in September 1939, he provided a series of strained rationalizations. The Soviet Union abhorred, above everything else, war and capitalism, he wrote. The Allied powers welcomed wars; they provided jobs to the jobless and made the powerful wealthy. But

Stalin, he wrote, faced no unemployment at all, so to dispatch his workers from factories to battlefields made little sense. He anticipated another argument. "We have an active labour movement. . . . The Communist party is not illegal in Britain or France, but Communists are beheaded in Germany." Why then, Haldane's hypothetical reader asked, should Stalin not stand with the democracies of the West? Bereft of any real response, Haldane feebly volleyed an accusation back upon the Allies:

> The answer is simple. What do you mean by *we*? Do you mean Britain or the British Empire, France or the French Empire? I would sooner be a Jew in Berlin than a Kaffir in Johannesburg or a negro in French Equatorial Africa. If the Czechs are treated as an inferior race, do Indians or Annamites enjoy complete equality? Until the British and French Empires become Commonwealths, they can only expect Soviet friendship if their foreign policy is a hundred per cent peace policy.

The next year, Haldane became the chairman of the *Daily Worker*'s editorial board. Under him, the paper followed the party line faithfully, and Haldane was cautioned at least twice, via letters from the Home Office, that the *Daily Worker* was undermining the war effort. To Haldane, these warnings made plain the illiberal instincts of his government. The Home Office was borrowing tricks from Hitler and Mussolini, Haldane wrote in a fundraising pamphlet titled *Hands Off the Daily Worker*. The only way to peace and security was not by bombing German towns "but by throwing out Chamberlain and his gang and putting in a People's Government."

On a Tuesday in January 1941, just after 5 p.m., police officers occupied the Cayton Street office of the *Daily Worker*, handed over an order banning its publication, and locked down the printing press. The order was signed by Herbert Morrison, Home Secretary. "No one likes the idea of the suppression of a newspaper even during a war," *The Guardian* wrote in an editorial. But the *Daily Worker*, day after day, "has vilified the British Government and its leaders to the exclusion of any condemnation of Hitler." Haldane issued another pamphlet, splenetic about this murder of democracy:

> The Editorial Board of the DAILY WORKER offered to stand trial in a court of law; it offered to discuss with the Home Secretary the question of its news presentation.
> These offers were rejected. The Government was determined to suppress the DAILY WORKER.

Leading the *Daily Worker*'s board during the years of the ban required not an editorial sensibility but a stubborn capacity to deal with disaster. In April 1941, the newspaper's premises were destroyed by an air raid. When the bombers finished, only the building's frame remained standing, its roof and windows blown out and its facade now a gap toothed grin. In any case, there was nothing to print. Haldane wrote many times to the Home Office for permission to resume publishing, arguing that the government had misunderstood the *Daily Worker*'s attitude. The Home Secretary declined even to meet Haldane and his board. In his speeches, Haldane held aloft the banner of the *Daily Worker*. This was a newspaper that had improved British morale, not sapped it, he said. The government was frightened of the paper's influence. It had been banned, he went on, not for what it had done but for what it might do.

In June, after Hitler invaded the Soviet Union, Stalin was transformed overnight into an ally. "Now at least we are on the same side; this ache I have had all the war because of the Soviet Union is now healed," Naomi wrote in her diary. The CPGB was free once again to bellow its support for the war. This was a turning point, Winston Churchill said on the radio, and the cause of Russia was now the cause of "free peoples in every quarter of the globe." The very next day, Haldane sent Churchill a telegram:

> IN VIEW OF YOUR STATEMENT RE TURNING POINT OF WAR AND TASKS OF ALL FREE PEOPLES ASK YOU CONSIDER IMMEDIATELY WITHDRAWAL OF ORDER SUPPRESSING DAILY WORKER.

A full year passed before the ban was lifted, in August 1942, and orders for the *Daily Worker* flooded in—half a million copies a day, its editorial board claimed. But no other newspaper was willing to run the *Worker* off its presses, so Haldane helped start a campaign to raise £50,000 (more than $2.5 million today) to buy a printing company. Paper was still in short supply; only 83,000 daily copies could be published. "THE BLACKOUT IS OVER!" a headline exulted in the September 7 issue, the first in 19 months. In a cartoon, a burly embodiment of the *Daily Worker* stalked toward his workplace, a wrench and a rifle gripped in his hands.

THE FIRST DAY OF JUNE, 1939: high summer, glassy seas, a perfect day for a submarine trial. The HMS *Thetis* swam out of Birkenhead that morning for a final lap of tests—down the Mersey, through Liverpool Bay, and into the Irish Sea just north of Wales. When she reached her position, the 103 men aboard ate their pies and cold cuts, drank their beer, and then prepared for the submarine's very first dive. Just before 2 p.m., the *Thetis* vented the air in her tanks and dropped below the waves.

But she didn't sink as she ought to have. It could only be because she hadn't taken on enough water, the captain knew, and so they audited the various tanks on the submarine to see if they were full. The main tanks were flooded; so were the auxiliaries. Then, in checking the torpedo tubes, an officer read a mislabeled dial, assumed that tube no. 5 was dry, and opened its rear door for an inspection. The sea hurtled in—2 tons of water every minute, quickly annexing the torpedo room and the torpedo stowage space. Its nose suddenly heavier, the *Thetis* plunged down into the seabed, striking it at an angle of between 30 and 40 degrees and wrecking her signaling systems in the process.

What was to be done now?

Send a man into the torpedo compartment, so he could shut the door to tube no. 5 and seal the ocean out again. Then they might attempt to pump the water out.

One sailor tried, around 4 p.m., wearing his Davis Submerged Escape Apparatus, a knapsack-like device strapped to the chest. The

Davis fed its bearer pure oxygen, which poured through the mouth-piece with a soothing *haw-haw-haw*. But the pressure of the water, 150 feet below, was immense. The volunteer, feeling faint and dizzy, asked to return. Another officer tried, and then a third, and they both retired midway as well, their ears screaming with pain. One felt his heart constrict, as if it was being squeezed for juice.

Another plan was needed. Ordinarily, they would evacuate one by one, through an escape chamber—like an air lock, but for water. A man would climb into the chamber wearing a Davis, wait for the chamber to be flooded, open the outer hatch, slip out, shut the hatch behind him, and float up. The chamber would then be drained, and the next candidate would climb in. But aboard the *Thetis*, there were problems with this standard procedure. Of the 122 Davis sets, 29 were in the torpedo compartment, out of reach. Of the 103 men, 43 were civilians, so they had never trained in fleeing a submarine. It was past 8 p.m., and even if they all managed to make it up, they weren't sure if a ship would be around to spot them and pluck them out of the water. Best to wait the night out and pop a few sailors to the surface in the morning, by which point a rescue ship might have arrived. The sailors could explain the situation, and the ship could send down a hose to push compressed air into the torpedo compartments. The air would expel the water back the way it came, the door to tube no. 5 could be closed, and the *Thetis* could rise once more, like a great armored whale making for the light.

The night inched past. The crew worked to lighten the stern, until the mouth of the escape chamber was just 20 feet from the surface. By dawn, 103 men had been inhaling and exhaling for nearly 20 hours, and the air was turning thick with carbon dioxide. The *Thetis* had no purification systems installed. Ordinarily, a crew of 53 men could breathe easily for 16½ hours—enough to last until darkness fell, when the submarine could come up and take in more air. On this dive, after the two forward compartments had been inundated, the volume of available air per pair of lungs shrank. The carbon dioxide rose by 0.4 percent every hour. The men felt sick; they were breathing fast, and they reported headaches. Their minds dulled. When an officer was asked for a gauge reading, he needed many seconds to comprehend the request, then many more to determine what he had to do. His thoughts had to take brief furloughs.

At 8 a.m., two officers put on their Davis sets, hiked up the inside of the near-vertical submarine, and locked themselves into the escape chamber. Just as it started to fill, they heard signals above. A ship had arrived. When they broke water, the HMS *Brazen* hovered nearby, waiting to retrieve them.

In the *Thetis*, when the crew drained the escape chamber and opened the door, water slopped over the threshold, ran down the incline, and splashed into the main motors. The electrics short-circuited, and white, acrid smoke billowed through the submarine. The fire consumed more oxygen still; rib cages now heaved and ached with every breath. In a fugue of panic, the officers decided to send four men out of the escape chamber at once, but after it filled with water, the outer hatch remained shut. No one swam out and up. The crew waited for 20 minutes and then emptied the chamber, to find that the hatch had jammed and that three of the four men had died. Foam flecked their lips. The undiluted oxygen had shocked their bodies poisoned by carbon dioxide, and they had vomited into their mouthpieces.

Two more men were able to leave, forcing the hatch open and rising fast. Rescuers in rowboats picked them up, wrapped blankets around them, stuck cigarettes in their mouths, and put them aboard the *Brazen*. Other ships had steamed into position around the *Thetis* by now, and a diver, after half an hour's investigation, failed to see how an air hose might be connected to the submarine. The tide was running swiftly, so the ships looped a wire around the submarine's stern, raising it out of the water so that a hole might be sliced into her flank. Before this surgery could commence, though, the wire broke. At 3:10 in the afternoon, having allowed just four of her 103 men to vacate her alive, the *Thetis* sank out of sight.

IN A COUNTRY GIRDING FOR WAR, the loss of a new submarine and the failure to save its crew set off tremors of consternation. "Did We Fail Them?" the *Sheffield Telegraph* wondered. "Someone Has Blundered," the *Daily Mirror* declared, and urged members of Parliament to demand answers of their ministers. The government formed a tribunal of inquiry, and on behalf of the Amalgamated Engineering Union, Haldane offered to run tests to ascertain how the men would have flagged and flailed in the *Thetis*'s decaying air. He secured the use of an escape chamber on the premises of Siebe Gorman— the manufacturer not only of the Davis apparatus but also of instruments and diving suits designed by J. S. Once again, Haldane was his father's son, turning over the problem of how human beings breathe.

The chamber was 6½ feet in both height and diameter, barely tall enough to fit a standing Haldane. He began his first experiment at 10 p.m. on a Thursday in July. Once the door shut, the air mixture within was changed. Haldane gave himself 2.3 percent carbon dioxide.

"As far as you could tell," a lawyer asked Haldane when he gave evidence before the tribunal, "the atmosphere was in substantially the same conditions as you would have expected it to be in the submarine at the same time on June 1st?"

"Roughly so," Haldane replied.

He slept in patches through the night and woke at 8:30 a.m. The carbon dioxide level read 4.7 percent; after another 3 hours, it climbed to 5.55 percent. Haldane was panting heavily, and someone was beating a heavy drum within his skull. When the light shone on his eyes, the pain grew worse: photophobia. He lay down with a handkerchief over his face. One of the survivors from the *Thetis*, who visited that morning, thought Haldane looked in worse shape than his crewmates. At 12:15 p.m., he was tottering when he stood to talk into the telephone, and half an hour later, he let himself leave.

"Then when you came out of the chamber, did you put on a Davis breathing apparatus?"

"I did."

For the first 2 minutes, Haldane felt oxygen washing blissfully into his cells. The air tasted hot; the Davis's purifier was scrubbing excessive amounts of carbon dioxide from his breath. Then he vomited with violence, pulled the Davis away, and continued to throw up.

Over the next few days, Haldane ran more tests. Once he entered the chamber with four volunteers, friends of his from the International

Brigade in Spain. ("I reasoned that men recently on battlefields would be unlikely to be affected by panic," he told a newspaper.) The carbon dioxide concentration was ratcheted up rapidly, and the men felt instantly ill—although less so, Haldane noted, than when the air had spoiled more gradually. They spoke to each other, only to find that all sense had leaked out of their sentences. They wrote things down, and Haldane read these notes later and thought them rather stupid. Brought out and strapped into Davises, the men all reacted differently. One threw up, as Haldane had; the others professed varying grades of headaches. No one was fully well.

"Did that suggest to you any conclusions with regard to the use of the Davis apparatus under these conditions?"

It did, Haldane said. "The correct course, if men are attempting to escape with the Davis apparatus from a very foul atmosphere, would be that before going into the escape chamber they should breathe oxygen either from the Davis apparatus itself, or from a manifold for a period of about three-quarters of an hour, in order that they should get over these symptoms which arise from the breathing of air containing no carbon dioxide."

Another day, he sat alone in the chamber, sickened from 7 percent carbon dioxide, and asked for compressed air to be piped in, as the rescue ships had wanted to do with the *Thetis*. His breathing grew even more minced; the new, cold air stayed at the bottom of the chamber, where it entered, so the pressure of the carbon dioxide, floating near the top, actually rose. If a submarine crew was to be saved in this manner, he realized, it was essential to mix the air. He repeated the test, but this time he ensured that the fresh air streamed to the top as well as the bottom of the chamber; he also used a fan to blend the atmosphere around him. He didn't feel better, but he didn't get any worse.

What if there was soda lime on the submarine, to absorb the carbon dioxide?

"I think I was the first person to use soda lime in a submarine, and that was, I think, in the year 1905, when I was 12 years old," Haldane said. "We stayed down for only three or four hours and did analyses."

How much soda lime would the *Thetis* have had to carry?

He needed a minute to calculate that; he had been in the Siebe Gorman chamber just 12 hours earlier, and he still wasn't at his best. He scratched some figures onto a piece of paper. "If you had something like 100 pounds of soda lime there, it would have absorbed a substantial quantity of the whole carbon dioxide produced by the men in the submarine."

The tribunal thanked Haldane. Did he have anything to add?

In the room, watching the proceedings, were the families of some of those who sailed and died on the *Thetis*. "I think it might cause some comfort to say that I believe that the men suffered absolutely no severe pain," Haldane said. The aching head and the taxed breathing were unpleasant, but "it is not an intolerable feeling of bursting of the lungs which you get if you are buried alive, for example. . . . No one could say it was pleasant, but it would not, I think, be called torture." In his awkward way, he was trying to be kind. He genuinely believed that, however wretched the facts were, there was solace to be gained by fully knowing them.

ONCE THE *THETIS* WAS LOST, the Admiralty recognized how much it still had to learn about the physiology of men under water. Britain had just built 15 U-class submarines, and after the war started, another 34 would be ordered. So the government asked Haldane if he would look into this some more and study how humans responded to the demands of the deep sea. Haldane accepted. Over the next five years, he invariably mentioned in his letters that he was engaged in some rather serious war duties and so didn't have the time to write at any length. He wasn't bragging, or only a little; more likely, he was treasuring the thought that this work had instant practical utility—the way his father's did—and that he had been drafted into the battle against fascism. And perhaps he was enjoying the irony: that he was working for a government he constantly and publicly criticized.

Haldane drafted his own little battalion: Helen Spurway and his secretary Elizabeth Jermyn; Hans Kalmus, whom he had helped settle as a refugee from Germany; Juan Negrin, the former Spanish prime minister; and at least a dozen others, including some accredited

Communists. Month after month, he introduced himself and his colleagues into Siebe Gorman's equipment—an unlit cylindrical chamber of steel, 8 feet by 4, from which a person could communicate only by tapping code on the walls or by holding written messages up to the glass portholes. Within this chamber, Haldane was able to vary the pressure and the composition of air. Separately, in a tall tank, one of his crew, wearing a Davis or a diving suit, could remain underwater for an hour or more. The memory of the *Thetis* shadowed these trials. All these British sailors encapsulated in submarines around the world—what would happen to their bodies in extreme cold, or under intense pressure, or in this mixture of gases or that? How would they survive? In all, Haldane ran more than 100 tests. It was the most exhaustive, sustained spell of research into underwater physiology ever conducted by a scientist.

In his systematic style, Haldane progressed through every scenario he could think of—every possible coalition of gas, temperature, and pressure, every possible way in which these elements could pillage the body. In the chamber of horrors, as they came to call it, his colleagues were compressed and decompressed. They breathed blends of helium and oxygen, or oxygen and hydrogen, or oxygen and nitrogen; they breathed pure oxygen; they breathed a variety of concentrations of carbon dioxide. They breathed these when they were warm and when they were plunged into ice baths, when they were dosing themselves with benzedrine and when they were in atmospheres of high pressure. They appraised their ability to do regular things in irregular conditions: eat a meal, solve arithmetic problems, exercise with weights, or just maintain an even state of mind. They watched each other roll through confusion, anger, stupefaction, and suffering.

From within the chamber, their senses muddling, they wrote out messages for the observers outside:

> Nose clip very painful
> I can't see anything
> Eyepieces so dimmed
> Why did you trick me?

Years later, when they could discuss this work more freely, Haldane and Helen dramatized one of their experiments for the BBC. In the script, Haldane remained outside the chamber, as an observer. Helen went in, to an atmospheric pressure 10 times higher than normal, so that she could be fed measures of carbon dioxide. Another colleague, Martin Case, sat with her, wearing a respirator and taking gas samples.

"Have you got everything?" Haldane asked.

"Pencils. Pads. Sweaters. Watches. Sampling tubes," Helen replied. "That's the lot."

The steel door clanged shut. Case uncorked a cylinder of carbon dioxide, and the air in the chamber took on a new, sharp smell. When the compressor started up, Helen's ears roared. Her voice became nasal, and she began to pant gently.

"Gosh, the heat," she said. "My fingers feel like bananas."

"If you'll open the next cylinder of carbon dioxide, I'll fan," Case offered.

Just 8 minutes in, Helen was in poor shape, fumbling for her thoughts. "Chase fanning me makes me feel hyper . . . I can't spell the wretched word. H Y T E R," she stopped, and then "O P," and paused again, "H E R A," and another pause, "T E R A," pause, "P," pause, "H E R E A. A length of word—then it's all over." Her voice faded.

"She's looking fairly bad now," Haldane noted. "But she's still trying to write. I should think the CO2 must be up to 8%."

Case undid his respirator, then examined Helen's pupils. "Chase has looked at my eye," she said. "Under my lid . . . to eye . . . If they have looked if I am still C O N (pause) C I O (pause) C O U S."

The level of carbon dioxide neared 10 percent, and both Helen and Case appeared to be on the brink of sleep. Her face twitched. Half an hour after the test started, they blacked out.

"Stand by to decompress," Haldane said. "Right. Go ahead." The pressure eased with a steady hiss, the temperature dropped, and faint wisps of fog formed within the chamber. The subjects came around.

"We have . . . reduced . . . pressure," Helen said. "But I haven't lost consciousness."

"You did," Case replied. "So did I."

The carbon dioxide dropped to 4 percent.

"It's cold," Case suggested. "Put your sweater on."

"I am afraid . . . to think . . . about being sick," Helen said. Licks of pain ran through her head. She checked her watch and wrote notes: "12.10. Feel much better. 12.24. Hear the talking outside for the first time. 12.44. Gradually waking up."

The experiments were always edged with danger. A valve could malfunction, trapping a test subject in a haze of noxious air; an individual could react unexpectedly badly to a gas or to a shift in pressure. Lungs collapsed, noses bled, teeth ached. Eardrums clamored for mercy, and one of Haldane's perforated. ("If a hole remains in it, although one is somewhat deaf, one can blow tobacco smoke out of the ear in question, which is a social accomplishment," he said.) On occasion, Haldane lost his memory for a span of time. Twice, in high-pressure atmospheres, he grew fidgety and anxious, then slipped into convulsions. They were, he wrote in a letter, "like ordinary epileptic ones, and two or three minutes of kicking are followed by about ten minutes of unconsciousness." He recovered into a state of high terror.

The convulsions were so violent that they crushed three of his vertebrae. In another session, he damaged himself further. Having breathed a mix of helium and oxygen, he was waiting for the chamber's pressure to bleed down when his body erupted with pain. Then he felt a patch of skin above the back of his hip itch and tingle, as if his nerves were dancing. Later, he learned that a bubble of helium had formed in the conus at the very tip of the spinal cord. The injuries brought on Kümmell's disease, where the bones in his lower back, starved of blood, turned necrotic. For the rest of his life, Haldane would need a cushion to intercede between himself and his chair.

FOR THE ADMIRALTY, he wrote compact reports detailing his results and giving practical advice: what the best oxygen-nitrogen mixture for divers was, or when carbon dioxide began to enhance the toxicity of high-pressure oxygen, or what the symptoms of oxygen poisoning were, or how best to use the Davis apparatus. In its specifics, Haldane's research was classified, but the nature of its ordeals came to be widely known. His glamour brightened; his repute as a scientist of courage and conviction enlarged further still. "J. B. S. has become, in the eyes of the younger generation, the most romantic figure in the biological world of today," a letter to *Nature* declared. A Labour Parliamentarian, attacking the ban on the *Daily Worker*, reminded his colleagues of Haldane's sacrifices: "Professor Haldane, for many months during the war and before it, has almost daily risked his life in dangerous experiments for the benefit of this country, and when the Home Secretary sent somebody to serve a notice on him I should not have been a bit surprised to have been told that he had been taken out of some appallingly dangerous poisonous tank to have the notice served upon him." A newspaper profile of Haldane opened with a rhyme:

What, teacher, can that object be, inside a plate-glass drum?
That is Prof. Haldane whom you see, testing a vacuum.
See, floating near the waterside, that buoy of strange design!
That is Professor Haldane, tied, decoying of a mine.
On sea, on land and in the air, protecting us from harm,
Prof. Haldane meets us everywhere, our scientific arm.

But Haldane didn't feel like any sort of invincible warrior. His body was mutinying. His back troubled him all the time; an injury to his arm, sustained in Harpenden, briefly turned septic; he had a gastric ulcer. Death dogged his thoughts. He drew up a will, leaving most of his estate to Helen and his scientific periodicals to libraries in Yugoslavia or the Soviet Union. (He made no provisions for Charlotte, he mentioned: "I have made her an allowance of thirty pounds a month until my death and I am of the opinion that I have already done sufficient for her.") He wrote a lengthy note to his provost, explaining what was to be done with his department if he died. He itemized the strengths and weaknesses of his colleagues and considered who could replace him as the chair.

His party comrades noticed how weary and morose he was. "He is in a really dreadful state just now, by the nature of his work," Harry

Pollitt told a caller in April 1944. "I was with him last night. . . . You see, every day his life is in his hands, and I think he is only keeping himself going by a bit of doping." Bob McIlhone, another party member, thought Haldane was working himself to death because he was "too bloody afraid of us losing the war." In 1945, when Haldane was nearly 53, Germany was finally defeated. It ought to have revived him. But he had seen too close the nature of civilization and of his own body, both blown as brittle as glass. The war bit into him, crunched through his spirit, and spat the shards back out.

THE UNDERWATER RESEARCH wasn't the only work Haldane did for Britain during the war. He consulted with the air ministry on the mathematics of air raids—on how best to spread bombs on and around a target to minimize error and inflict maximum damage. For the army, he wrote a paper on interpreting risk and casualty statistics, using some of the tools he had put to work in a 1938 paper on genetics. When someone suggested to General Neil Ritchie that, to guard against invasion, it might be a fine idea to coat the sea with oil and set it on fire, Ritchie sought Haldane's advice on the matter. With Helen, he wrote a paper explaining why aircraft on completely random patrols were less likely to spot U-boats than those on regular patrols. When Germany lobbed V-1s toward England, Haldane was among the scientists who were asked for ways in which the bombs might be shot down. After 1945, he served on government committees on underwater physiology and on the medical effects of atomic energy.

The war rendered Haldane a visible paradox. Here was a man who railed against the government's faults, who couldn't wait for it to be supplanted by a revolutionary socialist state, who edited a newspaper judged to be so destructive of British interests that it was shut down, and who counted as friends men and women in thrall to Moscow. Here, also, was a man who was employed by the government to conduct classified research and who was familiar with many of the military's worries, deficiencies, and stratagems. The question asked itself: Where did Haldane's loyalty lie? Was he ever a spy?

MI5 didn't think so. In a short memo in 1947, an analyst wrote:

With every Communist there is a risk of leakage to the
Communist Party. . . . Professor HALDANE was entrusted
during the war with secret information and there is no
evidence in our possession that he betrayed it to anyone.
That purely negative statement cannot take us very far. Risk
of leakage exists in Professor HALDANE's case as in those
of other Communists. Beyond that it is hardly possible to
go, except to say that the risk is greater than that of a mere
possibility: it is an appreciable risk. The chance of his passing
information direct to the Russians may be regarded as a good
deal slenderer.

But British intelligence didn't know everything. (For one thing,
among the officers kept apprised of Haldane's movements was Kim
Philby, who was certainly spying for Moscow.) The existence of
Gruppa Iks, or X Group, a small band of British agents passing information to Soviet military intelligence, was discovered only after the
war as part of an American program to decrypt cables sent home by
Soviet embassies. Venona, as the program was called, exposed the
Cambridge Five, the spy ring that included Philby, as well as a Soviet
effort to pry into the Manhattan Project. By comparison, X Group
was a much less momentous enterprise; in some of their cables, in
fact, X Group's handlers in the London embassy sounded positively
exasperated by how little its leader, code name Intelligentsia, was able
to do.

The first surmise that Intelligentsia was Haldane came in 1999, in
a book titled *Venona*, by Rupert Allason, a historian of espionage who
uses the pen name Nigel West. Before Venona read Soviet messages,
no one thought that Haldane had spied for the GRU, the Soviet
army's intelligence directorate, West wrote. "The embarrassment
factor of this disclosure is such that the texts identifying Haldane
as INTELLIGENTSIA have been deleted by the British government." For support, West offered only guilt by proximity. Haldane
was close to the CPGB, even subservient to it; he was friends with a
man who was friends with a man who turned out to be a Soviet spy;
he worked on submarine research, and one of the secrets transmitted

by the Soviet embassy to Moscow was Britain's pursuit of sonar to locate submarines.

The release of the unredacted Venona intercepts into the British National Archives showed that West was wrong about Intelligentsia. The earliest cable about X Group, sent by the Soviet embassy in July 1940, named not Haldane but his friend Ivor Montagu, the patrician filmmaker and table tennis player. Meeting his handler, a Soviet attaché, for the first time, Montagu offered some airy political gossip: what Britons were making of a speech by Hitler and of another by Lord Halifax and of the Germans—"the sausage-dealers"—in general. In the cable, the attaché steamed with impatience: "He had not yet obtained a single contact. I came to an agreement with him about the work and pointed out the importance of speed."

The next month, Montagu disappointed again: "INTELLIGENT-SIA has not yet found the people in the military finance department." Maybe the Soviets needed another man, a smarter man, the cable writer suggested. But on this occasion, Montagu pledged to bring more material soon: documents from "Professor Haldane, who is working on an Admiralty assignment concerned with submarines and their operation." In September, a new message claimed a table-spoon of triumph: "INTELLIGENTSIA has handed over a copy of Professor Haldane's report to the Admiralty on his experience relating to the length of time a man can stay underwater." The attaché had only a meager hold on the science of Haldane's research; the title of the eight-page paper, written jointly by Haldane and Case, was a summary of their first 50 experiments on the effects of high pressure, carbon dioxide, and cold.

The caliber of Intelligentsia's material improved with time. Montagu relayed British success in deciphering a Soviet code, details about where the government moved its troops and how many tanks it produced, the weaknesses of its coastal defense, and a top-secret operation to confuse the Luftwaffe's guidance systems. Haldane's name, though, doesn't occur again in the Venona intercepts.

Haldane knew Montagu well; they had first been incendiary radicals at Cambridge, then lost sight of each other, and regathered within the arms of the party. In 1942, in fact, Haldane agreed to a cameo in a film Montagu was producing for the Soviet War News Film Agency. Just a couple of sentences to introduce the Lazarus picture, Montagu specified, in a letter to Haldane—it was to be called, "if we dare, 'Experiments in Bringing the Dead to Life.'" The final title, *Experiments in the Revival of Organisms*, was less gothic but still dramatic. It promised Victor Frankenstein, although it was only delivering Sergei Brukhonenko, the Russian inventor of an early heart-lung machine. Brukhonenko's *autojektor* kept air and blood pumping mechanically through the lungs and the heart, but the Soviet Union exalted his work to suggest that he could lift fresh corpses out of death. Montagu's film, having explained the *autojektor*, claims to show a dog's severed head lying sideways on a dish, kept alert and sentient for hours by the machine. Later, a dog's blood is drained, and it lies dead for 10 minutes before the *autojektor* brings it back to full, frisky life. The experiments were re-creations, for the purposes of the film, and while some of Brukhonenko's work on the heart-lung machine was legitimate, there's no evidence to show that he reanimated dogs after many minutes of death or that he could decapitate them and still have their heads remain alive. But if Haldane had any doubts about lending his name to an artifact of Soviet propaganda, he didn't mention them in his letters to Montagu.

In the movie, Haldane sits at a table, round glasses on his nose, his collar breaking free of his dowdy jacket. He's referring to a sheet of paper just below the frame, so his eyes sink and rise periodically behind his lenses, like indecisive fish in a glass bowl. "I should like to tell you," he says in his aristocratic tone, "that I have seen some of the experiments shown in this film actually carried out." He pronounces it "eck-chually." "As you can imagine, technique is everything. Besides such work as you are about to see, Brukhonenko shares the credit for the methods of human blood transfusion which were first developed in the Soviet Union and are now practised in this country, which have saved so many lives during the war." He practically slumps with relief after the last word; throughout, he has been stiff and stilted, as if he were in a hostage video, reading a statement under the point of a gun held just out of sight. Before a camera, he was a ghost of the sprightly orator he was on a rostrum. He wasn't happy with his performance, and he blamed the recording machine. "I may be a lousy speaker," he wrote to Montagu, "but I don't think my voice is so like that of H. G. Wells at his squeakiest."

Had Montagu asked to read Haldane's report for the Admiralty, Haldane would likely have given it to him. He didn't attach any virtue to ironclad secrecy in scientific matters. Science couldn't be owned by one government or another; it was intrinsically universal. Of course, he was alert to the dangers of treason. He wouldn't have suggested giving Germany kind advice on how to safeguard its U-boat sailors. But in 1940, Britain wasn't at war with the Soviet Union, and the next year, the countries became allies. Haldane refused to believe that sharing science with Moscow was immoral. In 1946, when the British physicist Alan Nunn May was convicted of divulging atomic secrets to the Soviet Union, Haldane pleaded the case against scientific selfishness in the *Daily Worker*. His own work on underwater physiology, Haldane remarked, "was shown to naval officers of several navies, including American and Dutch, but not to Soviet naval representatives, though it would probably have saved the lives of Soviet sailors."

Despite this belief, though, nothing suggests that Haldane gave his report to Montagu knowing that it would, in turn, be slid into the hands of Soviet intelligence; nothing suggests that Haldane was wittingly an informant for a foreign power. If he had been intent on aiding the Soviets, he would have passed on more material, but the archives contain just the one incident of the Admiralty report shared unwittingly with Montagu. X Group seemed to have abandoned him altogether as a source thereafter. And while Haldane was adamant about the principle of openness in science, he was discreet in practice. His letters through the war never hinted at the details of his discoveries; his conversations in the party headquarters, all studiously tapped and transcribed by MI5, held no mention of his research. Even his sister, visiting him in 1942, was privy only to what was already broadly known to the public. "He seems to be going to do another series of pressure experiments," she wrote vaguely in her diary. "I hope he won't kill himself." Haldane didn't even discuss politics with her. They talked about genetics and poetry, and they played Halma.

If espionage involves the deliberate transfer of secrets to another state, nothing stains Haldane's innocence at all.

But even in Haldane's time, not everyone believed him reliable, and his presence on government committees fell under scrutiny. The wintry winds of the Cold War had started to blow, and Communists were instinctively regarded with suspicion. Haldane responded in impeccable form. Of course he was a Communist, he told the *Daily Express* in 1948—as good a Communist as anyone. He was on two scientific subcommittees. "They don't pay me anything, and they can throw me off them if they want to," he said. "But if they'd thrown me off six months ago, they might not have had certain increased efficiency in underwater craft." There were no instructions from Stalin, whispered in his ear, to do this or that. "Sometimes I wish we did get orders from Moscow. I would like to know what they are thinking." Then a final Haldaneian sting: "The only group of people in this country who get orders from foreign powers are Roman Catholics."

In Parliament, Prime Minister Clement Attlee was asked to clarify why Haldane, "an avowed Communist," was serving on committees at all, rather than being purged like the radicals in the civil service. Attlee resisted, but questioned again in 1950, he sought the counsel of MI5. An operative gave his opinion: "I . . . said that there was a risk to security in his access to classified material and that if he could be denied such access that would be entirely satisfactory from the Security Service point of view." Haldane couldn't even attend a discussion about radioactive effluent in the Thames without being viewed with circumspection. An MI5 official wrote in a note that he would be exposed to plenty of confidential information from the Department of Atomic Energy: "D. At. En. rather suspect that Haldane's motives are really directed towards obtaining this information rather than towards saving the Thames from pollution."

In January 1950, the subcommittee on the effects of atomic energy was rearranged, and Haldane was told that his services were no longer needed. In June, the Medical Research Council removed Haldane from the subcommittee on underwater physiology; the Admiralty wasn't willing to share any information with him anymore, a council officer wrote. A rare tone of hurt sounds in Haldane's letter of response:

> I have delayed answering your letter of June 1st to allow myself time for reflection. I have little doubt that the Board of Admiralty has been misinformed concerning me. It is a fact that I disapprove of some aspects of the foreign policy of His Majesty's present Government, and make no attempt to hide this disapproval. If the Board has any other reason to suppose that I would divulge secret information imparted to me, I should be interested to learn it. . . . I have, in the past, been particularly scrupulous in keeping my mouth shut on such matters, both as a matter of honour, and in my own interests.

He didn't say it, but he might have: he had locked himself into airless chambers for his country, he had punished his body, he had blacked out and thrown up, sweated and shivered, ached and suffered, all so British submariners and sailors could be kept safe. To be ejected on the grounds of disloyalty, of all things, demeaned his work and saddened him.

And the affair augured poorly for science, he thought; it placed researchers in leg irons, thrust them wholly under the control of companies or governments. In 1951, in a Science for Peace meeting organized in a small hall in Bloomsbury, Haldane outlined the future. An MI5 agent in the audience took notes.

Haldane knew of a scientist, he said, who watched fish all day and realized that some of them communicated through pulsations. But the navy clamped down on him; if fish could do it, so could submarines. He knew of another scientist who studied the sex life of extinct fungi—"no interest to anyone but a dirty-minded scientist," except that some secretions signaled the presence of oil in the neighborhood. That, too, became a corporate and then a state secret.

How could science possibly achieve anything under these conditions? How would scientists talk to each other, drink out of each other's wells of thought, if the threat of treachery followed them around? MI5 was wasting its time restricting the legitimate activity of research, he said. Restrictions would kill science, and when true science died, the world would also perish.

MAYBE THEY THOUGHT he was a spy because of the Lysenko affair. Maybe they thought that if he was able to betray his science for the Communist cause, betraying his country must be just as easy.

Three years after the war ended, and after he had returned from Harpenden to University College, Haldane made his fateful broadcast on the BBC. Let us not rush into judgment on Lysenko, he urged. There may be something to his science, he said of a man who denied that chromosomes bear the information of heredity, who insisted that organisms can acquire new physical traits and hand them down to their offspring, who concluded that heritable changes are predictable and controllable, who scoffed at statistics and random mutations. He had heard only rumors of purges of scientists like Vavilov, and no one knew what truth lay in them, he said. And if the Soviet Union declared that Lysenko had doctored the inheritance of his wheat and that it was now able to grow in the hostile soil of Siberia, he saw no reason to disbelieve it. His colleagues in genetics, bright and devoted scientists, pointed out how hollow and dangerous Lysenko's work was. He argued that he saw some merit in it. He raised criticisms of Lysenko, but they were mild; he disagreed with Lysenko on many points, he said, but he thought some of his most profound arguments were correct.

Even in 1948, it was apparent that Haldane had picked the wrong side of this battle. The question was: Why had he done it?

He couldn't have believed truly in the logic of Lysenko's scientific precepts; to do that, he would have to admit to the worthlessness of his own career. Lysenko erased Mendel from Soviet genetics; Haldane had braided Mendel into natural selection. Lysenko dismissed Morgan and his chromosome theory of heredity; Haldane had plotted the addresses of genes on the X chromosome, and he had based his work on the inheritance of hemophilia and color blindness on Morgan's principles. Consistently, he had derided the Lamarckian scheme of the inheritance of acquired traits, and yet here was Lysenko, celebrating that defunct idea. No one had been able to repeat Lysenko's experiments and obtain his fantastic results, and this was the heart of Haldane's cherished scientific method.

Haldane was aware of how little he actually knew about Lysenko's techniques. Late in 1944, Lysenko wrote to Haldane through the Soviet embassy in London. He had read Haldane's *New Paths in Genetics*, he said, and had ordered a Russian translation to be made available. Meanwhile, for Haldane, he enclosed a copy of his own book, "which in many points differs essentially from the orthodox Mendel-Morganistic genetics."

Haldane wrote back, saying that wartime Britain was short on translators who might be able to explain Lysenko's book to him. Then he added: "Unfortunately it is extremely difficult in this country to obtain accounts of the actual experimental results on which your views on genetics are based, and I should be very grateful if I could obtain some reprints giving accounts of the facts in question, as well as the general conclusions based on these facts."

To this there was no reply.

The best guides to Haldane's real views of Lysenkoist science are two pieces of writing—one a 1940 column for the journal *Science and Society*, the other a lengthy letter to a colleague in Prague, sent in October 1948, a month before the BBC broadcast. These were still mealymouthed arguments in part, but they included flares of forthright criticism. He permitted Lysenko only small, hedged victories: that in theory a plausible way to alter an organism and thereby modify its chromosomes might be imagined, even if no evidence of it had yet emerged; and that in plants a small number of cases had been recorded in which inheritance seemed to have nothing to do with nuclear chromosomes. (Subsequent research has shown that plants can pass down traits through chromosomes in the mitochondria and the chloroplast, and not just through the cellular nucleus, where chromosomes mainly reside.) In animals, such cases were rarer: "I think that nine times out of ten Lysenko is wrong, that is to say you cannot improve a breed of animals by improving its food." But in both his article and his letter, Haldane cited what he thought was a glaring exception: a 1933 study in which day-old mice, from a strain highly susceptible to mammary tumors, were suckled by females from a strain far more immune to those tumors. The chosen pups grew up to contract cancers at a lower rate, as if they had imbibed a partial resistance from the milk. It appeared to point to a novel form of inheritance—until 1942, when scientists found that this class of tumors was caused by a virus transmitted through mouse milk—nothing to do with heredity after all. Haldane couldn't have known this when he wrote his *Science and Society* article in 1940, but if he read of the virus after that, his letter of 1948 stayed silent about it.

When he did reproach Lysenko, Haldane picked at the very fundaments of his ideas. Lysenko's attack on the chromosome theory was greatly flawed. "His statement that 'any hereditary properties can be transmitted from one breed to another even without the immediate transmission of chromosomes' is, in my opinion, absolutely false," Haldane wrote. The charge that orthodox genetics was sterile, unconcerned with any real benefit, was not true. In Britain, a scientist had worked out a way to determine the sex of a cross-bred chick at birth, aiding poultry farmers immeasurably. "This was done with severely practical motives, and not to confute Lysenko." He didn't share Lysenko's contempt for statistics or for Mendel and Morgan. He stated all this tactfully but firmly, like a parent guiding a wayward child. Even his favored phrase for Lysenko had a paternal indulgence to it: Lysenko had, Haldane liked to say, "gone too far."

So he had in his hand all these cards, all these instruments of objection. He just didn't play them when he spoke to the world on the BBC.

Haldane wasn't some simpleton to be seduced by any easy vision of transforming heredity. To be sure, his father and then Engels had taught him that science ought to lift humankind into better conditions of life. In that regard, a full mastery of the gene was the key to all fortunes: it could keep people fed, rid them of disease, and tune them to produce a more harmonious society. Genetic modification was the kernel of every human destiny Haldane imagined. But he was hardheaded and pragmatic and too knowledgeable to think that Lysenko had tripped and fallen face first into the distant future.

So possibly Haldane saw some other value to Lysenko—his challenge of convention, for instance. In an interview, of which Haldane had seen an English translation, Lysenko said: "A sharp struggle of ideas is going on, and the new always meets resistance from the old. But with us, in the Soviet Union, the new always wins." He couldn't have furnished a better credo to appeal to Haldane. The old order must constantly be questioned; it was as important for Lysenko to force classical geneticists to reexamine their views as it was for the Soviet Union to force capitalist nations to reexamine theirs. "An argument I do remember him using was that we should take seriously any alternative to our favourite orthodoxy," John Maynard Smith, Haldane's student and Communist colleague, wrote many years later about Haldane's consideration of Lysenko's theories. Only rebels like himself bucked the establishment in this manner, Haldane believed, so Lysenko must have felt like a kindred soul, deserving of his support.

Or possibly he was just being stubborn in continuing to talk Lysenko up. In his own work, Haldane was forever ready to acknowledge a mistake; scientific accuracy was more precious than pride. "He was not over-concerned with always being right and often enjoyed pointing out in front of audiences and even students how he had been shown to have been wrong," Hans Kalmus later wrote. But the Lysenko affair had become about something other than science; it had become political, calling into question not Haldane's reasoning but his emotional attachment to the party. Admitting he was wrong about Lysenko would mean admitting he was wrong about Communism and the nature of Stalin's regime. He was prepared to hold this hill and fight upon it. There was a "fantastic amount of anti-Soviet propaganda by numbers of people who are carrying on the work of the late Dr. Goebbels," Haldane grumbled in a letter. He saw no cause to trust any of it. When another scientist, Sir Henry Dale, resigned from the Soviet Academy of Sciences to protest Lysenko's imperium, Haldane was asked if he would follow. "Certainly not," Haldane snapped. "It would have been more scientific if Sir Henry had waited until translations of Lysenko's work were published in London." When a group in Cambridge began rendering into English the fractious summer debates of the Lenin All-Union Academy of Agricultural Sciences—in which Lysenko was believed to have evicted his rivals from genetics—Haldane insisted on financing part of the work.

In his pathology of open-mindedness, he refused to credit any news of the purges of Soviet academia. That would comfort the enemies of Communism too much. He wanted first to learn more about why Lysenko had removed other biologists from their posts, telling his friend in Prague that Lysenko's "attacks on his Soviet colleagues may be justified. I don't know." It didn't seem to matter that Hermann Muller, once as radical as Haldane, had felt compelled to leave Moscow in 1937, his faith jarred by Lysenko's pseudoscience and the state's rough treatment of dissident geneticists. Or that Charlotte, returning from the Soviet Union in 1941, warned him that Vavilov may have met an ugly end. Or that Julian Huxley, who had visited Moscow in 1945 and thought Lysenko maddening, worried that some Soviet geneticists had been "liquidated."

If Haldane shrugged away these reports, it wasn't because he had accepted the erasure of scientists as collateral damage. He certainly had that streak in him—that willingness, in moments when his heart iced over, to look past a death or two in service of the larger ideal, as he had in Spain when the Republican police executed a foreigner without trial. But in this situation, he wanted to be certain, wanted to discover for himself what the truth was, before he committed to a criticism of Moscow.

The fate of Vavilov became a bellwether for Haldane. Repeatedly, through 1946 and 1947, the CPGB and its various allies urged Haldane to be more outspoken about the uproar in Soviet genetics, and repeatedly Haldane demurred. In September 1946, a Marxist journal called *Modern Quarterly*, where he sat on the editorial board, considered a new piece on the controversy. Haldane asked them to wait for his opinion on Vavilov. The next April, the editors asked him to write a piece. The minutes of the meeting record: "Prof. Haldane said he might be able to do something in September but he could not proceed till he knew what had happened to Vavilov."

On a letter from John Lewis, the editor of *Modern Quarterly*, Haldane scribbled, almost as a reminder to himself: "Try to find out how and where Vavilov died."

But then in the summer of 1948, before he had been able to satisfy himself with an answer, Lysenko held his conference in Moscow and rinsed genetics clean of his opponents. The storm broke, and Haldane, as the CPGB's foremost scientist, had to plant his feet and take his stand. He saw other biologists shrink from the party, aghast at how the Soviet state was enforcing a doctrinal perversion of genetics. "If the central committee of the Soviet communist party could be wrong about that, what else could they be wrong about?" Maynard Smith wondered at the time. He quit the party over Lysenko: "It was the crack in the dyke." This was just what Haldane feared: that doubt would vaporize the credibility of the Soviet Union, wobble the foundations of Communism. So when he went on the BBC to debate his fellow scientists, he hammered together his best defense of Lysenko, dusted it with as much criticism as it could safely bear, and presented it to the world.

FOR THE CPGB, that wasn't enough. Within science, Haldane was becoming a lonely advocate for Lysenko, but the party felt he was still too equivocal, too careful with his public statements on Soviet genetics. He was leaving himself loopholes, as one party member carped to another; old Haldane wasn't making his bloody mind up. In private, among his Communist comrades, Haldane was much more withering about the failings of Lysenkoism than he allowed himself to be in print or on the BBC. He was torn between two poles of loyalty, and that conflict spilled out of him eventually, first cracking his relationship with the party and then snapping it altogether.

The morning after the BBC broadcast, in the CPGB headquarters on King Street, Haldane's speech went down poorly. It sounded too liberal, a member named James Klugman complained, especially after three other fire-eating scientists had assailed Lysenko. More and more, people in the party were calling Haldane a silly old fool; there was a real danger he would be chased out of the CPGB.

Bill Wainwright, another member, said he had listened to the speech. It was a step in advance of any previous position Haldane had taken, he thought.

Klugman didn't agree. He grumbled about Haldane some more. If he had been in Haldane's position, he said, he would have either shut up or put forward the party's line wholeheartedly, however much he disagreed with it. Haldane hadn't done that.

Maybe Haldane wasn't sufficiently a Marxist yet, Wainwright said.

That was the trouble, Klugman agreed. And now that a well-known Communist like Haldane had voiced a few differences with Lysenko, reactionaries the world over would see an opportunity to attack them. Haldane must consult the party first before taking part in any further debates or making any more statements.

The CPGB ought to work more closely with Haldane, Wainwright said, to make sure his speeches worked better in the future. It was just that all this science was so new, not anything the party had ever experienced before.

But other Communist parties had dealt with the Lysenko issue well, Klugman pointed out. The French Communists—they had put up a terrific show—whereas Haldane—Haldane hadn't even talked about the importance of Lysenko to agriculture and collective farming, which was key. And he hadn't discussed the party's basic conception that you can change nature, which to the thinking man or woman was the greatest thing of all. He just hadn't convinced anyone to come down on the side of Communism.

The discord between Haldane and the party over Lysenko was visible to anyone looking on. In December 1948, the cartoonist David Low produced a panel for the *Evening Standard*, part of a series of prophecies for 1949. Haldane stands straight and haughty in a room, in front of a framed portrait of Lysenko. He has been indicted by a seated panel of five men, who point at him balefully; one of the men has a potted plant labeled "Communist Potato" next to him. A party member is divesting Haldane of his pipe and his coat buttons. "J. B. S. Haldane, refusing to recant utterly, is denounced by the DAILY WORKER as a bourgeois pseudo-scientist," Low's caption read.

The CPGB resorted to damage control. In January 1949, the

Daily Worker issued one of its periodic "educational commentaries," a leaflet on Lysenko and genetics. The *Worker*'s editorial board and its chairman, Haldane, weren't consulted on the content. Modern geneticists in the West, the commentary claimed, weren't interested in the causal connections by which inheritance was altered; they thought that genes were immortal and immutable and were only occupied by the mutational games of statistical probabilities. Haldane grew enraged. No modern geneticist held any of these views, he told Bill Rust, the editor of the *Daily Worker*. If the commentary wasn't withdrawn, Haldane said, he would resign from the editorial board. But the leaflet had already been dispatched, so Rust, trying to control the damage, asked Ivor Montagu to draft a correction. The document he produced was no less deficient, so in February, Haldane handed out his own mimeographed note within the party, responding to the *Daily Worker* commentary. Its statements, he wrote, "are so completely false that no constructive discussion is possible until they have been denied." Then, over four pages, he took the commentary—and, by extension, the party's support of Lysenko—apart.

A similar episode occurred with the Engels Society, a group of Marxist scientists. A draft of a discussion statement titled "In Support of Lysenko" reached Haldane around the same time, and he found plenty to disagree with. He marked up the draft with violence. More than once, in the margin, he wrote "No" and then, at the end, a series of numbered remarks. "Evidence!" "This is plain nonsense." "Utterly incomplete." "Have you read e.g. Knight on cotton breeding." He sent a letter to Maurice Cornforth, the society's secretary: "I am in absolute disagreement with the document sent round to me, and should it be sent in I should have no option but to resign from the Engels Society."

Two months later, Cornforth contacted Haldane again. The society was organizing a two-day conference on Michurinism, the name Lysenko gave to his version of genetics, after the Russian horticulturist he revered so much. Would Haldane cast his eye over a rough resolution? Haldane responded:

I cannot agree with the resolution in its present form. . . . The first duty of communist scientists in this country is to make data available on which Michurinism can be examined. This has not been done. The second is to examine these data, and to examine them critically. I certainly cannot support a resolution which calls for us to examine our own work critically, and not that of others.

Comrade Haldane was finally standing up for science—but being a staunch Communist, he conveyed his demurrals only in loud whispers. As he told the *Daily Worker*'s new editor, John Campbell: "During the past few months a number of statements have been made in Party organs with which I am in disagreement. As a Party member I have refrained from public criticism of them." In spite of his discretion, though, Haldane's obstinacy was a pounding headache for the party. It couldn't afford an insurrection by its most famous scientist.

For some true believers, the solution was to discredit Haldane. At one meeting of the Engels Society, the air turned sour with slander. Taking advantage of Haldane's absence, Cornforth argued that Nazi race theory came from the same wellspring as Mendelism. This was why, he said, Haldane had refrained from attacking Nazi science. Not everyone in Cornforth's audience swallowed this falsehood. One man, remembering Haldane's forthright indictments of Nazism before the war, got up and threw himself into a row, calling Cornforth a bloody liar. The party was tearing apart over Haldane.

Harry Pollitt wrote to him. Perhaps it would be best for everyone if he sloughed off his other public duties and restricted himself to the *Daily Worker*'s editorial board? Haldane took it to heart. At a rally, invited to speak, he cited Pollitt's instructions and declined. In December 1949, he sent a disheartened letter to Campbell. "From the two letters which I have received in the last week it is clear that I no longer enjoy the confidence of an important section of our readers," he wrote. "I am therefore no longer the right person to be chairman of the editorial board, and I must ask you to put my resignation in its hands at the next meeting."

HE STILL WANTED TO BE with the party, still wanted to belong. He just couldn't work out what their relationship ought to be.

He went to see Pollitt once, in the summer of 1949, climbing the stairs with difficulty.

Pollitt apologized for making him come up. The Academy of Sciences in Moscow had invited Haldane to the Soviet Union, for a sort of holiday. Would he like to go?

Haldane grew flustered. Not this summer, he said. He wouldn't be able to take a holiday this summer.

"You won't?" Pollitt replied, with surprise.

He couldn't, Haldane said, stammering now. He had to keep his laboratory going. The offer was very tempting, but he couldn't get away for more than three or four days.

Pollitt was disappointed. "You're a Party man, you know." How about September? But Haldane said he was booked through that month as well. Then, when Pollitt asked him how a recent visit to Czechoslovakia had been, Haldane admitted he hadn't made the trip. He was still not well, he said. This only irritated Pollitt further. You can't go on like this, he warned Haldane.

After April 1949, Haldane stopped figuring regularly in the minutes of the *Modern Quarterly*. He turned down invitations to speak or to write about his politics. That autumn, he was paid a visit by Ivan Glushchenko, part of a Soviet delegation touring Britain. Glushchenko was a plant geneticist; Maynard Smith called him Lysenko's hatchet man. To British scientists, Glushchenko seemed vapid and unimpressive, prattling on about how the Soviet Union had produced apples that looked like pears. He came to University College, Maynard Smith recalled later, with a couple of goons:

> And he and the goons . . . were actually closeted in Haldane's office for about three hours, trying to persuade Haldane how right Lysenko was. And Haldane was absolutely unapproachable for two or three days afterwards. He was pretty clearly shaken by what I take to be Glushchenko's ignorance.

In March 1950, he was back in Pollitt's office, attempting to wriggle free. He thought he was out of place now at the *Daily Worker* even as a columnist. They must think him hopeless and outdated, possibly even reactionary, he said. He had just been assigned a piece, but it was for the best if someone else did it.

Not at all, Pollitt said. Haldane's articles were unique. Then he read from a letter Haldane had sent him: "Dear Harry, I am in fundamental disagreement with the political line of the Party; it's not the slightest use you trying to make me change my mind and I hereby send in my resignation." He broke off and addressed Haldane: Didn't he see how difficult he was making things by writing like this? It wasn't the time to desert the party, he said. Stay on, take another shot at it. Puffing with reluctance, Haldane allowed himself to be swayed for the moment.

But he was pulling himself away from the party, like a stamp lifting off an envelope; the glue had dissolved, and they were linked to each other only by tenuous tendrils of adhesive. In July, terminating a 12-year association, he told the *Daily Worker* he was going to stop appearing in its pages. He couldn't write about Soviet science because he couldn't read Russian, he wrote, and he was getting older and had less time and energy to read widely. He wasn't being honest; just a few months later, he was investigating the possibility of becoming a columnist for *Reynolds News*, telling one of its contributors that he and the *Worker* "don't always see eye to eye."

Then, on a weekend in mid-November, the CPGB's phone line shimmied with activity. Journalists were calling. They had heard a rumor that Haldane had quit the party. Had he? The party claimed it knew nothing about it. After that, the reporters got hold of Haldane himself, crabby from the pain of an ankle he had broken six weeks earlier. Through his secretary, he made his statement:

> Professor Haldane considers that a very bad precedent would
> be established if university teachers were expected to state
> their membership or otherwise of political organisations
> when requested to do so by newspapers.

He also considers that, when suffering from a broken leg (Pott's fracture of malleolus) he was under no obligation to ascend a staircase to answer a telephone call from a newspaper (the Sunday Pictorial) which refused to state the subject under discussion.

He has not read the various statements made about him in newspapers and leaves it to their readers to speculate as to their truth or otherwise.

For the guidance of such readers, he remarks that in view of the experiences of 1914–1918 and 1939–1945, he regards the rearmament of Germany as a suicidal policy for Britain, and that, as a doubtless old-fashioned believer in such notions as honour, he regards the wholesale massacre of civilians, whether by atom bombs or otherwise, as unworthy of a civilised people.

In this respect he is in full agreement with the policy of the Communist Party.

It was vintage Haldane: willful attention to principle, a sense of being boundlessly hassled, a gobbet of scientific knowledge, and a minor seminar in political history, all dispensed in tones of utmost brusqueness. And in flagging just one point of agreement with the party, he had dodged any mention of genetics or his resignation, which would have embarrassed his former comrades. If he wrote and signed a letter formally renouncing the party, in that year or any year afterward, it never emerged. He had done the party one final kindness before shutting the door softly behind him.

FURTHER DOWNSTREAM, from time to time, Haldane would be asked about Lysenko. It would go too far to say that he recanted, but his once-steadfast support grew dilute. He was still prepared to accept Lysenko's views on wheat, he said not long after he left the party. A few years on, he gently chided Lysenko; even if he had discovered some things of considerable interest, he had been intolerant. The Soviet state had tried to make people toe the line, which was very silly.

Did he approve of science being controlled by the state? his interviewer asked him.

"Well, the alternatives are probably worse. I mean, I'm in the ridiculous position of having to support a lot of my own scientific work out of my own pocket, because I can't get enough money to do it otherwise."

After Stalin's death in 1953, the Soviet Union's autocracy thawed, and the state's fervid support of Lysenko dissolved slowly in the meltwater. It would still be another decade before Lysenko was dismissed from academic posts, but his influence weakened when Soviet scientists, finding they could denounce him at last, did so. Haldane slackened a little as well. To a correspondent in Berkeley, he claimed he didn't exactly remember what he had said in his BBC broadcast. In any case, he wrote, Lysenko's vaunted transformations of wheat had evaporated. "Lysenko was too dogmatic about his own results, and above all Stalin interfered unjustifiably with research." And some time after that, back on the BBC, Lysenko's name came up once more, and almost reflexively, Haldane started by saying: "In my opinion, Lysenko is a very fine biologist and some of his ideas were right." But some of them, he went on, were badly wrong, "and I think it was extremely unfortunate both for Soviet agriculture and Soviet biology that he was given the powers he got under Stalin."

To his disgrace, out of adamancy or defiance or a faithfulness to the cause in which he had steeped his soul, Haldane still refused to fix any blame on Stalin. Details of Stalin's massacres, starvations, and penal colonies were by now well known, and yet Haldane wanted to keep ammunition out of the hands of the Soviet Union's critics. "I thought, and think, that he was a very great man who did a very good job," he wrote to a friend. "And as I did not denounce him then, I am not going to do so now."

If Haldane reflected on his own role in the Lysenko controversy, and if he pinched a lesson for himself, he may have tucked it into an essay he wrote for the *Rationalist Annual* in 1958. In the essay, he told the story of the Pandavas, the five noble brothers who defeat the forces of evil in the *Mahabharata*, the great Hindu epic. After ruling wisely for years, the Pandavas renounce the world and embark on a pilgrimage around India. A stray dog attaches itself to their party, and they cannot shake it off, even when they climb up into the Himalayas to attain heaven. One by one, on their journey up the mountains, all the brothers die save one: Yudhishthira, the eldest. When he finally reaches the threshold of heaven, he is welcomed by Indra, the king of the gods. His family awaits him, Indra tells Yudhishthira— but he cannot bring his dog in with him.

"Yudhishthira, who, on the whole, had been rather a prig throughout the epic, redeemed his character at the last moment," Haldane noted. He refuses Indra's offer: It isn't right to abandon any creature that has placed its trust in him. At that, the *Mahabharata* reveals that the dog had all along been a disguise of Dharma, the Hindu god of justice and Yudhishthira's father. Having passed his trial, Yudhishthira takes his place in heaven. "The moral," Haldane wrote, "is the wholly admirable one that a man must not do an action which he regards as dishonourable even if ordered to by the chief of the gods in person."

6.

India

IN 1957, HALDANE WAS 65 YEARS OLD. It was early in the evening of his life—the sun still out, but the work winding down and the day feeling worn: a time to slow and to mellow, anyone else might have thought; a time to bask in the glow of concluded labors; certainly not a time to shift forever to another continent or to dig up and replant a career in unfamiliar soil.

So naturally, that was what Haldane did. In first-class seats on an Air India flight, Haldane and Helen flew to Calcutta on July 24 to take new positions at the Indian Statistical Institute (ISI). With them they took a dozen jars of live fish. He couldn't check them into the baggage hold, he explained to an ISI official; the temperature would swing too much and kill them. He needed them in the cabin: "They will be sealed up, and will <u>not</u> smell."

He announced his resignation from University College in November 1956, days after the United Kingdom, France, and Israel attacked Egypt to wrestle away its control of the Suez Canal. He didn't miss the opportunity to score a political point. "I do not want to be a citizen here any longer because Britain is a criminal state," he told *The Times*.

Was this because of the Suez affair? the reporter asked.

"If you mean the mass murders at Port Said, it is."

Just before he caught his flight, Haldane gave a reporter a second reason: the presence of US soldiers on bases in Britain. "I want to live in a free country where there are no foreign troops all over the place. Yes, I do mean Americans."

The Suez crisis had certainly maddened him and convinced him more deeply still that the Western powers were blinded by their impulses to make profits and war. For Haldane, the attack on Egypt was a criminal act for which Prime Minister Anthony Eden's government ought to pay. Once, in the midst of the Suez crisis, he summoned Brian Silver, an undergraduate at University College, to his room. Silver entered and absorbed the flotsam of Haldane's career: the piles of notepaper and books, the fossil bones, the box of sprouting potatoes under his desk.

"Sit down, Silver."

"Thank you, sir."

"I'm told you're a leader among the students," Haldane said. Then he proposed, in all seriousness, a plot to force Eden out. What if thousands of students, each with a few pennies, went into telephone booths and started ringing the offices in Whitehall? "When the phone rings, replace the receiver immediately. That way you can use the same coin repeatedly." The government, its phones endlessly jammed in this manner, would get so exasperated that it would resign. It was a preview of the kind of distributed denial-of-service attacks that hackers use to take down websites today.

Even in his official resignation letter, Haldane brought up Suez. He stressed that he no longer wanted to be "a subject of a state which has been found guilty of aggression by the overwhelming verdict of the human race. I believe that we shall find in India opportunities for research and teaching in a country whose Government, by its active work for peace, gives an example to the world."

But in a way, he was posturing, giving his country one last performance of the scientist as radical.

India had been on his mind for a while. He had visited there with Helen in the winter of 1951–1952, attending the Indian Science Congress in Calcutta and then hopscotching around the country, giving 35 lectures in the span of seven weeks. Sometimes he felt that people were intrigued more by him than by what he had to say, he groused in a letter after the trip, and he hated being "paraded about like a Rajah's elephant, as an object of interest." But he made a number of friends, including P. C. Mahalanobis, the director of ISI. An appointment with Prime Minister Jawaharlal Nehru in Delhi fell through; Nehru was too busy. But they had met once before, at a dinner party that Naomi hosted in 1934, so Haldane wrote to him after returning to London. The state of Indian biology was uneven, Haldane wrote, and in particular, he had seen very little research into animal genetics. If someone undertook to develop that, India would have more milk and more eggs for its people. Then, having described this vacancy, he applied to fill it:

> I could, I believe, assist in the development of human physiol-
> ogy and of the more academic side of genetics. . . . However,
> I fully realise that the time has ceased when an Englishman
> can claim any right to advise Indians. If such a view is taken,
> I can make no complaint. If it is not, perhaps I may be of
> some service to India.

Nehru wrote back. He had appointed a committee to consider Haldane's suggestion. "Your help will, of course, be welcome."

The Haldanes visited again in the summer of 1954, spending more time at ISI and staying at Mahalanobis's house. Haldane was besot-ted by the profusion of plant and animal life, their traits and char-acters yet to be analyzed. Around every corner, there was something to study and think about. "Dear Prasanta," he wrote to Mahalanobis, who was traveling at the time, "you have a tree in your garden which seems ideally adapted for the teaching of elementary statistics. . . . This sheds flowers every morning." He drew a table, showing his patient tally of petals on fallen flowers over two days. "A daily count might give interesting results—perhaps a change in mean or variance during the season."

Haldane wanted the chance to do some "real work" in India, he told Mahalanobis, rather than make brief visits to deliver inspir-ing lectures. The year before the Suez crisis, Haldane wrote again to Mahalanobis, with a frank account of his wishes. He was due to retire in 1958, and although he would probably gain a part-time position easily, he preferred to do his research in India, even in an unpaid role. Helen, he added, is "homesick for India." In January 1956, speaking to the Socialist Club at Oxford, Haldane praised

Nehru and his vision for his country: "I feel far more at home there than I do in some European and American cities. . . . It is entirely possible that I shall die as a citizen of the Indian Republic, if they will have me."

Mahalanobis made it happen: a position as reader for Helen and a professorship for Haldane. "Helen is overjoyed at the offer of a full-time post," Haldane wrote. "There is no question of her wanting to go anywhere else." The letter went out on November 1—five days before British marines charged onto the beaches of Port Said.

He had liked India as a country ever since his time there as a convalescent soldier during the First World War, but he loved it more as a notion. It was his exemplar of choice when he wrote about races and civilizations, when he was rooting around for a people to contrast against the societies of the West. To discuss how disease weakened a populace and allowed them to be conquered, he looked to India and its malaria; to ponder democracy in the midst of inequality, he looked to India and its illiteracy; to consider religious miracles, he looked to India and Hinduism, which "is more sophisticated, and less censorious of human psychological variation, than is Christianity."

After India gained its independence in 1947, it fascinated Haldane even more. It became a petri dish for scientific socialism, a vast experiment to see how a wise application of science could advance the lives of hundreds of millions of people. The Soviet Union had been just such an experiment, but it was letting him down; India gave him hope. Under Stalin, the Soviet Union suffered from an absence of democracy and of any openness to ideas. He asked himself: Where, in the world as it is at present, should I find the greatest amount of freedom of this kind? "I suspect that the answer is 'In India,'" he wrote in 1954, in an essay titled "A Rationalist with a Halo." He attributed this to the Hindu tradition of discussion, explained the *Āstika and Nāstika* systems of knowledge in Hinduism, and finally came to Nehru, the titular haloed rationalist. Nehru was determined to make India a secular state and to avoid being yanked into either side of the Cold War, and for these decisions Haldane admired him. "I have more or less retired from politics," he wrote in a letter a month after his 60th birthday. "However . . . my opinions as to world politics are probably as near to those of Nehru as of any other well-known politician."

That was all very well, this high-minded esteem for a leader and his ideals. But Haldane wouldn't have moved to India on the strength of that alone. He was discovering other reasons to leave London. For one thing, his university was frustrating him. Later, after he resigned, University College named him a professor emeritus; in a letter of reply, saying he would have declined the honor if they had cared to consult him, Haldane listed his dissatisfactions. He left because, for 20 years, he had been promised more space and more equipment, and nothing had materialized. Also, the university had behaved as if Haldane embarrassed them—which, he admitted, might well have been the case. "To take one example," he wrote, in a small mood, "a colleague junior to myself was chosen as its official representative at the last International Genetic Congress."

In the postwar years, University College was struggling. Its buildings had been damaged by bombs and its funds were low; its sports facilities had been so thoroughly wrecked that the nearest tennis courts were 14 miles away. The number of grants available had wilted. In at least one term, Haldane personally donated £300 —about an eighth of his annual salary—into his department's funds. At other times, he asked for the fees he earned, as an examiner or a writer, to be made over to the university. Once he bought a set of teaspoons for 10 shillings for the men's staffroom when he realized it had none.

This was no way to work, he thought, and maybe in India, these fiscal reins on his research would fall away. He would like, he said in an interview, to be "looking after a certain number of animals, which can be kept enormously more cheaply in India than here. Quite obviously, if I want to keep a thousand poultry, it's going to be a big job here. In India, well, labour is very cheap, and the food is fairly cheap, and so on."

He was brooding over his own finances as well. In the late 1940s, his bank statements—blue, square slips of paper from Westminster Bank—sometimes showed him overdrawn, figures in red inscribed into the rightmost column under "Balance." By the early 1950s, his account was healthier, but even then he only ever managed to retain a balance of £500 to £600. He was fighting Charlotte's solicitors over alimony, which a court had increased, in 1948, from £360 to £680 a year. He didn't have the money for it, he pleaded to the court; it had been a particularly sparse month when he filed his affidavit, and in his account, "the amount at present standing to my credit is approximately the sum of £20." He had gifted Roebuck House to Helen and sold her the copyright of the *Journal of Genetics*, which he owned, so that Charlotte could claim no share of them. Even after the court issued its order, Haldane set his jaw and dug his heels in. He did mulish defiance well. Charlotte's lawyers were forced, in the summer of 1950, to dangle the threat of penal action in front of him if he didn't pay, within seven days, an outstanding sum of £255.

A journalist sent a letter from Chicago: "Dear Mr. Haldane, would you mind telling me more about your life, to assist me in a book I'm writing about evolutionary theory and its prime scientists?"

"Dear Madam," Haldane replied, "You ask me a number of personal questions, expecting that I should devote some hours to the job. We Europeans are now learning from Americans the hard way. I shall be glad to answer your questions in return for a check for $100 payable to University College London. If I did not do so I should, to judge from the conduct of Americans in this context, be regarded as a 'sucker.'"

A zoologist in California asked to be included in the list of people who received reprints of any new papers that Haldane published. "Your request that I should put your name on my reprint list is fantastic," Haldane wrote bitterly. "If you honestly suppose that professors in Europe west of Poland and Czechoslovakia can afford such a luxury, you have been listening to propaganda."

Haldane was still a party member during these years, so he confided in Harry Pollitt. Money was a difficulty, he said on a visit to the King Street office in 1949. Things weren't as easy as they could be.

The party owed him money, Pollitt reminded him—a couple of hundred pounds, perhaps fees for his duties at the *Daily Worker*. Just sign for it, and it'll be in your account, Pollitt told Haldane.

That was the thing, Haldane replied. He didn't want too much money in his account while this alimony action was in progress.

Nine months later, he still felt harried. He wanted to get out of London now, he said to Pollitt, and to buy a cottage somewhere. But his money was tied up in a way that he couldn't make a purchase like that without Charlotte's consent.

Then he left the party, and that rupture was vital and decisive. All at once, it deprived him of a political center, of a network of friends and colleagues, and of an outlet for his writing. The party had been the anchor of his public life; without it, he was weightless, liable to drift. He had been roundly disillusioned by how the CPGB had paltered with scientific truth and had hushed him up in the interest of the policy line. Maybe he was upset with himself as well, for coddling the party on Lysenko and for speaking his mind less plainly than he might have. He remained an unwell man, his body still hurting and buckling from his wartime experiments; he was in hospital, in fact, when Charlotte's solicitors declared a deadline to pay her. All in all, the decade after the war was humid with stress and dejection. It was enough to make anyone want a fresh start.

Additionally, there was the question of footwear. "One of my reasons for settling in India was to avoid wearing socks," Haldane said, some years later. "Sixty years in socks is enough."

THE ISI CAMPUS LIES ON a trunk road stretching north, alongside the Hooghly River, out of Calcutta; on buses halting nearby, conductors shout "Statistical, Statistical!" to announce the stop. When Haldane arrived in 1957, public transport to this area of Baranagar was scant. You should have a car of your own, Mahalanobis warned Haldane in advance. In the neighborhood, the rural was slowly transforming into the urban. The campus comprised a few buildings set within a grove, trees of palm and mango nodding over ponds and flower beds. Thanks to Mahalanobis, the vegetation kept multiplying. Every time ISI hosted visitors of eminence, Mahalanobis asked them to plant a new mango seedling.

And ISI received dozens of such visitors. Many of them came because of Mahalanobis's energy, his mania for inviting the best minds in the world to his corner of eastern India. (Or even those not quite among the best. "Every second-rate economist from Europe or

America" must have come to ISI in the 1950s, the Swedish economist Gunnar Myrdal declared.) But many also came because in the years after India became free, the institute turned into a regular waystation for the politicians who came to court this new nation. Lyndon Johnson made the trip; so did Alexei Kosygin, Che Guevara, Ho Chi Minh, Henry Kissinger, and Zhou Enlai.

For Nehru, the institute sat at the focus of his centralized plan to build his country. The plan was everything. It moved ore from here to there, it determined where roads needed to go, it decided which industries deserved thrift and which deserved lavishness. It ordered the economy on the very Soviet premise that with enough data and enough thought, the fields would grow heavy with grain and the factories would turn out unending rivers of products. Mahalanobis promised to provide the data as well as the thought. In 1937, he had used sample surveys to gauge the performance of Bengal's jute crop, a technique that proved so effective that he expanded it into a national survey in 1950, covering as many aspects of Indian socioeconomic life as could be stuffed into it. These data could then be fed into a succession of statistical tools—to estimate, to forecast, to budget, to prepare. Three years before Haldane moved to Calcutta, Nehru asked Mahalanobis to be the architect of India's second five year plan. The national economy was being processed through the halls and corridors of Baranagar.

Since ISI enjoyed such prestige and importance, Mahalanobis had an unusual latitude in deciding whom he wanted to invite to work at the institute and how he lured them there. During Haldane's time, ISI was home to a number of foreign scientists from China, East Germany, Japan, the Netherlands, America, Iran, the Soviet Union, and Great Britain. Haldane told Mahalanobis that his annual salary at University College was £2,300 and Helen's around £800; in addition, the *Journal of Genetics* brought in £800 or so a year. None of these were ordinary sums for an institute in a young and poor country. Not to worry, Mahalanobis wrote back; he would not only meet those figures but also pay for medical care and the occasional vacation in the hills.

The Haldanes first lived in a fourth-floor flat on the campus,

Haldane speaking at the Indian Statistical Institute, Calcutta,
August 1960.

which they promptly cluttered with their papers and books—"a mess
by European standards," he wrote in a letter, "though not, I think, by
Indian standards." A year later, they moved into a bungalow not far
away, sharing it with a student, Haldane's secretary, and the secretary's mother. Haldane, who knew no half measures, immersed himself in India. His attire consisted of a white dhoti and a loose white
kurta—a fine, practical fit for the clammy heat of Calcutta, except
that he also wore them on trips to Europe, determined to signal his
break from the West. He taught himself Bengali and Sanskrit; he
even bought Hindi glossaries of botanical terms, although no one
recorded whether he began referring to germ plasm as "*janitra dra-
vya.*" He read Hindu philosophy until he was able to fling quotations
out with ease. In the evenings, as the skies faded to black, Haldane
gave impromptu lectures to his friends and students on the planets
and the constellations, citing their Greek as well as their Sanskrit
names. "You should probably avoid being on the roof with me at
night," he wrote mischievously to his secretary's fiancée. "This is not
for the reason which you might guess, for I am sixty-five years old,

and love my wife; but because I am liable to start talking about the stars and many people find this very boring." He turned vegetarian and made it a point to mention his diet whenever he was invited anywhere. Baranagar had seen white men come and go, but none like Haldane, wrote Krishna Rao Dronamraju, one of his students. He went with the Haldanes often on their excursions through the countryside nearby, Haldane smoking his cheap cheroots and discoursing on the flora, Helen collecting food for her silkworms:

> As our small group continued walking, others—total strangers—used to join us. Our numbers swelled, getting bigger and bigger. Some of those people who followed us had no clue who Haldane was. He was impressive, tall and bald headed, wearing saffron or white robes and surrounded by disciples, he was occasionally mistaken for a "holy" man or a sage of the kind one encounters frequently in India. I recall one air hostess approached me on a plane, asking, "Sir, this holy man, which Ashram is he from?"

In 1960, after the Haldanes had been accelerated into Indian citizenship, they received new passports. Helen, looking fierce, wore short hair and a shirt. Haldane, his moustache tamed and shrunken, his spectacle lenses darkling, wore a kurta. His face was creased, as if in petulance, as if he were wishing they would take the damn photo quickly and stop wasting his time.

WHEN HE HAD a vacant evening, Haldane would go to a pond on the campus, clad in a pair of orange trunks, and wade in. It didn't make for very vigorous exercise; once he wrote to Mihir Sen, the Calcutta hero who had swum the English Channel, inviting him to lunch and a dip, but he warned that the 80-by-30-yard pond might be too domestic for Sen's taste. For Haldane in his late 60s, though, it was just right. He would flip onto his back, a pipe or a cheroot gripped between his teeth, and lose himself in the stars as they arrived to hang over the Indian night. Every so often, a carp might nudge his leg, its skin cool and slick. He reveled in being this voluptuously close to nature, so close he could hear it breathe and feel it move. The question put perennially to Haldane was: "Why did you settle in India?" And this was the answer he usually gave, Helen remembered: "Wouldn't any biologist like to live in a country where one can find chameleons in the garden?"

The kind of work Haldane thought possible in India derived from its proximity to a luxuriance of plants and animals. The labs at ISI couldn't afford electron microscopes or cyclotrons. But Darwin hadn't needed any expensive instruments to observe the finches on the Galápagos Islands or the earthworms in his lawn. His primary research relied on his patient respect for other organisms and on his sense of kinship with them—qualities that, to Haldane, were analogous to Hindu philosophy's nonviolence against nature. Christian theologians insisted that humans and animals were distinct, he wrote; in the Hindu, Buddhist, and Jain faiths, "animals have rights and duties." Sometimes Helen liked to say that Darwin had converted Europe to Hinduism—an exaggeration, Haldane held, but not terribly distant from the truth.

Under his guidance, his colleagues and students watched, counted, and measured. Helen bred guppies and silk moths; the fish did poorly, but the moths provided a paper on the two kinds of cocoons they produced and the varying quantity of silk from each kind. One associate planted combinations of rice varieties together to see if they affected each other's yield. Dronamraju, whose first interview with Haldane consisted of a problem set sent by post, studied lantana plants and the butterflies that preferred lantana flowers of one color over another. The more choosy the butterflies were, Dronamraju realized, the less probable it was that plants of one sort would be cross-pollinated with plants of the other—and so, over time, the likelihood that the two variants became distinct species increased. Another student, whose initiative had impressed Haldane during a visit to the Central Coconut Research Station in Kerala, was brought to Calcutta and put to work examining the leaf spirals of palm trees.

His teaching became elemental. In Britain, no student came to him without a sense of how to formulate research; in India, it happened often. So he had to first train his pupils in the scientific method and in the power of originality. He didn't want them to follow a process capably and reach an anticipated conclusion, he remarked in an essay for the *Rationalist Annual*. He wanted them to know how to proceed after encountering an unforeseen event. "It is an essential part of scientific method to expect the unexpected," he wrote. "In most university teaching of science it is, I fear, discouraged. The student who notices the unexpected has generally made a mistake in his technical work. After a few such mistakes have been pointed out to him he realizes that he had better keep them dark in future." To be open to surprises made for an exciting life. The men and women of this temperament, he noticed,

> do not take the laws of gods or men very seriously and are more or less incapable of the dull but, for all I know, satisfying form of existence which is now called social adjustment or integration into the group, and used to be called doing one's duty to God and one's neighbor. There may, of course, be "laws of Nature." But if so, I don't know what they are, and I also know that nobody else knows them.

It was a remarkable statement to make after a lifetime spent picking the locks of nature: that despite all he had learned, there was still so much to discover. In a letter to his old friend, the poet Robert Graves, he wrote: "The world is not only queerer than anyone has imagined but queerer than anyone can imagine."

Of his own labors in Calcutta, he wrote to Mahalanobis that he had only managed to complete some rather trivial theoretical papers. He was wasting time, he worried, and the heat tended to daze him. He would be more efficient soon, he promised.

EVEN WHILE THE LYSENKO AFFAIR was lashing about him, Haldane had worked away on genetics, but his energies had abated. In 1947, he refined his famous calculation, from 12 years earlier, of the mutation rate for hemophilia. Someone had published a sturdy set of 63 pedigrees of hemophiliacs in Denmark. With these data, he improved his estimate from one mutated gene in every 50,000 new

X chromosomes to one in 31,250. He also gauged that the rate in women was less than a tenth of that in men—the first-ever discovery of sex-dependent mutation frequency, of which several other examples would later be uncovered. Again with more fresh data, he revised his estimate of the recombination frequency for the genes for color blindness and hemophilia, from around 5 percent to around 9.8 percent. Past work could never rest; like old silver, it had to be brought out from time to time, to be held to the light, reexamined, and polished.

He still fizzed with ideas, but now he left these in his talks and conversations, rather than taking them home to develop them. Attending a symposium in Milan in 1949, he spoke of how infectious diseases shape natural selection. He had been the first to suggest this, back in 1932, when he wrote, in his book *The Causes of Evolution*, that immunity was a character that held "survival value." Since then, one major study of flax had indicated that Haldane was right and that disease could be an agent of evolution. In a discussion after his talk in Milan, he added an offhand curl of thought. Blood disorders like thalassemia, so often fatal at a young age, ought to have been rubbed out by natural selection. Instead, they persisted—and they appeared at particularly elevated rates in populations living in malarial surroundings. The correlation couldn't be an accident. It must be, Haldane speculated, that people who are thalassemia heterozygotes—who have only one copy of the faulty gene, rather than the full, dire complement of two—have some resistance to malaria. That limited immunity acts as an evolutionary agent, providing a reason for natural selection to keep the thalassemia gene in circulation.

At the time of the symposium, unaware of its proceedings, a South African geneticist named Anthony Allison was collecting blood specimens in East Africa, and he noticed that sickle cell anemia, another blood disease, was more common in tribes living in the hot, humid regions near the Kenyan coast or Lake Victoria, but less frequent in drier, higher terrain. In 1954, having tested the blood of nearly 5,000 people and having found lower counts of the malarial parasite in children with the sickle cell trait, Allison published his research—only to be told afterward that Haldane had preempted him. At University College, Allison wrote, "Haldane invited me up to his room and we had a long and interesting discussion, in which I remember trying to keep him from following too many tangential arguments—on thrushes and worms and palmtrees, on everything but cabbages and kings and what I wanted to discuss!" But Haldane was kind enough, Allison noted, to not mention his own anticipation of the idea.

Later researchers proved Haldane and Allison right, finding that the sickle cell mutation emerged independently in different African populations and then stuck around, selected for the protection against malaria that it provided. Studies mapped correlations between malaria and other blood disorders, including a kind of thalassemia, which was Haldane's original formulation. A 2011 paper in *Science* explained the mechanism of this perverse immunity. The malarial parasite ordinarily breaks into a healthy red blood cell, then scrambles the proteins of the cell's wall. But in red blood cells rendered unsound by sickle cell anemia, the parasite struggles to latch on to the cell wall to sabotage it. Heterozygotes for these blood disorders have enough dysfunctional red blood cells to throttle the parasite's rampage through the body, but also more than enough normal red blood cells to avoid the disorder's most severe effects. If the disease doesn't kill you, it turns out, it makes you stronger.

Allison didn't contest Haldane's prior claim on the hypothesis. In any case, it was Allison who went to Africa, ran blood audits, and verified it; that kind of field study would never have been on Haldane's horizon. The epiphany in the Orange Tree, though, was a much thornier story.

The Orange Tree was a pub near University College to which Haldane and his colleagues often repaired. One evening, probably in the 1950s—the legend is undated—someone asked Haldane: Would he sacrifice himself to save someone's life? Haldane's response was first to churn through a few sums on the back of an envelope and then declare his conclusion. He would jump into a river to rescue either two of his brothers or eight of his cousins. To ensure the survival of as much as possible of his own genetic material, these two selections of relatives were mathematically the most logical; a brother or a cousin less, and they were better left to drown.

This was how Maynard Smith narrated it in 1975, years after Haldane's death, in a *New Scientist* article. It certainly carried the flavor of a Haldane tale: the bold pronouncement, the numerate ethic, the clever-pants answer to a casual question. But like Darwin and Fisher, Haldane had done a little thinking about altruism. In *The Causes of Evolution*, he had wondered in passing if saints and military martyrs aided the survival of the group even as they impaired their own. It was the precursor to a fiercely contested theory of group selection, advanced by other scientists, who ascribed altruism and social cooperation to an innate desire to preserve the wider species. Haldane's comment in the Orange Tree, if he made it, had a more selfish scope, restricting the benefits of an organism's altruism only to those who share a chunk of its genetic material. Haldane had, Maynard Smith wrote, not only presaged the theory of kin selection; between beers, he had even worked out a piece of its basic arithmetic.

The *New Scientist* article dismayed a biologist named William Hamilton, who had independently arrived at a model of kin selection in the mid-1960s. He had termed it *inclusive fitness* to distinguish it from the individual fitness that otherwise lay at the center of evolutionary thought. In one paper, Hamilton had even written that "everyone would sacrifice [his life] when he can thereby save more than two brothers or four half-brothers or eight cousins," the language echoing Maynard Smith's anecdote a little too perfectly.

The *New Scientist*'s Letters pages grew roiled in squabble. He had been a student at University College himself, Hamilton wrote, but no one there seemed to be interested in altruism and kinship. Was there any evidence that Haldane really had this insight at the Orange Tree?

Maynard Smith replied. Hamilton certainly deserved the credit for kin selection; he had been the one to develop the science, and that's what mattered with any theory, "to see its relevance and to work out its consequences." But the principle had occurred to Haldane; he had even written up a version of it in a 1955 essay titled "Population Genetics." The essay included this snatch of a scenario:

Let us suppose that you carry a rare gene which affects your behaviour so that you jump into a flooded river and save a child, but you have one chance in ten of being drowned, while I do not possess the gene, and stand on the bank and watch the child drown. If the child is your own child or your brother or sister, there is an even chance that the child will also have this gene, so five such genes will be saved in children for one lost in an adult. If you save a grandchild or nephew the advantage is only two and a half to one. If you only save a first cousin, the effect is very slight.

In a beehive or an ant colony, Haldane added, every member was a sibling in some way or another. These pervasive ties of kinship presumably gave rise to the high degree of altruism and social behavior witnessed in these communities.

No clarity ever emerged about Haldane's moment of inspiration in the Orange Tree. Hamilton, who fell out with Maynard Smith, didn't recover from his sense of injury. Still, the theory came to be associated most strongly with Hamilton; discussions of altruism and social behavior begin with his papers, and an equation sitting at the heart of the kin selection model is known as Hamilton's rule. It has been used to account for a multitude of kindnesses and collaborations in the living world: squirrels calling out predator alarms, red howler monkeys forming coalitions with their relatives, and even our own tenderness toward our children.

Criticisms of inclusive fitness have accrued to Hamilton as well. The naturalist E. O. Wilson, who has spent a career studying ants and was initially entranced by Hamilton's explanation of their social customs, later reversed his position, suspecting that Hamilton's equation was too simple for the messy complexities of evolution. Other biologists, including Richard Dawkins, defend the idea. In 2010, when *Nature* published a mathematically dense refutation of Hamilton's theory, 137 scientists signed a letter of response, arguing the case for inclusive fitness.

In this, also, the poetic imagination longs to ascribe the origin of kin selection to Haldane. He was the kind of man who would have been delighted to ignite half a century of argument with a wisecrack in a pub.

WHILE JUGGLING HIS FINAL DUTIES at University College and the arrangements of his move to India, Haldane wrote and revised his last significant paper. Published in 1957 and titled "The Cost of Natural Selection," it retrieved an example he had previously used: *Biston betularia*, the peppered moth. A change in the environment, during the Industrial Revolution, had selected against the pale-winged members of the species until only dark-winged moths remained. What if the environment had altered even more radically, acting on not the single trait of color but 10 independent traits? Haldane calculated that only one in 1,024 moths could have lived through such an intense rigor of selection. The cost was too high; the species would have vanished.

The cost is part of the compact of evolution. A variant of a gene makes some individuals better adapted to an environment—more "fit"—and as those individuals reproduce and spread the variant gene, the population evolves toward that fitter state. But the less fit variants must die along the way. Their numbers have to dwindle— but they have to dwindle at the right speed. Too fast, and the population goes extinct before the fitter trait can become fixed within it; too slow, and the fitter trait is outnumbered, never to be fixed at all. Further, independently inherited characters vie for the attentions of natural selection, like swains around the woman they desire. If one character is intensely selected, that is detrimental to the other suitors, Haldane wrote. As a result, if an environmental shift ever requires a quick replacement of several genes by their variants, that demand proves impossible to meet, and the organism is bound to go extinct. This limit on the speed of evolution came to be called Haldane's dilemma.

In his paper, Haldane set up a basic scenario of low, rather than highly intense, pressure of natural selection, to work out how many generations it would take for a fitter trait to fully replace its alternatives. About 300 generations, he calculated: 300 generations' worth of genetic death—that was the cost of selection. He could imagine some scenarios in which natural selection might act faster: when the first vertebrates colonized the land, for instance, and found themselves without competitors. But these were exceptional circumstances. The speed limit he proposed "accords with the observed slowness of evolution." He included a rider, though: "I am quite aware that my conclusions will probably need drastic revision."

Haldane's framework of the cost of selection is still valuable, but others have contested its details and have argued that a gene variant can spread through a population well within 300 generations. The cost of selection, in other words, must be lower than Haldane thought. The most important doubts were cast by molecular biologists, who in the 1960s began to sequence the vast variety of proteins in the living world. By comparing a protein like hemoglobin across species, scientists could discover how its differing versions must have diverged from a common precursor—and then, by extension, how fast the underlying genes, too, must have evolved. The rates thus deduced in mammals smashed Haldane's speed limit comfortably; one estimate expected a nucleotide substitution every couple of years, and another every 50 years, compared with Haldane's estimate of one every 300 generations. If natural selection were acting at these intensities, winnowing gene variants out of the population with such fury, an individual mammal would have to produce millions of offspring so that even a single one could survive. The cost became intolerable.

A possible solution was proposed in 1968 by a scientist named Motoo Kimura: perhaps, at the molecular level, most changes in genes are neutral, not improving or degrading an organism's fitness, and so not subject to the pressures and costs of natural selection. These neutral changes are fixed, instead, by chance—by genetic drift—and so don't need to pay Haldane's cost of natural selection. Though Kimura's solution continues to be counterargued, it signaled a crucial, unmistakable transition. From the statistical scrutiny of population genetics, the study of evolution and natural selection had passed into the miniature focus of molecular genetics. The notepad and pencil gave way to the apparatus of the lab. Haldane had migrated to India hoping to seed a culture of frugal genetics. But he arrived just when genetics started to require banks of modern equipment and deep wells of funds and just when his own methods were beginning to be pushed to the side.

FROM CALCUTTA, HALDANE URGED his friends to visit him. Stay with us, he wrote to Isaiah Berlin. "Our food is probably too Dravidian for your taste . . . . Don't worry, we will kill a fatted goat for you." Have a drink at ISI, he wrote to Aldous Huxley. "If you visit me, you will meet, among other denizens of my house . . . four tortoises, to remind me that if I cannot be good, I should be careful." To the geneticist Theodosius Dobzhansky, flying to Bombay for an Indian Science Congress, Haldane recommended playing hooky so that they could tour a complex of rock temples instead. "It would be grand if you could come with us, for we could talk biology after dark." Despite himself, he missed the company of the people he had known in England, and when they flew in, he made it a point to meet them at the airport. A science writer, landing after a long and tiring flight, was informed that "Jagdish Haldar" was waiting for him. It turned out to be Haldane, unrecognizable in his dhoti. Most guests were bundled into a car and taken to the museum and the zoo. With Haldane, the education always began right away.

One winter, Haldane was visited by someone he hadn't previously known, a Canadian biologist named Gary Botting, who worked, like Helen, on moths. Botting, now a legal scholar, was a devout Jehovah's Witness at the time, and he talked with the Haldanes about moths for hours. Then he made his mistake. He believed, Botting said, that the various species of moths and butterflies had all descended from the single pair of Lepidoptera that had traveled with Noah on his ark and landed on Ararat. It was the same for every kind of animal: birds, reptiles, monkeys, cows, cats.

What about kangaroos? Haldane asked.

"And kangaroos."

"And platypuses?"

Platypuses were aquatic, Botting explained kindly. They just swam through the flood. But the similarities between various other species couldn't be explained any alternative way. "Look at the emu of Australia, the ostrich of South Africa, and the rhea of Argentina.

Look at the snow leopard of China and the jaguar of Brazil. Look at the anteater and the aardvark and the armadillo. Look at—"

Haldane interrupted. He understood, he said. Now, would Botting do him the favor of meeting him at the Calcutta Zoo at 5 a.m. on Saturday? "I want to show you something. Something very important."

On Saturday, when the zoo was drowsing and still, Haldane took Botting to a footbridge over a stream. For an hour or more, they watched as the day shook itself awake and birds started to rend the silence and plow the sky. Haldane knew them all, Botting wrote later,

> from sarus cranes to florikens, from chukors to bee-eaters, from barbets to sunbirds, from drongos to shrikes. Some, such as the terns and ibises, sounded familiar. Others, like the rusty-cheeked scimitar-babblers and the red-billed leio thrixes, sounded like a put-on, but when I challenged him on his identification of a flock of blossom-headed parakeets, he merely intoned in Latin, "*Psitticula cyanocephala bengalensis*," with enough flair and flourish to allay any doubts.

There were 310 species of birds to be found in India, Haldane said: 238 genera, 62 families, 19 different orders. "All of them on the ark. And this is only India, and only the birds—"

But some of the birds, Botting feebly responded, were aquatic.

Haldane laughed the laugh of a man who had won an argument. They walked back to their cars.

He traveled frequently, in spite of his age: through India, to see the Ajanta cave art near Aurangabad, to Kashmir, to Allahabad and Agra, to Varanasi and Madras, to the wildlife sanctuaries of Kaziranga and Periyar. For a conference, he flew to Malaysia and Singapore. In Sri Lanka, he threatened to cancel his speech unless he was moved to a less opulent hotel with vegetarian food; later on the trip, touring a mine shaft, he fell down and broke two ribs and a foot. When he was invited to the United States in 1960, though, he declared that a visit was impossible. He had been to the American consulate in Calcutta to apply for a visa, he explained a friend, a slime mold biologist at Princeton:

I was asked . . . to give a list of all associations to which I
have belonged since my 16th birthday (in 1908) with date of
joining and leaving, with a threat of jail or fine if I get one
wrong. I don't know if I joined the Oxford University Liberal
Club in 1912 or 1913. Having a professional regard for truth
I am not going to guess. If President Kennedy has the guts
to tear down this Iron Curtain I will come when next asked,
if I can manage. But I think there are too many officials who
have a vested interest in that sort of nonsense.

So you had better come here. There are plenty of molds,
especially in the monsoon.

Being the most celebrated biologist in the country had its pit-
falls. India loved its public functions, loves them still. A person has
not been honored or welcomed unless a ceremony has been arranged
expressly for the purpose. Haldane was always turning down requests
to be a chief guest or to deliver a vote of thanks, or to speak on some
other pretext. Having hired a secretary, Haldane wrote him a letter, describing what his most important duty would be: to guard his
employer from the people trying to distract him from his work: "I
mean people who want me to appear on public platforms, or make a
20 min speech after travelling 200 miles." When he answered invitations himself, he didn't flinch from curtly pointing out that this
endless chief-guestery and vote-of-thanksifying would be a waste
of his time. Why couldn't he give lectures on biology instead? And
why, when he and Helen visited other universities, were they trotted
around laboratories doing work bearing no connection to their own?

When Queen Elizabeth II came to Calcutta, Haldane was asked
to the governor's mansion for a reception. He oiled out of the obligation. He had never wanted to go; he thought, in particular, that
the Duke of Edinburgh pasted a permanent smile onto his face at
such functions, making him look too much like a shark for Haldane's
liking. But fortunately he had a sturdy roster of excuses. He had just
returned from Sri Lanka, and his broken leg was still in plaster, he
wrote. He didn't own a car. He didn't even own the right attire, he
explained to the governor's office. "My European tail coat was bought

before 1939, and looks it. My Indian clothes come from Gram Khadi
Udyog, and look it. My wife has also no formal evening dress, jewel-
lery, and the like." Their appearance was sure to insult Her Majesty,
he wrote, and they had no wish to insult her. Wickedly, he had struck
the right note; the only way to deter them, he realized, was to promise
to appall their refined tastes.

India's genius for bureaucratic tedium brought out the hellion in
Haldane. It was as if an edifice of rules and customs had been built
around the practice of science, holding it imprisoned. Why should
the labs have to be shut at 5:30 p.m., for instance? Why did university libraries lock away their best journals? Why couldn't he fly out
of the country without obtaining the consent of India's central bank?
His profession was riddled with hierarchies, as if it had imported a
kind of caste system from the society around it. He couldn't have a
meal with a technician; it was frowned upon. He wasn't allowed to
hire an assistant with anything less than a first-class degree in the
sciences, even though he himself possessed not a single scientific
degree of any class. The niceties of rank were hobbling Indian science like nails in its feet. The lethargy of progress, Haldane wrote in
an article, was "not because Indians are stupid or lazy. It is because
they are too polite. They spend hours daily in conversation with others not on professional matters, but on personal topics." His colleagues were just too civil to criticize his work a most inefficient
way to do science. "I hope that as Indian science grows up, it will
become less acute."

For Haldane to fall out with Mahalanobis was inevitable.
Mahalanobis ran ISI with a feudal hand, distributing favors gener-
ously but expecting in return to be permitted his autocratic ways. He
wanted to control everything. If travel arrangements had to be made
for a visitor, he made them; when the campus needed new buildings, he insisted on designing them, even though he had no training
as an architect. Haldane could never have tolerated such meddling
for long. It jabbed and roused his instinct to rebel against authority,
which had not left him since he was an Eton schoolboy fulminating
against his masters. For all his declamations against hierarchies, he
had also never fully expelled the aristocrat within him. Visiting India

as a young soldier, he had observed scathingly: "The Hindu caste system is the greatest glorification of snobbery that the world has ever known." Now, living in the country as an older man, he still censured the indignities of caste regularly; even if affirmative action enrolled lower-caste students at the expense of deserving Brahmins, he wrote, for example, it was "a rough kind of justice. It certainly does not outweigh the huge injustices which the ancestors of Brahmins inflicted on their fellow Indians in the past." But Haldane also admitted to a "sneaking sympathy" for the Brahmin class. He thought he shared their ethic of prizing knowledge over wealth and their stature at the top of a pyramid of intellect. And perhaps, too, despite himself, there was a vestige of entitlement of the Briton living in India. He was happy, even eager, to live as his colleagues lived, but he didn't see why he should work as they worked, tied down and constrained in a thousand different ways. He expected the rules, the silly as well as the sensible ones, not to apply to him.

The Indian Statistical Institute held two caged egos, spoiling for a fight.

Haldane encountered a stream of small vexations. He had been told, when he came, that his classes would be part of a unified science degree, but no one consulted him on how to frame a curriculum. When some higher authority sent him a proposed syllabus, Haldane said he was horrified at how inept it was, "a rather poor copy of British teaching 30 years ago." An institute circular asked researchers to sign in to an attendance ledger every day; Haldane thought this ridiculous, especially since his students spent most of their time outdoors. When he traveled for work and wanted a student to accompany him, ISI said that it couldn't fund an extra air ticket. Annoyed, Haldane paid for the ticket himself, but this, too, was a problem. Consider the effect on other professors, Mahalanobis told him, "who have a smaller salary than yours [and] cannot afford to pay for the travel expenses of their juniors by air." He was deputed to a library committee, and then the committee was abolished. He complained about the librarians, their incompetence and their unwillingness to improve; they complained about him, his rudeness and his rage and his accusation that they had removed some of his papers from a locked cupboard.

Minor breezes whipped into storms. In the autumn of 1958, Haldane rejected some curtains tailored for his flat, arguing that neither he nor Helen had approved them. They were likely to be of the wrong width, he wrote, since no one had stopped by to measure the windows. Now that the set of curtains had been delivered, he would "arrange for its destruction, if as is possible, it proves useless. . . . I also venture to hope that you will send a representative to the bonfire if it proves necessary to destroy the material." This prompted a 13-point letter from an institute official to Mahalanobis himself, the text squealing with alarm. "I find this idea of destroying the curtains by bonfire most distressing. . . . He could give it away to someone who would be able to use them. . . . Why is he thinking of making a bonfire? I am feeling apprehensive that this shows a certain lack of balance of mind which is alarming in a scientist of his eminence."

Mahalanobis wrote to Haldane often, to comfort him. The institute was still short of money and trained staff, and it couldn't move as quickly and competently as the universities of the West. They were trying their best. But once a feeling of unjust persecution took over Haldane, it stopped up his reasoning; he became immune to any appeal. He had been at ISI for three years, and still his position was undefined, he wrote to Mahalanobis: "For some purposes I am treated as a head of a department, for others as a subordinate." Another time, in a sulk, he remarked that his had become a "menial position."

In February 1961, the Haldanes quit the institute in tandem. His note to Mahalanobis was full of grievances: the absence of practical physics and chemistry courses to make up a well-rounded science degree, as well as Mahalanobis's string of broken promises about the curriculum. Only in a short paragraph did he refer to the triggering incident at all. Alexei Kosygin, the Soviet politician, was scheduled to tour ISI, and Mahalanobis had chosen to exhibit the work of another British scientist at the institute rather than that of the Haldanes' junior Indian researchers. Helen held Mahalanobis guilty of discriminating against Indians; he pleaded that he was only showing a customary courtesy to his visiting scientist. Within this altercation, Helen perceived somehow that Mahalanobis was accusing her of negligence. Smoldering from the insult, the Haldanes wrote out their letters of resignation that very week. Mahalanobis replied to each of them, explaining that he had leveled no charges of negligence and that the curriculum would yet come together. He tried to persuade them to stay. But he did not try very hard.

Once again, Haldane had forced people to ask themselves: What is to be done about J. B. S.? The matter went all the way up to the prime minister. The Council of Scientific and Industrial Research (CSIR), a government body, offered Haldane the chance to start his own genetics and biometry unit in Calcutta. This was a good move, Nehru wrote to the director of CSIR:

> There can be no doubt about Dr. Haldane's eminence as a
> scientist and more especially as a biologist. To have him with
> us will be an acquisition. But he is not an easy person to get
> on with and does not normally fit in with rigid Government
> rules. If we want to keep him, as presumably we do, we shall
> have to be flexible in some matters. These are not likely to be
> money matters as he does not care much for money.

But then, for a year, this proposal disappeared into the innards of the Indian bureaucracy. When his new offices failed at first to materialize, Haldane started his unit himself, turning his flat into a laboratory and his bathroom into a place to breed wasps. His volumes of books and journals, 60,000 of them, spilled out of cupboards and shelves. Signing contract after contract, Haldane learned the nature of his indenture to the government: that he couldn't speak out on politics, that he couldn't buy an item priced more than 10 rupees without sanction from New Delhi, that he could be transferred to any other part of India without warning, that he had to pay his researchers solely on the basis of their degrees. He came to call CSIR the Council for the Suppression of Independent Research. He was champing at his bit, anxious to get on with the science, to be of some practical value to India. "There are several hungry chaps in this country, and I should like to do something to fill their bellies," he wrote to a friend in the spring of 1962. "The CSIR thinks otherwise. In fact," he added, quoting a James Elroy Flecker couplet, "by now they would like: 'To fire and fell the monstrous fort of fools / Who dream that men may dare the deathless Rules.'" (He capitalized "Rules" as if to underscore his irritation.)

That June, one of Nehru's ministers gave the Indian Parliament an update: office accommodation had been provided for the new biometry unit. Haldane considered this a lie. "This accommodation," he told a student, "consists of half a table 10 km away for my administrative officer." The day after the minister's statement, Haldane resigned—from a job that had not even begun. "This is entirely typical of the official treatment of scientists in India," he wrote. "It is the intolerable conditions imposed by bureaucrats, and not the low salaries or the lack of equipment, which cause so many Indians to take up posts abroad." This time Nehru, occupied with the tense preludes to war with China, was not on hand to intervene.

Slyly, though, Haldane had been negotiating all the while with the government of the state of Orissa. The state's chief minister, Biju Patnaik, had told Haldane how much he admired him and offered land and buildings for a new genetics laboratory in the capital of Bhubaneswar. The situation seemed too similar, at first, to ISI. Like the institute, Haldane worried, Orissa was one man's fiefdom. But Patnaik was younger than Mahalanobis, and he did not, at least to Haldane, pretend to know how scientific research ought to be conducted. By the end of the summer, the Haldanes had shifted once more. He sent a German naturalist a letter: "I am now settling down here, as I hope, for the remainder of my life."

YOU COULD CALL BHUBANESWAR A garden city, although the only things that grew in the garden of Haldane's new house were shrubs of crown flowers, their leaves waxy and their flowers tinted a pale purple and shaped like stars. Strictly speaking, it wasn't even a city yet. The capital had been founded 14 years earlier, swallowing an older town of sandstone temples and the villages that clung to its borders. Only 38,000 people lived there in 1962—so few, by Indian standards, that an airmail envelope addressed to "Sri J. B. S. Haldane, Scientist, Bhubaneswar" found him easily. It was still a half-formed city when he arrived. The roads were new and as yet unfilled with traffic; many of its plots of land gaped vacantly, and others held just-constructed buildings, designed to be simple and austere. The state was not rich, and it was planning Bhubaneswar as a purely administrative center.

Haldane's house, his "one-storied ivory tower," lay a mile from his lab, close enough that he could walk to work. He lived in astonishing luxury, he told Case: "fluorescent tube lamps, fans, whisky, and gin, all made in India," although his principal vice was *bhang*, a cannabis-milk concoction that made him giggle and loosened his grasp of time. Bhubaneswar was still so small, so bucolic, that the bungalow felt like an interloper in the natural world. Every evening, hundreds of egrets flapped over the Haldanes' residence. Cows and buffaloes stalked through the grass, and elephants pottered around in the woods nearby. Haldane sat on his veranda and watched yellow-wattled lapwings go about their domestic affairs. Once, in his bathroom, he met an intruder, a green, 4-foot-long lizard.

No one asked him to serve on a library committee or meddled with his budget or installed attendance registers in his lab. No bureaucrats came by with their spools of red tape. Everyone was "highly civilized," Haldane wrote to Case, "though so easy-going that I can be a nuisance by working a little too hard for them at the age of 70." For the first time in India, Haldane was content. Bhubaneswar, he thought, was paradise on earth.

The Sino-Indian War began, and the government's plan for a new building to hold Haldane's lab stalled. Temporarily, he was given quarters on the campus of an agricultural college: four rooms, containing just the Haldanes and two of their students, and a bare minimum of apparatus. Their ambitions were modest as well. Once Haldane had manned the front line of genetics, thinking always about how to move the line further, doing the kind of vital work that would win new territory. Now he had fallen behind the line; with his students, he was stocking the storehouse, building provisions of knowledge for others to draw on. They charted color blindness in local tribes, deaf-mutism within an inbred caste, and the frequency of a shorter fourth toe in the region's men; they watched, minute by minute, the habits of wasps and geckos and birds. It was robust basic research, but its horizon was limited.

"I think we can still learn a great deal from observation," he wrote to his friend S. Radhakrishnan, who had become the president of India in 1962. He described how Helen had inspected a wasp's nest for 8 hours daily, over a span of three months. "Nobody has ever done this before. We shall have a history of this little animal community better than that of many human communities." He was right, in his way, and his happiness in gaining knowledge for its own sake recalled all the studious observations he had made himself throughout his life: the notebooks he kept as a boy, recording the plants he encountered on his hikes, or the pocket-sized diary he had as an adult, with its notated descriptions of the cats he saw. He had never stopped being fascinated by the variety of the natural world around him, and now, in the absence of funds and facilities, he fell back on this most primal aspects of his love for science. Given his age and the constraints of his lab, and given how the pursuit of genetics itself had jumped onto other tracks, it was no surprise that the vital work once done by Haldane was now being done elsewhere. The lab received a succession of visiting scientists—from Berkeley and Columbia, from Paris and London—but they were drawn by Haldane's reputation, not by the nests of wasps he maintained in Bhubaneswar.

Haldane did as much for the lab as he could. When his mother died in 1961, at the age of 98, he inherited some money and planned immediately to spend it: a fund for the Royal Society, but also hundreds of pounds' worth of equipment for his students. He invited his old student Dronamraju to join him in Bhubaneswar, promising: "If I can't get you laboratory accommodation with some sort of guarantee here, I shall give you £2,000, and tell you to seek your fortune in Europe or North America." In 1961, Haldane was awarded the Feltrinelli Prize by Italy's Accademia dei Lincei: a purse of £11,500, around 150,000 rupees then and a third of a million dollars today. He decided that 100,000 rupees of that money would go into financing genetics in Bhubaneswar. To another student, he wrote: "The sum seems to be very large. . . . I will be able to leave enough in Italy to pay for a visit by yourself and your family to Rome." He was leaving for Italy to pick up his prize, so he let his student know where his will was: the bottom drawer of a cupboard in the bedroom. "Don't open it unless both my wife and I are killed, in which case you will benefit financially, if not in other ways." If Haldane believed in a cause or in people, he would drain his treasury for them.

He traveled less, although he still made some trips overseas. He went to Israel and rode in a bus through the Valley of Elah, where David had once toppled Goliath. Nothing in Israel moved him more; later, he asked his hosts for a stone from a brook in the valley. David had chosen five stones for his battle. But even one, Haldane wrote, "would today make me a better man, for I am getting old, and to-day giants are armoured in files and fight with red tape." He knew himself well enough to realize that the story of David and Goliath, of the powerless against the tyrant, had formed his character. In Bhubaneswar, he chose "Elah" as his telegraphic address, and he always carried that small, round stone from the brook in his pocket.

He did go to Italy to collect his Feltrinelli award. The last time he had been in the country, eight years earlier, he was attending a small workshop on population genetics. On the last day, as dinner began, his colleagues asked Haldane, as one of the senior scientists, to deliver a speech of thanks to their hosts. Haldane pulled a piece of paper out of his bag, and through the dinner, every now and then, he made a quick note to himself. His speech, after the meal, consisted of extracts from Dante and Lucretius picked for their hints of ideas in genetics—Dante's foreshadowing of mutations, for instance:

Natura generata il suo cammino
Simil farebbe sempre ai generanti
Se non vincesse il provveder divino.

Or, loosely: "Generated nature would make its way / Just like its progenitors / If providence did not intervene." Haldane had sieved these quotations straight out of his memory.

He went back to America, at long last. This time, the consul in Calcutta didn't demand a watertight history of his various associations; Haldane was asked no questions, and he told no lies. Still, his notoriety needed no introduction. He was scheduled to visit the University of North Carolina, but university officials, invoking a new state law that barred Communists from speaking on campuses, badgered him about his links with the party. He refused to respond. Instead, he traveled to the University of Wisconsin–Madison, where he gave a seminar with Sewall Wright. At a reception, a professor and some students approached him to talk about his theories of relationships in biochemical kinetics. "We've been extending the Haldane Relations," the professor said.

"So have I," Haldane replied, with a snuffling *hrnff-hrnff-hrnff* laugh. "Been married, you know!" *Hrnff-hrnff-hrnff.*

He went to Rochester, where he told the evolutionary biologist Richard Lewontin that his stomach was troubling him. They went, for some reason, to a pharmacy counter in a Sears, Roebuck store, where Haldane bought antacid.

Where are you from? the sales clerk asked him.

India, Haldane said.

"That's funny," the sales clerk replied. "I thought you were one of those high-class Englishmen."

Late in the autumn, he flew to Florida, where the National Aeronautics and Space Administration was hosting a conference on the origins of life. He served as the chair, and Alexander Oparin, the simultaneous creator of the Oparin-Haldane hypothesis, was the first speaker. Haldane was unstinting in his introduction of his colleague. "I suppose that Oparin and I may be regarded as ancient monuments in this branch of science, but there is a very considerable difference, that whereas I know nothing serious about it, Dr. Oparin has devoted his life to the subject."

A discussion followed Oparin's speech. The astronomer Carl Sagan remarked that the first paper in his field to describe the methane-ammonia atmospheres of the Jovian planets was published in 1931. Prebiological Earth was likely held within a similar envelope of air. A reducing atmosphere—an atmosphere devoid of oxygen but stocked with gases such as hydrogen and carbon dioxide, as in the theories of Oparin and Haldane—could act on methane and ammonia as these two scientists had described, to begin the first anaerobic processes leading to life. "If I am not mistaken, Haldane's argument was published before the astronomical papers were," Sagan said.

But Oparin was there first, Haldane insisted. "I am ashamed that I haven't read his early work," he said. "I didn't publish until 1927. . . . I think if his first book was in 1924, the question of priority doesn't arise. The question of plagiarism might."

Not long afterward, Haldane told one of the conference's organizers that he needed to see a doctor. He had been experiencing some rectal bleeding. Perhaps it was a recurrence of an earlier attack of dysentery? But at the hospital, the doctors warned him that it could be cancer.

From America Haldane went to London, where he was diagnosed with colorectal cancer. In a surgery in December, they removed a length of his gut. While he was still recovering, he agreed to see Jonathan Howard, a young student who hoped to come to Bhubaneswar to work in his lab. There he was, Howard thought when he entered the hospital room, lying in bed, looking unmistakably like J. B. S. Haldane. "He was reading the poems of John Donne when I arrived," Howard said, "perhaps the sonnet 'Death Be Not Proud.'" But the doctors had told him that the operation had been a success, Haldane said to Howard. He was no longer a cancer patient.

He composed an essay while he was on the mend: on the first few pages of a crisp new notebook he had brought from India, an overview of human genetics, a contribution to an anthology titled *Society and Science*. Fate and mortality were on his mind. Genetics hadn't yet given people control over their physiology, he wrote, but that certainly lay in the future; in the meantime, it had at least dispelled superstitions and contradicted Shakespeare's thesis that man is "servile to all the skyey influences."

He sent the *New Statesman* a poem titled "Cancer's a Funny Thing," such a wild and comic creation that it must always be recited in full:

I wish I had the voice of Homer
To sing of rectal carcinoma,
Which kills a lot more chaps, in fact,
Than were bumped off when Troy was sacked.

I noticed I was passing blood
(Only a few drops, not a flood).
So pausing on my homeward way
From Tallahassee to Bombay

I asked a doctor, now my friend,
To peer into my hinder end,
To prove or to disprove the rumour
That I had a malignant tumour.

They pumped in BaSO4.
Till I could really stand no more,
And, when sufficient had been pressed in,
They photographed my large intestine,

In order to decide the issue
They next scraped out some bits of tissue.
(Before they did so, some good pal
Had knocked me out with pentothal,

Whose action is extremely quick,
And does not leave me feeling sick.)
The microscope returned the answer
That I had certainly got cancer,

So I was wheeled into the theatre
Where holes were made to make me better.
One set is in my perineum
Where I can feel, but can't yet see 'em.

Another made me like a kipper
Or female prey of Jack the Ripper,
Through this incision, I don't doubt,
The neoplasm was taken out,

Along with colon, and lymph nodes
Where cancer cells might find abodes.
A third much smaller hole is meant
To function as a ventral vent:

So now I am like two-faced Janus
The only* god who sees his anus.
*In India there are several more
With extra faces, up to four,
But both in Brahma and in Shiva
I own myself an unbeliever.

I'll swear, without the risk of perjury,
It was a snappy bit of surgery.
My rectum is a serious loss to me,
But I've a very neat colostomy,
And hope, as soon as I am able,
To make it keep a fixed time-table.

So do not wait for aches and pains
To have a surgeon mend your drains;
If he says "cancer" you're a dunce
Unless you have it out at once,
For if you wait it's sure to swell,
And may have progeny as well.

My final word, before I'm done,
Is "Cancer can be rather fun."
Thanks to the nurses and Nye Bevan
The NHS is quite like heaven

Provided one confronts the tumour
With a sufficient sense of humour.
I know that cancer often kills,
But so do cars and sleeping pills;
And it can hurt one till one sweats,
So can bad teeth and unpaid debts.

A spot of laughter, I am sure,
Often accelerates one's cure;
So let us patients do our bit
To help the surgeons make us fit.

In January 1964, there was a second, minor surgery; in February, a third. "I am told that my chance of a metastasis in the next five years is under five per cent," he wrote to Arthur C. Clarke. The wound was healing too slowly. He shouldn't have censored himself in his poem, he said. His original couplet about the holes in his perineum was:

One is an artificial c**t
Much shorter than the gash in front.

For another two weeks, he recuperated in Maynard Smith's house. His young son, Julian, was tasked with burying Haldane's used colostomy bags in the garden. "He did my Greek homework for me," Julian remembered later. "He could just do a sight translation of these texts that he probably hadn't seen in forever." When Haldane was cleared to fly home, there was a late fall of snow in London, and the city was paling as John Maynard Smith drove Haldane to the airport. They crossed Hampstead Heath, but then Maynard Smith stopped the car so that they could climb out and gaze upon the park upholstered in white.

"I'm afraid I'll never see this again," Haldane said.

"No, I guess you won't, Prof."

ONCE HE WAS BACK in Bhubaneswar, he had plenty to do, as he told a friend in a letter: that study of the shorter fourth toe and a new project in poultry genetics and plans to preside over the International Congress of Human Genetics in Chicago in a couple of years. As soon as he was elected a member of the National Academy of Sciences, he began brooding on what he could publish in its journal. "I have a real snorter of a paper nearly ready for them," he told Dobzhansky. Every two months, he went to Calcutta to get a checkup. His body creaked all over now, as he neared 72, and the effects of his surgeries slowed him down, particularly during the summer. But nothing seemed to alarm his doctor, and he took courage from that.

"I never want to leave this place again on any pretext," he told Dobzhansky. "I know that I shall sometimes have to. Perhaps if I recover fully I may even want to. But I don't at present."

Haldane in India, with Marcello Siniscalco, a visiting geneticist, in August 1964.

Death stuck around in his thoughts, an unbidden guest refusing to vacate the premises. He knew how cancer worked, after all, how it might have sent a troop of malignant cells to another part of his body. His last essay for the *Rationalist Annual*, written for the following year's issue, was called "On Being Finite." What a strange thing it is to recognize, as humans, that we must die, he mused: how certain that prospect is, and yet how unimaginable. He considered the various tales that religions had come up with to fill the nothingness after death, to pretend as if life could be prolonged beyond the lived existence. But this was not the way to live; the way to live was to accept finitude and act upon it.

"I doubt if happiness is possible unless one works fairly hard and enjoys one's work," he wrote. "The most satisfying work . . . is probably that which brings obvious and immediate benefits to a number of other people." That would make the most of life; eluding death was another question altogether. To ally yourself with a great cause is one partial alleviation; another is to be able to believe that your work "will be of use or of ornament" after you. He quoted Catullus on Cinna, the author of *Smyrna*. (He had already quoted the Upanishads, Bertrand Russell, Epicurus, Shakespeare, Algernon Swinburne, Lenin, and the *Bhagavad Gita*.) "Smyrna cana invarua alino menih pri unohe ni," Catullus had written: "The hoary centuries will long unroll *Smyrna*." Except that *Smyrna* never survived. The advances of technology, on the other hand, even one as old as fire, were still in existence. Time was ruthless with the works of humans, but it was kindest on science.

"I should find the prospect of death annoying if I had not had a very full experience mainly stemming from my work," he wrote. He hadn't walked on the seafloor from England to France, as he'd have liked, but he had been wounded in a war, known the love of two women, tried heroin and *bhang*, eaten 60 grams of hexahydrated strontium chloride, and spent 48 hours in a miniature submarine. "I doubt whether, given my psychological make-up, I should have found many greater thrills in a hundred lives," he wrote. He quoted Omar Khayyam's *Rubaiyat*: "So when the angel with the darker drink at last shall find me by the river's brink, and offering his cup, invite my soul forth to my lips to quaff, I shall not shrink."

In August 1964, a new batch of tests revealed cancer in his liver, occupying the organ so thoroughly that he had only a few months left. Helen read the letter from Calcutta first, then went to his room, with his students, to hand it to her husband. He registered no emotion on reading it, but he did have words for his students: a verse from the *Bhagavad Gita*:

jaatasya hi dhruvo mrtyuh
dhruvam janma mrtasya ca
tasmad apariharye 'rthe
na tvam socitum arhasi

For one who has taken birth, death is certain.
For one who is dead, birth is certain.
Therefore, for what is unavoidable,
Do not grieve.

Then he wondered why Helen hadn't taken his pulse before and after giving him the news.

What did affect him was the information that his doctors in London must have realized the cancer had spread even as they operated him. Out of misplaced kindness, they had chosen not to relay that fact. Maybe they were frightened of him, Naomi thought. The deliberate masking of truth enraged Haldane. If he had known, he would have tried some alternate cures, he wrote to his friend Peter Medawar, and he certainly wouldn't have planned studies that he wouldn't see through to their end: "I don't mind dying. I mind dying with unkept promises." He had to sort out the future of his lab in Bhubaneswar, and of his students, and of the papers he had wished to publish. Rest formed no intended part of his schedule: "If I can do even an hour's work in the last twenty-four hours of my life, I shan't feel too ashamed."

Through the autumn, he diminished, spending hours on the porch of his house, his students faithfully telling him about their research. On December 1, he died on that porch. Until the end, he kept in his hand the stone from the Valley of Elah. He never stopped being the boy bullied at Eton, the scientist harassed by his employers, a David in search of Goliaths.

He had left strict instructions about what was to be done with his body. "I hope that I have been of some use to my fellow creatures while alive," he once wrote, "and see no reason why I should not continue to be so when dead." His organs went to the Rangaraya Medical College, in the neighboring state of Andhra Pradesh, to be preserved and studied. Bones from his ear and from the skull just above it went to the University of Chicago. The rest of his bones were articulated into a skeleton, to be displayed in an anatomy museum. He wanted to give every particle of himself to science.

7.

Ten
Thousand
Years

FORTY-ONE PORTLAND PLACE, a brief walk from University College, used to be the home of the Ciba Foundation. Set up by the chemicals manufacturer Ciba, the foundation hosted scientists and their meetings for half a century, beginning in 1947. Not long after it moved into its premises, there was first a fire and then a series of small floods, one of them the result of a scientist forgetting to turn off his bath. After the foundation reopened in November 1962, a conference room was inaugurated with a symposium called, grandly, "Man and His Future." Twenty-seven scientists—all men, all European or American—gathered and, between sherries, lunches, teas, and cocktails, wondered and fretted about the fortunes of humanity.

Haldane's speech closed out the symposium. It was titled "Biological Possibilities for the Human Species in the Next Ten Thousand Years," and in its initial minutes, it wasn't one of his best. He shuffled in desultory manner through scenarios of atomic annihilation and world-conquering tyranny, and he yielded too easily to his taste for contrarian shock. "A few centuries of Stalinism or technocracy might be a cheap price to pay for the unification of mankind," he said, even though he knew himself to never relinquish even a grain of his freedom. He hoped, he said, for some kind of planetary government, some system of organizing people everywhere. Within that system, they could respond to the important questions that faced them. What were the biological capacities and limits of human beings? And in what directions would humanity evolve, especially if it chose to direct its change?

Having framed these inquiries, Haldane treated them as launch pads for his pet futuristic themes; for most of his speech, he was merely being fanciful. "I have sketched my own utopia, or as some readers may think, my own private hell," he said. "My excuse must be that the description of utopias has influenced the course of history." He envisioned aseptic human beings, freed of diseases; then the elimination of congenital defects; then the selection of young men and women into occupations that best suited them. He thought that with enough education, a man with rectal polyposis and a woman heterozygous for hemophilia would decide for themselves not to have children. He predicted that people would clone themselves in middle age, so that they could bring their younger doubles up to speed. He imagined men and women engineered for maximum efficiency for life in space, ridding themselves of their pelvis and legs or perhaps regaining the prehensile feet of their mid-Pliocene ancestors. Or disastrously, perhaps humankind would split into two or more species, each mistrustful of the other. On general principle, he said, humans will make every possible mistake before choosing the right path.

We might imagine that the murderous, pseudo-genetic programs of the Nazis froze all talk of eugenics after the end of the Second World War, but that didn't happen. Watchfully breeding a better human race seemed to be politically and psychologically impossible at the time, Julian Huxley said immediately after the war. But the notion had to stay under examination, he said, "so that much that is now unthinkable may at least become thinkable." Over the course of the symposium, the scientists returned time after time to the subject. They had learned, of course, the American geneticist Joshua Lederberg said euphemistically, that the "worst anomalies of biological powers" must be anticipated and avoided. But to refrain from reaching for these powers at all was, for most of the participants, beyond discussion. Science doesn't work in that fashion. Once a form of technology is inside the castle, you cannot eject it, refill the moat, and nail shut the doors. You must sort through its retinue of evils and benefits, guided by your moral wisdom to quash the one and foster the other.

The house couldn't agree on how hard it would be to master methods of eugenic control. Some speakers thought that destination wasn't very distant. Surely very soon, Lederberg said, "we can expect to learn tricks of immeasurable advantage." He listed some of these tricks: manipulating chromosomes, switching genetic segments in and out, and maybe even commanding the sequences of nucleotides in our DNA. Hermann Muller predicted using nano-needles to tinker with genes. And certainly the husbandry of humans—breeding new babies by stud, so to speak, using artificial insemination and banks of high-quality sperm—was already possible. But others argued that the mechanisms of inheritance were still murky. Even to fix the traits of a good laying hen was difficult, Haldane countered, and they knew nothing at all about the genetics of desirable human characters like intelligence or strength. Two scientists thought that the improvement of humanity was best left to nature, which had already been working on the problem for a few million years. When Huxley mentioned that eugenicists were only planning a gradual improvement of the species, Peter Medawar shot back: "But you don't know how to do it!"

Medawar would still be justified in saying that today. The decades since the symposium have both clarified and complicated the status of our knowledge of inheritance and evolution. If we thought earlier that genetic material went only from one generation to the next, for instance, we are now aware of horizontal gene transfers, in which the material moves from one organism to another. A bacterium can build into itself a gene—for antibiotic resistance, say—purloined from another bacterium. In fact, 8 percent of the human genome consists of DNA acquired from retroviruses. The gene on chromosome 6 that helps construct a membrane between a placenta and a fetus was, in its original form, a retrovirus gene that stitched an envelope around the virus's contents.

Similarly, the theory that the gene is the sole mode of inheritance has ramified. Biologists are studying a suite of molecules in our bodies that affects how genes are expressed, regulating and modulating the effect they have on our physical forms. These molecules can be influenced by the environment—by diet or aging or chemicals or acute stressors. As a result, the expression of the underlying gene transforms as well. Its sequence of nucleotides is unaltered; only the dials controlling it are turned. Studies have uncovered rare cases in which a parent, having had the dials turned by one environmental factor or another, passed on these new settings to an offspring—a mode of inheritance that is not genetic but epigenetic. In an experiment, a batch of male mice was conditioned to associate electric shocks with the cherrylike odor of acetophenone. When these mice became fathers, their pups were removed from them. They never had the chance to enact for their children their fear of the smell of cherries. Yet the young mice turned out to be hypersensitive to acetophenone; they even had an enlarged set of frontal brain neurons, the sort their fathers developed after their traumas.

This experiment and others like it make for combustible debates in biology. We have proof that an organism's epigenetic molecules are vulnerable to its environment, but we haven't been able to form any specific causal relationship between environmental factors and epigenetic change. Evidence for epigenetic heritability is also still slight. It contradicts the long-held theory that every embryo undergoes a sort of epigenetic reprogramming, so that the markers around parental genetic material are erased and then rebuilt anew. Worse still, to the resistant ear, the principal tune of epigenetics can sound too Lamarckian for comfort, even though it is wholly different from the kind of inheritance of acquired characters that Lysenko once sought.

Epigenetics is only one of the wrinkles to have appeared in our understanding of Darwinian evolution since Haldane, along with Fisher and Wright, poured the foundations of the modern synthesis. Another is niche construction, by which an organism modifies its immediate environment, the way a beaver builds a dam. These modifications alter the selective pressures on the organism, often to its advantage. (Niche construction is a new version of an older idea; Darwin studied earthworms with keen interest and described how they reworked the soil.) A third wrinkle is phenotypic plasticity, the ability of a single species to create multiple kinds of physical phenotypes, so as to best thrive according to the environment. A smartweed plant grown in bright sunlight will have thick but narrow leaves, to maximize photosynthesis and to remain cool; a genetically identical plant grown in dimmer conditions will sprout thin, broad leaves to capture as much light as possible.

All these various mechanisms have prompted a small, vocal faction of scientists to call for an augmentation of the modern synthesis. Through its reign, modern synthesis gave too much importance to natural selection as the engine of evolution, these scientists argue; a new model, capable of fitting niche construction and epigenetics and several other engines, is required. The augmented model is known, by its champions, as the extended evolutionary synthesis.

It isn't to everyone's taste. In the larger opposing camp, biologists believe that the modern synthesis can already accommodate all these channels. More pertinently, these channels have not yet been proved to be even remotely as important to evolution as the forces acting on the gene: natural selection, mutation, genetic drift, and recombination. The modern synthesis is still effective, these biologists insist. It still explains nearly all of what we have found to be true in evolution.

But even amid all this doubt about how exactly an organism comes by its traits, scientists have gained the instruments to study individual genes and even to meddle clumsily with them. Prenatal tests tell parents if a fetus carries certain genetic defects, giving them the choice of an abortion if a baby is likely to be born impaired. The test can directly affect the genetic health of the population; in Iceland, so many women terminate their pregnancies after receiving positive fetal tests for Down syndrome that the disorder is fading out of the population. Whether that is a pattern to be sought is another matter altogether. Disability rights activists have argued that a life with Down syndrome is still a life worth living—particularly in the twenty-first century, when parents, doctors, and teachers have learned to make the world more inclusive of such children.

These questions of ethics must be debated seriously and soon; technology will not wait for them to be resolved before it takes its next steps. Already, CRISPR, an enzymatic tool, enables scientists to edit the genetic code—to add to it, to subtract from it, to adjust its sequences of nucleotides. University labs and companies have started to deploy CRISPR on human embryos, trying to modify their genomes, and in November 2018, a Chinese scientist claimed he had used CRISPR to create the world's first genetically modified babies. China declared his work illegal, and other scientists looking into his work showed how imprecise and dangerous it was. In practice, decades will pass before these techniques are foolproof enough to make safe, predictable changes to the genome; in concept, though, the distance from gene-editing tool to bespoke gene is the span of a single step. And while we wait, with hope or dread, for genetic couturiers to work on our DNA with fine precision, scientists are refitting the genomes of pigs so that we can harvest their organs for ourselves, and they are refitting the genomes of mosquitoes to eliminate malaria.

Although these advances augur all sorts of possibilities to improve human beings, rarely, if ever, are they described with the label "eugenics." That word is still discomfiting; it's still burdened with memories of tyranny, coercion, and racism from the middle of the twentieth century. But the word existed before the Nazis, and the types of genetic engineering being dreamed about now chime neatly with those early schemes for positive eugenics. To eradicate diseases, to hone the genomes of individuals, to screen prospective parents for defects they might hand to their children—all of these figured in Haldane's utopian visions for genetics. Ensuring that a new baby is born healthy is, in the most literal sense of the word, the aspiration of eugenics. And although these technologies will plan to work only at the level of individuals, they will collectively make for a transformed species. However scientists choose to dress it up, an irreversible march of positive eugenics is under way.

The accompanying problems are already evident. The tricks of gene editing will at first be expensive, available only to the wealthy—which will, in our already unequal societies, tear an even wider gap between the elites and the remaining billions. Our banks of genetic information will be besieged by concerns of privacy and control. The revelations of genetics, as they pour out, are sure to be manipulated and misrepresented by racists and nationalists. Long before the genetic natures of complicated traits like intelligence or physical fitness are resolved, whispers of eugenic methods to amplify these traits, or to sort people based on them, have begun to circulate. Every year from 2014 to 2018, for instance, a lecturer at University College London—Haldane's old employer—hosted an invite-only conference on intelligence, without informing the university's officials. Participants included white supremacists, academics funded by a notoriously racist foundation, and ideologues masquerading as scientists. Twenty-nine out of 75 talks obsessed over IQ differences between races and ethnicities, and two speakers raised the prospect of eugenic interventions in intelligence. Among bigots, the idea that their bigotry is justified by genetics has never died away, and it stands ready to be fertilized by their willful distortions and misjudgments of new research.

Of all the sciences, the study of how people differ from each other is the most instantly political. When we consider what human beings will be in the future, we implicitly address the nature of the society to come as well. Haldane always knew this, always recognized that genetics, ethics, and politics were manacled together. "In the long run," he wrote in 1938, "the application of biology to social problems must depend on the ideals of the community, and the possibility which its structure offers of realizing these ideals." Even in 1924, in *Daedalus*, he pressed humankind to search for a morality suited to its new powers.

And it is the duty of scientists themselves to think politically, Haldane believed. In his time, as the atomic age was dawning, this was not such a radical notion. The threat of swift destruction was all too present for scientists to ignore the implications of their work or to remain silent about the ideologies they thought would best protect the world. Since then, however, scientists have retreated from the public realm, even as their science itself has been continually politicized. They have hesitated to stake out positions associated with politics; the singular exception, perhaps, is the case of the planet's changing climate, but that still leaves many spheres—artificial intelligence, health care, genetics, weaponry, food policy, education—in which scientists' political opinions are only murmured out of range of the public ear.

Some of the reasons for this shift are clear. Scientists have cut themselves off from the humanities, so they have neglected to think enough about the knotted messes of social affairs. The fields of physics, chemistry, and biology are all so technical now that they compel their students to specialize early; it's impossible to think today of a classics major moving easily and quickly into a career in genetics, as Haldane did. Academic positions have become precarious, and research depends too much on funding from the government or from corporations. The science itself has grown too complex to communicate with ease to a lay audience. Moreover, being apolitical has come to be regarded as a corollary to being scientifically objective. In the past few years, as we've witnessed deliberate assaults on fact and truth and as we've realized the failures of the calm weight of scientific evidence to influence government policy, the need for scientists to find their voice has grown even more urgent.

HALDANE HAD A VOICE, and he used it—used it all the time, in fact, used it so much that it won him trouble, used it so much that it seemed he was put on Earth expressly for the purpose. He hadn't at first been any sort of believer in the need for scientists to communicate. But he came to see the value of conveying the facts to people. If the residents of London knew which type of air raid shelter was the safest, they could choose it; if voters knew what was happening in Spain during the civil war, they could force their government to act; if a woman knew what it meant for her to be heterozygous for hemophilia, she could decide if she wanted children.

Even the facts were secondary, however, to the faculties of processing them. The scientific mind—ah, that was the thing! He didn't just want to tell his readers *what* to think; he wanted to show them *how* to think. To assess a theory, to pit it against a new one, to seek out evidence, to weigh it, to be open to the unknown strangenesses of the world. These were tools, just like fire and the stone ax. Giving people these tools prepared them to make decisions about whom to vote for or which medical treatment to trust or what to make of race propaganda—and decades later, presumably, about whether to clone themselves or to genetically modify their babies. As citizens, these people became armed. Haldane's outspokenness not only made use of democracy, but helped to form and hold it.

Haldane ushered people into the deep, elegant mysteries of science. Before him, Darwin had broken the supposition that humans were separate from nature; he had shown that the secrets of human existence had to be sought not in theology but in biology. Haldane became a Darwinian preacher, leading his tribe through this new search. He set down the virtues—of patience, reason, experiment, and clarity. And he promised rewards: the slow revelations of nature, the joy of discovery, of unspooling the fine, invisible threads of life.

Once, during his research for the Admiralty, he found he could taste oxygen. At a pressure of 5 or 6 atmospheres, the gas grew thicker and more viscous, easier to swirl about the tongue. Sweetly acidic, he decided—it reminded him of "dilute ink with a little sugar in it." It was a trivial detail, but it pleased him greatly just to pin it down. One more truth of nature truffled out, one more unknown thing made known.

Acknowledgments

MY FIRST MEASURE OF THANKS must go to the Mitchison family: to Terence, Clare, and Tabitha, for permitting me to draw on the Haldane papers. They received an email out of the blue one day, but they were most gracious toward this writer whom they didn't know at all. I hope this book captures their granduncle in all his complexity and all his brilliance.

This project started during a two-month fellowship at the Center for the Advanced Study of India at the University of Pennsylvania, and I'm grateful to Devesh Kapur for inviting me there and being such a wise adviser and to Juliana Di Giustini for providing a warm, happy place to work. The project reached completion during another fellowship: at the Leon Levy Center for Biography, at the City University of New York. For a year, I relied on the guidance of Kai Bird (endnotes, Kai!) and Thad Ziolkowski, as well as my fellow fellows: Jennifer Homans, Rebecca Donner, and Stephen Heyman.

Through the course of my research, I drew on the resources of some incomparable archives and archivists. (A full list of archives is available at the beginning of the bibliography.) In particular, I'd like to thank Alison Metcalfe at the National Library of Scotland, Eleanor Hoare at Eton, Sarah Wilmot at the John Innes Centre, Rose Lock at The Keep in Brighton, Lynette Cawthra at the Working Class Movement Library, Kishor Satpathy at the Indian Statistical Institute, and V. V. Prasad at the Centre for Cellular & Molecular Biology. Their patience with my repeated queries has been remarkable, and I'm so grateful for their work—and for the work of all archivists everywhere.

The book grew through a bunch of rich conversations: with Ramachandra Guha, Sunil Khilnani, Katherine Boo, Keshav Desiraju, Srinath Raghavan, Rasika Kalamegham, Projit Mukharji, Pankaj

Mishra, Nils Roll-Hansen, Mark Adams, Manasi Subramaniam, Ram Ramaswamy, Ornit Shani, and Jairam Ramesh. I was touched by how they thought of my Haldane project during the course of their own work: to send me snippets they found in the archives, as Sunil and Keshav did; or to plunder other books in search of Haldane, as Jairam did; or to dispatch a special copy of the *Journal of Genetics*, as Ram did. With unbelievable resourcefulness, Simran Kapur found me books and journal articles when no one else could.

During my travels for research, I abused the hospitality of several friends: Vinod Ganesh and Soumya Subramanian; Sujatha Krishnan-Barman and Shayak Barman; and J. Krishnamurthi. I hope this book is worth the pain I inflicted on them.

A number of scientists—some of whom I still haven't met!—were incredibly generous with their time and attention, agreeing to read the manuscript, to recommend corrections, and to offer advice. This book would be riddled with flaws and faults, I'm sure, if they hadn't given me such detailed and painstaking feedback, and I'm hugely indebted to them: Shyamal Lakshminarayanan; Amitabh Joshi; Mukund Thattai; Massimo Pigliucci; Joe Felsenstein, especially for his primers on the cost of natural selection; Rasika Kalamegham, for her bracing and incisive critiques; Vijay Ramesh; and Saumya Ramanathan. Profound thanks to William deJong-Lambert for the long afternoons spent discussing bygone biologist, and for the dozens and dozens of emails exchanged about the history of genetics.

I owe an equal debt of gratitude to others who read the manuscript, in part or in full, and offered their feedback: Ravi Mundoli, who was blunt enough to tell me that he disliked an early version of the first section; Varun Rajiv, who courageously participated in reams of WhatsApp chat about Haldane; and Padmaparna Ghosh, who probed and questioned the material in fascinating ways, forcing me to think carefully through my conclusions and preconceptions.

Many thanks to John Glusman and Helen Thomaides at W. W. Norton in New York; to James Nightingale at Atlantic Books in London; and to Himanjali Sankar and Rahul Srivastava at Simon & Schuster in New Delhi.

I felt like I had a small posse of my own in my corner, propping me up throughout this endeavor: Shruti Debi at The Debi Agency and Anna Stein at ICM, my formidable agents, who championed this book and my idea for it; Svetlana Chervonnaya, briefly my researcher in Russia; and Sarah Ruth Jacobs, my researcher in New York, whose fierce energies and efficiency improved this book beyond measure.

And, as ever, my most indispensable team: Vidya, K. R. S., Harini, and Padma. Nothing would ever, ever get done without you.

Notes

Chapter 1: The Scientific Method

1   **A retired chemist in Surrey:** Haldane to H. E. Holtorp, HAL-DANE/5/2/2/130, University College London archive.

2   **"large woolly rhinoceros":** The journalist Laurence Thompson, writing in the *Evening Standard* in February 1953, as part of a series titled "Britain's Spotlight Scientists."

3   **"It ought not to be published":** J. B. S. Haldane to Hans Kalmus, May 1953, HALDANE/5/2/2/169, University College London archive.

4   **"the most brilliant scientific popularizer":** Arthur C. Clarke started his essay "Haldane and Space" in this way. The essay appears in the compendium *Haldane and Modern Biology*, edited by K. R. Dronamraju.

4   **"Start from a known fact":** See Haldane's essay "How to Write a Popular Scientific Article," first published in the *Journal of the Association of Scientific Workers* and subsequently collected in *A Banned Broadcast and Other Essays*.

5   **"the last man who might know all there was to be known":** Who actually said this? Ronald Clark, whose biography of Haldane is commonly cited for this quote, does not identify the speaker, who is merely described as "a younger colleague." The philosopher A. J. Ayer is a candidate; he was at University College as well, and once in a speech talked about how the complexity of science was growing beyond the reach of those who stood outside it. The last man to grasp every scientific discipline, he said, was Haldane. But it isn't clear if the phrase "the last man who might know all there was to be known" occurred in Ayer's speech.

6   **Vavilov was born to a bourgeois family:** A fine account of Vavilov's life is found in *The Murder of Nikolai Vavilov*, by Peter Pringle.

7   **In Moscow and Leningrad:** Much of this account of the Haldanes' time in the Soviet Union is drawn from Charlotte's memories of the trip, as laid out in her memoir *Truth Will Out*.

7   **parquet floors and marble mantelpieces:** See the essay "Vavilov," in Haldane's *Science Advances.*

8   **"One must spare a great scientist":** Lenin said this to A. V. Lunacharskii, quoted in *Revoliutsiia i Kultura*, No. 1, 1927. I found the translation and source in Bailes, *Technology and Society under Lenin and Stalin.*

9   **"My practice as a scientist is atheistic":** Haldane, *Fact and Faith.*

10   **"When applied science has created":** From a Conway Memorial Lecture titled "Science and Ethics," delivered at Essex Hall in London in 1928.

10   **"Diamat is a tank":** Milosz wrote this to a friend in 1951. My source for this was Andrzej Franaszek, *Milosz: A Biography* (Cambridge, MA: Harvard University Press, 2017), 310.

11   **40 pounds of bread:** Ipatieff, *The Life of a Chemist.*

11   **over one combative week in 1948:** In Moscow, the Foreign Languages Publishing House put out an English translation of the proceedings of that fateful week, titled *The Situation in Biological Science: Proceedings of the Lenin Academy of Agricultural Sciences of the U.S.S.R., July 31–August 7, 1948.* All dialogue and description of the sessions are drawn from that volume.

11   **"Stingy of words":** Quoted in Medvedev, *The Rise and Fall of T. D. Lysenko*, 11.

12   **"He is the peasants' demagogue":** Ashby, *Scientist in Russia*, 115.

13   **The son of a farmer:** The following account of Lysenko's life and career is indebted to a number of excellent sources: Medvedev's *The Rise and Fall of T. D. Lysenko*; Roll-Hansen, *The Lysenko Effect: The Politics of Science*; Ashby, *Scientist in Russia*; Joravsky, *The Lysenko Affair*; and my friend William deJong-Lambert's numerous writings on the subject, including the two volumes he edited with Nikolai Krementsov, titled *The Lysenko Controversy as a Global Phenomenon* (Cham, Switzerland: Palgrave Macmillan, 2017).

14   **43,000 hectares to be planted:** Roll-Hansen, *The Lysenko Effect.*

16   **"to saying that dogs give birth to foxes when raised in the woods":** Joravsky, *The Lysenko Affair.*

16   **At least 22 geneticists:** Estimate compiled from Joravsky, *The Lysenko Affair*, and Diane B. Paul's "A War on Two Fronts: J. B. S. Haldane and the Response to Lysenkoism in Britain," *Journal of the History of Biology* 16, no. 1 (Spring 1983): 1–37.

16 **"We shall go to the pyre":** From a speech made by Vavilov in May 1939, as quoted by Medvedev, *The Rise and Fall of T. D. Lysenko*, 58.

20 **"Merrily play on, accordion":** Quoted in translation by Medvedev, *The Rise and Fall of T. D. Lysenko*.

21 **the BBC broadcast a symposium:** Transcripts of the four speeches were published in *The Listener*, December 9, 1948.

25 **"I remember you saying":** Mitchison to Haldane, HAL-DANE/5/1/2/8/40, University College London archive.

26 **"the cleverest man I ever knew":** From Medawar's preface to Ronald Clark, *J. B. S.: The Life and Work of J. B. S. Haldane*, 3.

Chapter 2: *The Deep End*

27 **the outfit weighed 155 pounds:** Details of diving attire and excerpts from a log of J. S.'s diving experiments off Scotland can be found in Leonard Hill, *Caisson Sickness and the Physiology of Work in Compressed Air* (London: Arnold, 1912).

27 **In the diving logs:** From A. E. Boycott, G. C. C. Damant, and J. S. Haldane, "The Prevention of Compressed-Air Illness," *Journal of Hygiene (London)* 8, no. 3 (June 1908): 342–443. I also consulted "A Diary of the Deep Diving Experiments Carried Out off Rothesay, Isle of Bute, from H. M. S. Spanker, August, 1906," published as Appendix II in *Collected Papers of the Lister Institute of Preventive Medicine*, Vol. 5. Additionally, I'm grateful to Martin Goodman's lively biography *Suffer and Survive*.

27 **spent the days outdoors:** Details of Jack and Naomi's time in Scotland are drawn from Mitchison's memoir *As It Was* and from one of Jack's letters to his mother, part of the National Library of Scotland's collection of J. B. S. Haldane's papers, in a box numbered Acc. 4549.

29 **At the Lister Institute:** Details of these experiments are in *The Prevention of Compressed-Air Illness* and in Appendix I of the *Collected Papers of the Lister Institute of Preventive Medicine*, Vol. 5: "Details of the Experiments Made on Lieutenant Damant and Mr. A. Y. Catto, Gunner, R. N., in the Pressure Chamber at the Lister Institute." Also see Sir Robert Davis, *Deep Diving and Submarine Operations* (London: Siebe, Gorman & Co., 1909).

30 **By night, he slept:** Mitchison, *As It Was*. The tone of envy at Jack's access to these adventures is almost painful to read at times.

31 **leading from a 50-pound weight:** J. B. S. Haldane to Louisa Haldane, part of Acc. 4549, National Library of Scotland.

31 **dosed Jack with whisky:** Louisa Haldane, *Friends and Kindred*.

32 **he wrote in a pamphlet:** See *A Letter to Edinburgh Professors*, part of J. S.'s papers at the National Library of Scotland, Edinburgh, as part of Acc. 4549.

32 **In Dundee, in eastern Scotland:** See Thomas Carnelly, J. S. Haldane, and A. M. Anderson, "The Carbonic Acid, Organic Matter, and Micro-organisms in Air, more especially of Dwellings and Schools," *Proceedings of the Royal Society of London* (1886): 61–111.

33 **troubled by richly offensive smells:** See Thomas Carnelly and J. S. Haldane, "The Air of Sewers," *Proceedings of the Royal Society of London* (1887): 394–96.

34 **Many years after he left Dundee:** See Haldane's essay "Some Adventures of a Physiologist," in *Keeping Cool and Other Essays*.

34 **Through the Middle Ages:** A detailed history of the Haldane clan can be found in "Memoranda Relating to the Family of Haldane of Gleneagles," available as a digitized document with the National Library of Scotland. Haldane himself also wrote briefly of the history of his family in an essay titled "Some Reflections on Non-Violence," in *On Being the Right Size*.

34 **"He was very devout":** Richard Haldane, *Richard Burdon Haldane: An Autobiography*.

34 **there was church every Sunday:** Goodman, *Suffer and Survive*.

35 **J. S. followed Richard to the Edinburgh Academy:** I am immensely grateful to Steve Sturdy, at the University of Edinburgh, who first wrote a clear and informative PhD thesis on J. S.'s life and work, back in 1987, and then shared a copy of the thesis with me three decades later. Sturdy's thesis, "A Co-ordinated Whole," is a marvelous analysis of the philosophy whirring through J. S.'s life and science.

35 **But life was no mere machine:** J. S. Haldane, *The Philosophy of a Biologist*.

36 **he wrote anonymous letters:** J. S. wrote to his friend James Thomas Wilson about these letters. Quoted in Morison, Patricia, *J. T. Wilson and the Fraternity of Duckmaloi*, 34.

36 **He developed a system:** J. S. Haldane, *Methods of Air Analysis*.

37 **He had a wooden box built:** See J. S. Haldane and J. Lorrain Smith,

"The Physiological Effects of Air Vitiated by Respiration," *Journal of Pathology and Bacteriology* 1, no. 2 (October 1892): 168–86.

38   **On an afternoon late in 1883:** L. Haldane, *Friends and Kindred.*

41   **"The Haldane family was":** L. Haldane, *Friends and Kindred*, 195.

41   **When he was 10:** Mitchison, *As It Was.*

42   **"Will anyone have any of this":** Mitchison, *As It Was*, 55.

43   **When Louisa first came to Oxford:** Details of the two residences are drawn from the memoirs of Louisa Haldane and Mitchison.

45   **"I'm the overland mail":** L. Haldane, *Friends and Kindred*, 215.

46   **"I got some fairly good specimens":** From Haldane's diary, as quoted by Mitchison, *As It Was*, 27.

46   **a special notebook:** The notebook is in the collections of the National Library of Scotland, numbered Acc. 10306/10.

47   **In 1893, nearly 30½ million working days:** Henry Pelling, *A History of British Trade Unionism*, 5th ed. (New York: Palgrave Macmillan, 1992), 324.

48   **In the tunnels, he collected samples of air:** See J. S.'s report to the government, "Causes of Death in Colliery Explosions and Underground Fires" (1896), and a book he edited, *The Investigation of Mine Air*, published by Charles Griffin & Company (London) in 1905. Goodman's biography of J. S. also discusses his work in mines in fine detail.

49   **Reeling home from the university:** Goodman, *Suffer and Survive.*

50   **Once, on a new vessel:** L. Haldane, *Friends and Kindred.*

51   **"My dear Boydie":** J. S. Haldane to J. B. S. Haldane, part of Acc. 4549, National Library of Scotland.

52   **"You come in here":** L. Haldane, *Friends and Kindred*, 217

54   **"Well, that would simplify matters":** L. Haldane, *Friends and Kindred*, 176.

54   **"Friends, Romans, countrymen":** See Haldane's essay "Some Adventures of a Physiologist," in *Keeping Cool and Other Essays.*

57   **At the age of 11:** Young Jack's poem is part of Acc. 4549 at the National Library of Scotland.

59   **Mendel's paper remained unrecognized:** The rediscovery of Mendel in 1900 is often hailed as a triumph of early twentieth-century science, but it was also used to pour unfair scorn upon the scientists of the previous decades. "The publication of Mendel's paper was the throwing of pearls before swine," Darbishire said in 1911.

59   **Mendel certainly read:** The debate over whether Darwin owned a
      copy of Mendel's paper is a hot and involved one, but on balance, we
      have no conclusive evidence that he did. I'm grateful to Joe Felsen-
      stein for our email conversations about this.

59   **terms from these paragraphs:** *Lebensbedingungen*, for instance, the
      translation of Darwin's "conditions of life," or *Elemente*, for the mor-
      sels of heredity that Darwin used to refer to the reproductive mate-
      rial of the sex cells.

61   **"an intelligent, designing author":** Although Darwin found him-
      self unable to believe in an intelligent author, he thought highly of
      Paley. "I hardly ever admired a book more than Paley's *Natural The-
      ology*: I could almost formally have said it by heart," he wrote to a
      friend in 1859.

62   **"all corporeal and mental endowments":** From *Origin of Species*,
      4th ed. (London: John Murray, 1866), 577.

62   **"a little dose of reason and judgement":** Samuel Butler, *Life
      and Habit*.

63   **"while its friends are solicitous":** Eberhard Dennert, *At the Death-
      bed of Darwinism: A Series of Papers* (Burlington, IA: German Liter-
      ary Board, 1904), 28.

63   **"I confess to not attaching":** From a speech the Skipper delivered to
      parents in 1901, as quoted in *The Skipper: A Memoir of C. C. Lynam*,
      published by the Dragon School in 1940, 41.

63   **"the falseness of all the gods":** From a 1908 speech to the Associa-
      tion of Preparatory Schools, as quoted in *A Dragon Centenary, 1877–
      1977*, by C. H. Jacques, 84.

64   **"He must remember":** Jack's reports from the Dragon School are
      held in the collections of the National Library of Scotland.

66   **in an autobiography that remained:** Haldane's unpublished, half-
      finished memoir is titled *Why I Am a Cooperator*. A typewritten
      manuscript of this memoir is in the archives of University College,
      London, numbered HALDANE/1/2/63.

66   **"I was ill":** The diary entry isn't a contemporaneous one; rather,
      it was written in 1910, looking back on Eton. The diary is at the
      National Library of Scotland, in Acc. 10306.

66   **One of his teachers:** Jack's report cards and parent-teacher cor-
      respondence from Eton reside in the school's meticulously main-
      tained archive.

67   **The year Jack joined Eton:** Simon Ball, *The Guardsmen: Harold*

Macmillan, Three Friends, and the World They Made. The headmaster in question was Edward Lyttelton.

67 **"Where there was much disparity":** Haldane, *Why I Am a Cooperator.*

67 **"rampant and horrible":** Huxley, *Memories*, 45.

70 **"They are sometimes so thick":** J. B. S. Haldane to Mary Haldane, 1906, part of Acc. 4549, National Library of Scotland.

70 **Jack liked his science teachers:** Vivid portraits of these teachers can be found in J. L. Heilbron, *H. G. J. Moseley: The Life and Letters of an English Physicist, 1887–1915* (Berkeley: University of California Press, 1974). Moseley and Haldane had several teachers in common.

72 **"So sure his touch, that fame affirms":** Quoted in C. A. Alington, *Eton Faces Old and Young* (London: John Murray, 1933), 77.

72 **"It never crossed Porter's mind":** The physicist Thomas Ralph Merton, as quoted in Heilbron, *H. G. J. Moseley: The Life and Letters of an English Physicist, 1887–1915*, 25.

73 **deemed it irrational:** This worried Louisa so much that she wrote to his headmaster, asking him to pull Jack back onto track. He would do what he could, the headmaster promised. To a boy who'd been brought up in a household resounding with religious faith, "I can say England is a country called by God to fulfill a certain plain vocation," he wrote back. But without this anchor of Christianity, he felt hard-pressed to "give any reason whatsoever to a thoughtful boy why he should bother himself about his country. . . . He hears that the good of the country is to be the subject of his aspirations, but suppose he replies that his notion of good is different from mine, where are we?"

73 **wrote their arguments into a ledger:** The ledger is a part of the Haldane collection at the Eton archive.

76 **scholastic sausage machine:** The Old Boy is the composer Philip Heseltine. See Tim Rayborn, *A New English Music: Composers and Folk Traditions in England's Musical Renaissance from the Late 19th to the Mid-20th Century* (Jefferson, NC: McFarland, 2016).

80 **The results were published:** The paper was titled "Reduplication in Mice."

81 **"a great shaggy bear":** Gervas Huxley, *Both Hands* (London: Chatto & Windus, 1970), 69.

82 **One night in June 1914:** See Haldane, *Why I Am a Cooperator.*

82 **"E.g. today":** J. B. S. Haldane to J. S. Haldane, part of MS 20655, National Library of Scotland.

83   **"To judge from letters"**: J. B. S. Haldane to Louisa Haldane, part
     of MS 20655, National Library of Scotland.

84   **"too far into some parts of me"**: Haldane to Mitchison, Imperial
     War Museum, London.

85   *pup-pup-pup . . . whiwww*: Haldane described the sounds in this
     way in letters to his sister and his parents.

85   **A sergeant in the 1st Battalion**: From Victoria Schofield's terrific
     and comprehensive *The Black Watch: Fighting in the Front Line, 1899–
     2006* (London: Head of Zeus, 2017).

86   **"Somehow high explosives seem"**: Haldane to Gilbert Mitchison,
     part of MS 20655, National Library of Scotland. Mitchison later
     married Naomi.

86   **one of the happiest of his life:** Through these months, Haldane
     described the grand time he was having again and again in let-
     ters home. I tried first to search these letters for a sign that he was
     merely allaying the anxieties of his family. But Haldane was not the
     man to think of such deceptive kindness; in the moment, he genu-
     inely seemed to be enjoying himself.

87   **"would walk along a knife edge"**: L. Haldane, *Friends and Kindred*,
     236.

87   **"It is possible that"**: Haldane to Mitchison, Imperial War Museum.

88   **he wrote short poems:** Part of the National Library of Scotland's
     collection, in a notebook numbered MS 20579.

90   **"It reached the parapet"**: From the diary of A. T. Hunter, as quoted
     in *Canada in the Great World War* (Ann Arbor: University of Michi
     gan Library, 1918), 70.

90   **"What we saw was total death"**: Siebert's account is part of the col-
     lection of In Flanders Field Museum in Ypres, Belgium.

90   **Four days after the first attack:** Goodman, *Suffer and Sur-
     vive.* J. S. wrote an account of his time as well, located in the
     National Library of Scotland collection, as part of Acc. 9589.

91   **"Piss on your handkerchiefs"**: Quoted in Tim Cook, *No Place to
     Run*, 25.

91   **Jack was reunited with his father:** See Haldane, *Why I Am a Coop-
     erator*, and Haldane's essay *Callinicus*.

91   **The uproar engulfed him:** Haldane wrote about the injury in detail
     in an unpublished set of notes, now held at the National Library
     of Scotland as part of Acc. 10306. He captioned the notes "On no
     account to be copied or printed during the war."

94   **The Black Watch was helping:** See Victoria Schofield's *The Black Watch*; an anonymously written account called *With a Highland Regiment in Mesopotamia* (London: The Times Press, 1918); and Major-General A. G. Wauchope, *A History of the Black Watch*.

95   **"B is the biscuit":** See Haldane, *Why I Am a Cooperator*.

98   **"young officer who used to open":** See Haldane, *Callinicus*, 71.

99   **Duke of Plaza Toro:** See Haldane's essay "Meroz" in *Possible Worlds*.

99   **"so many fellows together":** Quoted by Jessica Meyer, *Men of War*, 146.

99   **"The soldier is working":** See Haldane, *Why I Am a Cooperator*.

100   **"Jack had been killing people":** From an untitled manuscript by Mitchison, held at the National Library of Scotland.

101   **"He disclaimed it so fiercely":** From "Beginnings," an essay by Mitchison published as part of *Haldane and Modern Biology*, an anthology edited by K. R. Dronamraju, 304.

101   **"It has always seemed to me":** Haldane to Graves, part of MS 20542, 1961, National Library of Scotland.

Chapter 3: Synthesis

102   **What happened when the blood acidified:** See H. W. Davies, J. B. S. Haldane, and E. L. Kennaway, "Experiments on the Regulation of the Blood's Alkalinity: I," *Journal of Physiology* 54, no. 1–2 (August 1920): 32–45; and "Experiments on the Regulation of the Blood's Alkalinity: II," *Journal of Physiology* 55, no. 3–4 (August 1921): 265–75.

104   **"Unfortunately, one could hardly try":** See Haldane's essay "On Being One's Own Rabbit," in *Possible Worlds*, 115.

105   **"He dropped in whenever he liked":** Huxley, *Memories*, 137.

105   **"How big do you think my liver is":** James Luck, *Reminiscences*, 23.

106   **newly knighted colleague:** Luck tells this story in his book. The distressed colleague was Haldane's supervisor, Frederick Gowland Hopkins.

107   **the most romantic figure of his day:** Haldane expresses this opinion in *Daedalus*.

108   **"If you're faced with a difficulty":** As quoted by John Maynard Smith in his obituary of Haldane, published in *Nature* (206: 239–40) in April 1965.

108   **"for five consecutive minutes":** See Haldane's preface to the revised

edition of his book *Enzymes* (Cambridge, MA: The MIT Press, 1965), vi.

109 **"all that algebra that Jack":** Quoted in Sarkar, *The Founders of Evolutionary Genetics.*

111 **Four years earlier, Ronald Fisher:** See "The Correlation between Relatives on the Supposition of Mendelian Inheritance," *Transactions of the Royal Society of Edinburgh* 206 (1918): 239–40.

112 **Haldane's first major paper:** Many of Haldane's most important papers have been helpfully collected by Dronamraju in *Selected Genetic Papers of J. B. S. Haldane.*

114 **"The paler ones the birds eat":** James William Tutt, *British Moths*, (London: Routledge, 1896), 307.

116 **an Oxford researcher named Bernard Kettlewell:** Biologists know all too well the subsequent discoveries of some of the flaws of Kettlewell's experiments and the kind of ammunition it gave creationists. Kettlewell planted his moths on the trunks of trees, for example, where they really ought to have been higher up in the canopy, set on small branches, as is their natural tendency. Between 2001 and 2007, though, the geneticist Michael Majerus ran more careful experiments in his garden in Cambridge and confirmed the differential predation of light and dark moths by birds—the spine of Kettlewell's work. See L. M. Cook, B. S. Grant, I. J. Saccheri, and J. Mallet, "Selective Bird Predation on the Peppered Moth: The Last Experiment of Michael Majerus," *Biology Letters* 8, no. 4 (February 2012): 609–12.

118 **"Pangloss' theorem":** John Maynard Smith, "J. B. S. Haldane," in Sarkar's *The Founders of Evolutionary Genetics*, 37–51.

119 **Maynard Smith once remarked:** Quoted in Kohn, *A Reason for Everything*, 19.

121 **"which I was never brought enough to fully digest":** Haldane to R. R. Race, of the Lister Institute, 1947, HALDANE/5/2/4/50, University College London archive.

121 **"we were always careful":** Recounted by James Crow in a contribution to *J. B. S. Haldane: A Tribute*, published by the Indian Statistical Institute in 1992.

122 **"For about seven years":** See Cyril Darlington's *Family History*, a 1946 account that is part of the Darlington Papers at the Bodleian Library, Oxford.

122 **"have a look round each day":** From Sarah Wilmot, "J. B. S. Hal-

dane: The John Innes Years," *Journal of Genetics* 96, no. 5 (November 2017): 815–26.

122   **"John Boredom Wanderson"**: The paper is part of the Haldane collection at the John Innes Centre in Norwich. The mock-exam and the note to his secretary are also part of that collection.

124   **"We have very strong evidence"**: See Haldane's "Some Recent Work on Heredity," published in *Transactions of the Oxford University Junior Scientific Club*, Series 3, No. 1.

125   **"The enzyme is a product of the gene"**: The unpublished "A Biochemical View of Some Genetical Problems" is part of the Haldane papers at the National Library of Scotland, numbered MS 20580. He originally titled his article "What Genes Do"—which was somehow lovelier and more direct.

125   **"suitably submissive and even adoring females"**: From Darlington's diary, F.10, part of the Darlington Papers at Oxford.

126   **"It is mere rubbish thinking"**: Darwin to J. D. Hooker, March 1863. The letter is part of the Darwin Correspondence Project, which hosts all his letters online.

126   **"some warm little pond"**: Darwin to Hooker, February 1871.

127   **"I have very little doubt"**: Haldane's speech, delivered at a conference in Florida in 1963, was published in *The Origins of Prebiological Systems and of Their Molecular Matrices*, 98.

128   **"If God did not do it this way"**: Quoted in Stanley Miller's obituary in the *Los Angeles Times*, May 24, 2007. The anecdote also appears in *Planet of Microbes: The Perils and Potential of Earth's Essential Life Forms*, by Ted Anton, published by the University of Chicago Press in 2017.

129   **a spiral of nucleic acids:** Haldane used the word *spiral* in an essay in an anthology volume called *The Planet Earth* (London: Pergamon Press, 1957)—four years after Crick and Watson discovered the double-helix structure of the DNA molecule. He was being inexact; he ought to have used the word *helical* to be perfectly accurate.

129   **"Whenever you can, count"**: The literature often refers to this as Galton's motto, but it should be said that we can ascribe that fact only to Karl Pearson, who mentioned it in his book *The Life, Letters and Labours of Francis Galton*, Vol. 2 (Cambridge: Cambridge University Press, 1924), 340. Pearson was a protégé of Galton, so presumably he heard Galton say this often.

130 **"As the ladies turned themselves about":** Francis Galton to his brother Darwin Galton, February 1851.

130 **"Could not the race of men":** Galton aired this question in a paper read to the Eugenics Education Society in June 1908. Quoted in Karl Pearson, *The Life, Letters and Labours of Francis Galton,* Vol. 3a (Cambridge: Cambridge University Press, 1930), 348.

131 **"men of strong and beautiful bodies":** Paul Cohn, writing in the *New Age* in 1913.

131 **Fitter Families contests:** See Daniel Kevles's terrific book *In the Name of Eugenics: Genetics and the Uses of Human Heredity*, first published by Alfred A. Knopf in 1985. An account of these contests is also found in "Registering Human Pedigrees," by Arthur Capper, a Kansas senator, writing in *Popular Science Monthly* in August 1923.

132 **"Yea, I have a goodly heritage":** A photograph of the medal is in the collections of the American Philosophical Society and has been uploaded by the Cold Spring Harbor Laboratory.

132 **"remarkable intelligence and already has":** The article appeared in syndication in several publications. I found it in a newspaper called *The Day Book of Chicago*, in the edition dated October 29, 1913.

132 **During the Boer War:** One of the most remarkable documents of this time is the "Report of the Inter-Departmental Committee on Physical Deterioration," submitted to Parliament in 1904. It lays bare all the anxieties of the new eugenics craze.

133 **A prominent psychologist:** G. Stanley Hall, president of Clark University.

133 **"'Civilization's going to pieces'":** F. Scott Fitzgerald, *The Great Gatsby* (New York: Simon & Schuster, 2003), 12.

134 **"Far too many live":** From Friedrich Nietzsche, *Thus Spake Zarathustra* (1911).

135 **in an 1875 report:** Dugdale extended his report and published it as a book titled *The Jukes: A Study in Crime, Pauperism, Disease and Heredity* (New York: G. P. Putnam's Sons, 1877).

135 **a researcher from the Eugenics Record Office:** Estabrook's study was titled *The Jukes in 1915*.

136 **"lesser sacrifices . . .":** The Supreme Court decision, written by Oliver Wendell Holmes, is known as *Buck v. Bell.*

137 **"So poor devil":** From a diary entry, written in 1910. The diary is part of Acc. 10306/8 at the National Library of Scotland.

137 **"It is true that in England":** The speech was the Conway Memorial Lecture, published subsequently as *Science and Ethics.*

138 **another scientist's rough computation:** R. C. Punnett calculated this in 1917.

138 **"Many of the deeds":** From Haldane's essay "The Future of Biology," in *On Being the Right Size*, 17.

139 **"in persuading a certain number":** From Haldane's essay "Eugenics and Social Reform," in *Possible Worlds*, 193.

139 **human diversity was not only desirable:** See Haldane's essay "Human Evolution: Past and Future," in *Everything Has a History.*

139 **"The races of the past":** From Haldane's essay "What Is Race?" in *Keeping Cool and Other Essays*, 33.

140 **"the black races of Africa":** From Haldane's "What Is Race?" 35.

140 **"they huddle around fires":** From Haldane's "What Is Race?" 35.

140 **be taken for a quack:** Huxley, *Memories.*

143 **Haldane's mother sent Huxley a letter:** Huxley, *Memories*, 126.

146 **During the Second World War:** Susan L. Smith, "Mustard Gas and American Race-Based Human Experimentation in World War II," *Journal of Law, Medicine & Ethics* 36, no. 3 (Fall 2008): 517–21.

146 **Among those who wrote to Haldane:** Charlotte gives a full, vivid account of her first meeting with Haldane in *Truth Will Out.*

148 **"We must have a shot":** Charlotte's letters to Haldane are part of the National Library of Scotland collection, as part of Dep. 300. They're individually unnumbered.

148 **The family had lived in London:** Details of Charlotte's background are drawn from *Truth Will Out* and from Judith Adamson's biography *Charlotte Haldane: Woman Writer in a Man's World* (Basingstoke, UK: Palgrave Macmillan, 1998).

151 **"I tried very hard to be pleasant":** Mitchison, *You May Well Ask*, 77.

153 **"Apparently it is impossible":** Haynes to Haldane, February 1925, part of Dep. 300, National Library of Scotland.

154 **he granted a divorce:** Divorce Court File 7699, J 77/2161/7699, National Archives, Kew.

154 **"The old-fashioned notion":** Haynes to Haldane, 1925, part of Dep. 300, National Library of Scotland.

154 **"I am going to sleep":** Ivor Montagu, *The Youngest Son*, 234.

154 **"was never a man not to blow":** Howarth, *Cambridge between Two Wars*, 54.

154 **"I am not a prude"**: Darlington, "Recollections of J. B. S. Haldane," draft in the Darlington Papers, Bodleian Library.

155 **"It is only fair to you"**: Quoted in Howarth, *Cambridge between Two Wars*, 54.

156 **"I am not sure whether"**: From Haldane, *Why I Am a Cooperator*.

Chapter 4: Red Haldane

157 **"and while pondering vaguely"**: Wallace published an article titled "The Dawn of a Great Discovery" in the magazine *Black and White* in January 1903. Even in this, he was fated to be yoked to Darwin. The article's subtitle was "My Relations with Darwin in Reference to the Theory of Natural Selection."

158 **"for amusement"**: Darwin, Charles, *The Autobiography of Charles Darwin* (London: Collins, 1958), 120.

158 **"not be considered as comfortable asylums"**: Thomas Malthus, *An Essay on the Principle of Population* (London: J. Johnson, 1798), 37.

158 **"our population is more largely renewed"**: Alfred Russel Wallace, "Human Selection," *Fortnightly Review* 48 (September 1890).

159 **"progress is merely accidental"**: Marx to Engels, August 7, 1866.

160 **volunteered to act as a bouncer**: From Haldane, *Why I Am a Cooperator*.

161 **"I consider the present distribution"**: From Haldane's essay "What I Think About," in *The Inequality of Man*, 218–19.

161 **"I would trust Shakespeare"**: From Haldane's essay "Possibilities of Human Evolution," in *The Inequality of Man*, 91–92.

161 **"How, in a society based on hierarchies"**: Gary Werskey, *The Visible College*, 108.

162 **"for the stupid to inherit"**: Julian Huxley, *What Dare I Think?* (New Yotk: Harper & Brothers, 1931), 112.

162 **"learned to appreciate sides"**: Haldane to Dronamraju, 1962, 162 in Dronamraju's *Popularizing Science*.

162 **"of all British intellectuals"**: Quoted in Patricia Cockburn, *The Years of the Week* (London: Macdonald & Co., 1968), 39.

162 **"Any political movements which diminish"**: From Haldane's essay "Possibilities of Human Evolution," in *The Inequality of Man*, 90.

163 **"Less than a million years hence"**: From Haldane's essay "Man's Destiny," in *The Inequality of Man*, 144.

163 **"It is hard . . . for an educated, decent, intelligent son"**: John May-

nard Keynes, "A Short View of Russia," 1925, in *Essays in Persuasion* (London: Palgrave MacMillan, 2010), 258.

164 **Haldane learned that the students:** From a lecture titled "The Place of Science in Western Civilization," collected in *The Inequality of Man*.

164 **"a spirit of festival":** The journalist Eugene Lyons worked for United Press and covered the Shakhty trial. Details of the trial have been taken from Lyons's *Assignment in Utopia* (Piscataway, NJ: Transaction Publishers, 1938), 114. See also Elizabeth Wood, *Performing Justice: Agitation Trials in Early Soviet Russia*.

165 **"glorious accuser":** Aleksandr Solzhenitsyn, *The Gulag Archipelago, 1918-1956*, Vol. 1 (New York: Basic Books, 1997), 306.

165 **"a shop window display of books":** From Haldane's lecture "The Place of Science in Western Civilization," in *The Inequality of Man*, 134–35.

165 **"the sweet principle of fat":** As described by Carl Wilhelm Scheele, the discoverer of glycerine.

166 **"It is only in so far as we renounce":** Bertrand Russell, *The Scientific Outlook* (London: Allen and Unwin, 1931), 264.

166 **"Until I took to scientific":** From Haldane's essay "A Mathematician Looks at Science," in *The Inequality of Man*, 242.

166 **"standardized people".** From Haldane's essay "What I Think About," in *The Inequality of Man*, 217.

166 **"bold and constructive":** Quoted in C. W. Guillebaud, "Politics and the Undergraduate in Oxford and Cambridge," *The Cambridge Review* (January 26, 1934).

166 **After a debate in 1925:** Howarth, *Cambridge between Two Wars*.

167 **One student, who had briefly spent time:** The student was the mathematician David Haden-Guest. He would later be killed in Spain, having volunteered to fight the fascists in the Civil War.

168 **"The peasants were starving everywhere":** C. Haldane, *Truth Will Out*, 83.

169 **"except . . . the spiritual luxury of open quarrelling":** Mitchison, *You May Well Ask*, 191.

169 **"In its endeavour, science is Communism":** Bernal, *The Social Function of Science*, 415.

169 **"seemed almost to dominate":** Neal Wood, *Communism and British Intellectuals* (New York: Columbia University Press, 1959), 121.

170 **"My old life was broken to bits":** Bernal, diary, part of the Ber-

nal papers at Cambridge University Library, document numbered O.23.1.

170   **"in the entire British Empire only two"**: From "The Place of Science in Western Civilization," in *The Inequality of Man*, 127.

171   **Trawling through Marx and Engels**: Haldane, *The Marxist Philosophy and the Sciences* (London: Allen & Unwin, 1938).

171   **"I have only one possible serious criticism"**: Haldane to L. Kislova, August 1944, HALDANE/4/4/22, University College of London archive.

173   **"Classical Mendelian genetics was damned undialectical"**: John Maynard Smith, 2004, interviewed for Web of Stories, https://www.webofstories.com/play/john.maynard.smith/33;jsessionid=366C7B8A225B1FE31E223AC57FB85528.

174   **"Like anyone on his first acquaintance"**: Andrew Rothstein, "Vindicating Marxism," *Modern Quarterly* 3 (1939), 290.

175   **"I began to realize"**: Quoted in Clark, *J. B. S*, 72.

175   **a brief typewritten note**: The MI5 files are numbered KV-2-1832_1 through KV-2-1832_6 and are held in the National Archives. Each file holds dozens of pages of documents, so I will only indicate which of the six files I am quoting from. The first-ever note MI5 makes about Haldane is in KV-2-1832_6.

177   **"Professor J. B. S. HALDANE"**: MI5, KV-2-1832_3, National Archives.

177   **"A meeting was held"**: MI5, KV-2-1832_5, National Archives.

178   **"the policeman who arrests"**: George Orwell, *The Lion and the Unicorn: Socialism and the English Genius* (London: Secker & Warburg, 1941), 42.

178   **so many wild undergraduates**: "Books and Authors," *New York Times*, March 18, 1928.

178   **She wasn't happy in Cambridge**: C. Haldane, *Truth Will Out*.

178   **"In those days"**: Quoted in Adamson, *Charlotte Haldane: Woman Writer in a Man's World*, 60.

179   **"a dangerous and unpleasant person"**: John Haffenden, *William Empson: Volume 1: Among the Mandarins* (Oxford: Oxford University Press, 2009), 595.

179   **"was mainly a refusal to be bored"**: A draft typescript by Empson, as quoted in Haffenden, *William Empson*.

179   **light dalliances with drugs**: See Haldane's essays "Pain-killers," in

On Being the Right Size; "Stimulants," in *Science and Everyday Life*; and "A Substitute for Morphine," in *Science Advances*.

179 **drive 500 miles through the night:** Haldane to S. P. Bhatia, January 1959, part of MS 20536, National Library of Scotland.

180 **Once Martin Case, leaving the Red Cow:** Adamson, *Charlotte Haldane: Woman Writer in a Man's World*, 79.

180 **"I think . . . the public has a right to know":** See the preface to Haldane's *Possible Worlds*.

180 **"Question: Who is the most widely read biologist":** See Crow's essay "Haldane, Marxism, and Popular Science," published as the preface to Haldane, *What I Require from Life*.

180 **"to make science plain for you":** *Daily Worker*, December 1, 1937.

181 **"Compare the production of hot gas":** See Haldane's essay "How to Write a Popular Scientific Article," in *On Being the Right Size*, 156.

184 **"We reserve the word instinct":** See Haldane's essay "Instinct," in *Science Advances*, 55.

185 **"On one occasion":** Interview of Case by Clark, part of the Clark papers, Acc. 9589/176, National Library of Scotland.

186 **"for fifteen years":** See Haldane's essay "Pain-killers," in *On Being the Right Size*, 151.

186 **"There is no indubitable evidence":** See Haldane's essay "The Future of Biology," in *On Being the Right Size*, 17.

187 **Harland was an old friend:** Sydney Cross Harland, *Nine Lives: The Autobiography of a Yorkshire Scientist* (Raleigh, NC: Boson Books, 2001).

189 **"From now on and until further notice":** May Lamberton Becker, *The Saturday Review of Literature*, November 19, 1938.

189 **"Dear Professor Haldane":** Tunstall to Haldane, HALDANE/5/1/2/3/5, University College London archive.

190 **truck driver in Australia:** Haldane to H. Rawlinson, December 1953, HALDANE/5/1/4/197, University College London archive.

190 **"If you want some calculations to do":** Haldane to Muriel Finn, May 1946, HALDANE/5/1/3/19, University College London archive.

190 **Once, on a trip to Europe:** Unnumbered diary, part of HALDANE/6/5, University College London archive.

191 **this mysterious fog in Belgium:** "Belgium's Poison Fog Cases Likened to the 'Black Death,'" *New York Times*, December 6, 1930.

191   **nearly 200 people turned up:** *Left News*, December 1937, No. 20.

192   **"A strange feeling began to oppress me":** See Haldane's essay "Lord Birkinhead Improves His Mind," in *A Banned Broadcast*, 13.

193   **at Columbia, he was in such a hurry:** Clark, *J. B. S.*, 55.

193   **Ronald Fisher read a gloomy message:** "Major Darwin Predicts Civilization's Doom Unless Century Brings Wide Eugenic Reforms," *New York Times*, August 23, 1932.

193   **The ideal society, he said:** "Not a 'Perfect Man' in Haldane's Utopia," *New York Times*, August 29, 1932. Also see "In Classroom and On Campus," *New York Times*, September 4, 1932.

194   **"as full of bloody Communists as Cambridge":** Howarth, *Cambridge between Two Wars*, 189.

194   **Haldane's patience was nearing tatters:** See Sarah Wilmot, "J. B. S. Haldane: The John Innes years," *Journal of Genetics* 96, no. 5 (November 2017): 815–26. Haldane's complaints about how the institute was being run were also expressed in letters held at John Innes, in Norwich.

195   **"But I have you on toast":** B. Schafer, 1937, "Prof. H——'s Report to the Council," quoted in Wilmot.

195   **After doctors placed him:** Mitchison, *You May Well Ask*, 211–12.

195   **At breakfast, something in the newspaper:** Charlotte Haldane, "My Husband the Professor," BBC Third Programme, September 11, 1965, as quoted by Adamson, *Charlotte Haldane: Woman Writer in a Man's World*, 68.

196   **the license plate of Haldane's car:** MI5, KV-2-1832_5, National Archives.

196   **His sister speculated:** Naomi's opinion is expressed in her memoir *You May Well Ask*, and Ronnie's in Adamson's biography of his mother, *Charlotte Haldane: Woman Writer in a Man's World*. Adamson quotes Charlotte as having told her daughter-in-law that Haldane was impotent. At least in public, Charlotte made no comment about this. In a letter to *The Listener* in 1969, responding to an article that speculated about Haldane's sexual abilities, Charlotte wrote: "Only myself and Dr Helen Spurway Haldane know the facts about J.B.S.'s sex life and these we would hardly discuss in your columns."

197   **Naomi and her husband lived:** Mitchison, *You May Well Ask*.

197   **"My personal character is such":** Haldane to Lionel Elvin, February 1939, HALDANE/5/1/2/11/3, University College London archive.

198   **At what rate did new mutations:** Haldane, "The Rate of Sponta-

neous Mutation of a Human Gene," *Journal of Genetics* 31, no. 3 (October 1935): 317–26.

200 **"No greater compliment can be paid"**: See Haldane's essay "On Being Finite," in *Science and Life*, 201.

200 **one mutated nucleotide per 30 million**: Y. Xue et al., "Human Y Chromosome Base-Substitution Mutation Rate Measured by Direct Sequencing in a Deep-Rooting Pedigree," *Current Biology* 19, no. 17 (September 15, 2009): 1453–7. Other studies have come up with a range from one in 55 million to one in 100 million.

202 **experiments didn't bear out its calculations**: The American scientist Bruce Wallace conducted studies on fruit fly populations in the 1950s and published two major papers in the second half of that decade. In a speech in 1986, speaking about genetic load half a century after Haldane's paper, Wallace said that, in his studies, "genetic load calculations failed to predict events that actually transpired in the *Drosophila* populations."

202 **not a burden but an asset**: Wallace makes this point in his speech, titled "Fifty Years of Genetic Load," delivered as the Wilhelmine E. Key Invitational Lecture in 1986.

202 **at least two, on average**: Yann Lesecque, Peter D. Keightley, and Adam Eyre-Walker, "A Resolution of the Mutation Load Paradox in Humans," *Genetics* 191, no. 4 (August 2012): 1321–30. See also Peter D. Keightley, "Rates and Fitness Consequences of New Mutations in Humans," *Genetics* 190, no. 2 (February 2012): 295–304.

202 **"Why have we not died"**: A. S. Kondrashov, "Contamination of the Genome by Very Slightly Deleterious Mutations: Why Have We Not Died 100 Times Over?" *Journal of Theoretical Biology* 175, no. 4 (August 1995): 583–94.

203 **observations of the great tit**: A. Husby, M. E. Visser, and L. E. Kruuk, "Speeding Up Microevolution: The Effects of Increasing Temperature on Selection and Genetic Variance in a Wild Bird Population," *PLoS Biology* 9, no. 2 (February 2011). The observed changes in breeding cycles were confirmed by T. E. Reed et al., "Population Growth in a Wild Bird Is Buffered against Phenological Mismatch," *Science* 340, no. 6131 (April 2013): 488–91.

204 **there is evidence that evolution has accelerated**: J. Hawks et al., "Recent Acceleration of Human Adaptive Evolution," *Proceedings of the National Academy of Sciences of the United States of America* 104, no. 52 (December 2007): 20753–58.

204 **"hence, number is of the highest"**: Darwin, *Origin of Species*.

204   **the first measurement of genetic linkage:** Julia Bell and J. B. S.
      Haldane, "Linkage in Man," *Nature* 138 (1936): 759–60; Bell and
      Haldane, "The Linkage between the Genes for Colour-Blindness
      and Haemophilia in Man," *Proceedings of the Royal Society of London*
      123, no. 831 (July 1937): 119–50.

205   **By gaining a numerical grip:** Milo Keynes, A. W. F. Edwards,
      and Robert Peel, *A Century of Mendelism in Human Genetics* (Boca
      Raton, FL: CRC Press, 2004); P. P. Majumder, *Human Population
      Genetics* (New York: Springer, 2012).

205   **the first partial map:** J. B. S. Haldane, "A Provisional Map of a
      Human Chromosome," *Nature* 137 (1936): 398–400.

205   **a 1936 paper on linkage:** J. B. S. Haldane, "A Search for Incomplete
      Sex-Linkage in Man," *Annals of Eugenics*, 7 (1936): 28–57.

206   **when Haldane was visiting Glasgow:** Dronamraju, *Popularizing
      Science.*

206   **"a pious hope included in the final paragraph":** E. B. Robson and
      Kay E. Davies, "The Human Gene Map [and Discussion]," *Philo-
      sophical Transactions of the Royal Society of London* 319, no. 1194 (June
      1988): 229–37.

206   **Scientists found a swatch of DNA:** Gilliam, T. C., et al., "Local-
      ization of the Huntington's Disease Gene to a Small Segment of
      Chromosome 4 Flanked by D4S10 and the Telomere," *Cell* 50, no. 4
      (August 1987): 565–71; Lindsay A. Farrer, R. H. Myers, L. A. Cup-
      ples, and P. M. Conneally, "Considerations in Using Linkage Anal-
      ysis as a Presymptomatic Test for Huntington's Disease," *Journal of
      Medical Genetics* 25, no. 9 (September 1988): 577–88; The Hunting-
      ton's Disease Collaborative Research Group, "A Novel Gene Con-
      taining a Trinucleotide Repeat That Is Expanded and Unstable on
      Huntington's Disease Chromosomes," *Cell* 72, no. 6 (March 1993):
      971–83.

209   **600 political meetings in London:** Richard Baxell, *Unlikely War-
      riors: The British in the Spanish Civil War and the Struggle against
      Fascism.*

210   **"You silly little fool":** C. Haldane, *Truth Will Out*, 93.

211   **Don't go getting killed:** Interview of Bell by Clark, part of the
      Clark papers, Acc. 9589/176, National Library of Scotland.

211   **On the train to Alicante:** J. B. S. Haldane, *A. R. P.*

211   **"To a man like myself":** Haldane, *A. R. P.*, 183.

212   **like great black pears:** Roderick Stewart, *Bethune in Spain.* Also see

Stewart, *Phoenix: The Life of Norman Bethune* (Montreal: McGill-Queen's University Press, 2011).

212   **screened for malaria and syphilis:** J. B. S. Haldane, *Science in Peace and War.*

213   **"A Spanish comrade was brought in":** Haldane, *Science and Everyday Life*, 170.

213   **Between the bombing runs:** Details of Madrid during the war are drawn from Haldane's and Bethune's dispatches from Spain, published as "Listen In! This Is Station EAQ" by the Committee to Aid Spanish Democracy, 1937; Virginia Cowles, *Looking for Trouble*; Richard Rhodes, *Hell and Good Company*; Amanda Vaill, *Hotel Florida: Truth, Love, and Death in the Spanish Civil War* (New York: Farrar, Straus and Giroux, 2014); Ken Bradley, ed., *International Brigades in Spain: 1936–39* (London: Reed International Books, 1994); Helen Graham, *The Spanish Republic at War 1936–1939* (Cambridge: Cambridge University Press, 2002).

213   **One biologist was investigating:** J. B. S. Haldane, "Genetics in Madrid," *Nature* 139 (1937): 331.

214   **Haldane found a fume cabinet:** Victor Howard, *The MacKenzie-Papineau Battalion: The Canadian Contingent in the Spanish Civil War.* Also see "Prof. Haldane Ill after Gas Test in Spain," *Daily Mail*, January 1, 1937.

215   **"At least ten times as much gas":** Haldane, *A. R. P.*, 24.

216   **"Here perhaps is what they feel":** Radio broadcast by Haldane, January 4, 1937, relayed over Station UGT, Madrid.

216   **"It was impossible not to be affected":** Fred Copeman, *Reason in Revolt*, 105.

217   **Once, at the Gran Via:** Cowles, *Looking for Trouble.*

217   **"like college kids on an outing":** Quoted in Vaill, *Hotel Florida*, 157.

218   **"Air raids are not only wrong":** Haldane, *A. R. P.*, 11.

218   **Once, Haldane told Julia Bell:** Interview of Bell by Clark, part of the Clark papers, Acc. 9589/176, National Library of Scotland.

219   **Pollitt wrote of a Christmas feast:** William Rust, *Britons in Spain* (London: Lawrence and Wishart, 1939); Harry Pollitt, *Pollitt Visits Spain* (London: International Brigade Wounded and Dependant's Aid Fund, 1938); Baxell, *Unlikely Warriors*; Bill Alexander, *British Volunteers from Liberty* (London: Lawrence and Wishart, 1982). Frank West is quoted in *Unlikely Warriors*, 278.

219 **"half-truths and hesitations"**: Laurie Lee, *A Moment of War* (London: Penguin, 1991), 46.

219 **"The only choice"**: Eric Hobsbawm, "War of Ideas," *The Guardian*, February 17, 2007.

220 **"Professor HALDANE has been urged"**: MI5, KV-2-1832_5, National Archives.

220 **"In fact, compared with a limpet"**: See Haldane's essay "No Caterpillars by Request," in *Science Advances*, 66.

220 **"Dear Sir, I have to thank you for"**: Haldane to unknown, 1938, HALDANE/5/1/2/15/1, University College London archive.

221 **"I don't think I can help you much"**: Haldane's self-obituary, recorded for the BBC in February 1964, broadcast on December 1, 1964.

221 **"The whole of my career"**: Quoted in Ronald Clark, *The Life of Ernst Chain: Penicillin and Beyond* (London: Bloomsbury, 2011), 14.

221 **"Perhaps all my discoveries will be forgotten"**: See Haldane's essay "An Autobiography in Brief," in *What I Require from Life*, xxxii.

222 **"the children of the Osagi [Osage] tribe"**: Untitled essay on racism, mid-twentieth century, HALDANE/1/2/79, University College London archive.

222 **"Let loose your motion"**: Haldane to Richardson, 1934, MS/792/5, Royal Society archive.

222 **The soil of Friesland**: See Haldane's essay "Blubo," in *Science Advances*.

224 **"We couldn't get a hall big enough"**: Lord Ashby interviewed by Gary Werskey in *The Visible College*, 256.

224 **In March 1938, during a speech**: MI5, KV-2-1832_5, National Archives.

225 **"In order to get into the Biological sections"**: Haldane to Garnett, David, November 1939, HALDANE/5/1/1/96, University College London archive.

226 **Naomi visited him in Harpenden**: Naomi Mitchison, *Among You Taking Notes: The Wartime Diary of Naomi Mitchison*, 116.

227 **John Maynard Smith remembered**: Maynard Smith, 2004, interviewed for Web of Stories, https://www.youtube.com/watch?v=qf1lsDsC-J0.

227 **"suspicious of all kindness"**: Mitchison, *Among You Taking Notes*, 206.

228 **"One day they came in to the cottage":** Montagu, *The Youngest Son*, 236.

228 **Soon after they settled:** Hans Kalmus, *Odyssey of a Scientist* (London: Orion, 1988).

228 **When Charlotte was back in London:** C. Haldane, *Truth Will Out*.

229 **"Dear Mr. Provost":** Haldane to D. R. Pye, 1945, HALDANE/3/5/1/4/16, University College London archive.

Chapter 5: The War at Home

232 **"far too optimistic":** "Haldane Offers to be Bombed to Test Refuges," *New York Herald Tribune*, August 2, 1939.

233 **In Valencia, they had hollowed vaults:** Haldane, *A. R. P.*

234 **"troglodyte existence deep underground":** Sir John Anderson, in the House of Commons, Hansard record of December 21, 1938.

234 **"You Have Received Your Death Sentence":** *Daily Worker*, April 22, 1939.

234 **"His views about the possible future":** Mitchison, *Among You Taking Notes*, 87.

234 **He professed himself:** Haldane wrote for Mass Observation as Directive Respondent 1972. The records are available at The Keep, an archive in Brighton.

235 **a memorandum for Anderson's department:** Haldane, "Memorandum on Shelter Policy," HALDANE/4/2/4, University College London archive.

236 **"As however this has so far":** Haldane to E. P. Cawston, August 1940, HALDANE/4/3/2/13, University College London archive.

236 **A *Daily Worker* cartoon:** *Daily Worker*, April 22, 1939.

237 **"I don't like to steal":** "Haldane Unearths a Bradford Scandal," *ILP News*, November 1, 1940.

237 **A section on evacuating pets:** MI5, KV-2-1832_5, National Archives.

237 **A man wrote from Shanghai:** S. C. Gregory to Haldane, November 1938, HALDANE/5/1/2/3/8, University College London archive.

237 **A man wrote from Charlottenlund:** Stig Veibel to Haldane, August 1938, HALDANE/5/1/2/3/6, University College London archive.

237 **A man wrote from the Australian Association:** Lindsay Bryant, July 1941, HALDANE/5/6/1/8, University College London archive.

238   **nearly 100 East Enders trooped into:** Angus Calder, *The People's War: Britain, 1939–1945* (London: Panther Books, 1969), 193.

239   **"Political schemers sailing under":** Herbert Morrison, "We Have Won the First Round," speech on the BBC, published in *The Listener*, November 4, 1940.

239   **Perhaps he would be arrested:** Kalmus, *Odyssey of a Scientist*.

240   **"The answer is simple":** J. B. S. Haldane, "Is There a Russian Enigma?" *New Statesman*, September 28, 1939.

240   **"but by throwing Chamberlain":** J. B. S. Haldane, *Hands Off the Daily Worker*, pamphlet published by the *Daily Worker* (Manchester: Working Class Movement Library, 1940), 5.

241   **Haldane issued another pamphlet:** J. B. S. Haldane, Sean O'Casey, John Owen, and R. Page Arnot *The Case for the Daily Worker*, pamphlet published by the *Daily Worker* (Manchester: Working Class Movement Library, 1941).

241   **"IN VIEW OF YOUR STATEMENT":** Telegram from Haldane to Winston Churchill, June 1941, HALDANE/4/8/3/4, University College London archive.

242   **The first day of June:** This account of the sinking of the *Thetis* is compiled from: the report of the Tribunal of Inquiry into the loss of H. M. Submarine "Thetis," HALDANE/1/5/2/23, University College London archive; C. Warren and J. Benson., *Thetis: The Admiralty Regrets: The Disaster in Liverpool Bay*; J. B. S. Haldane, "Human Life and Death at High Pressures," *Nature* 148 (1941): 458–60; J. B. S. Haldane, "Rough Summary of Analysis of Escapes," notes, HALDANE/4/16/1/44, University College London archive; "Some Afterthoughts on a Submarine Disaster," report prepared for the Underwater Physiology Sub-Committee of the Royal Naval Personnel Research Committee, May 1956, National Archives.

245   **"As far as you could tell":** Haldane's deposition, HALDANE/1/5/2/23, University College London archive.

246   **"I reasoned that men":** "International Brigade Men Who Fought in Spain Helped in Experiments," *The Telegraph*, date unknown, clipping numbered HALDANE/1/5/2/22, University College London archive.

248   **"Nose clip very painful":** These scrawled notes are still part of MS 20568, National Library of Scotland.

249   **Haldane and Helen dramatized:** HALDANE/5/8/1/17, University College London archive.

250 **"like ordinary epileptic ones":** Haldane to Warren McCulloch, September 1947, HALDANE/5/2/3/80, University College London archive.

251 **A Labour Parliamentarian:** Denis Pritt, MP for Hammersmith North, in the House of Commons, Hansard record of January 28, 1941.

251 **"What, teacher, can that object be":** "Civil Defence," *New Statesman and Nation*, November 12, 1942.

251 **He drew up a will:** Haldane's wills are part of Acc. 9589/167, National Library of Scotland.

251 **"He is in a really dreadful state":** MI5, KV-2-1832_5, National Archives.

253 **"With every Communist":** MI5, KV-2-1832_2, National Archives.

253 **The existence of *Gruppa Iks*:** The declassified Venona intercepts are tagged HW 15/43 in the National Archives.

255 **Just a couple of sentences:** Ivor Montagu to Haldane, July 1942, HALDANE/4/4/20, University College London archive.

255 **Haldane sits at a table:** *Experiments in the Revival of Organisms*, Techfilm Studios, Moscow, now in the Prelinger Archives: https://archive.org/details/0226_Experiments_in_the_Revival_of_Organisms_20_36_46_00.

256 **"I may be a lousy speaker":** Haldane to Ivor Montagu, September 1942, HALDANE/4/21/6/7, University College London archive.

256 **British physicist Alan Nunn May:** Robin Jardine to Haldane, May 1946, HALDANE/4/20/1/36, University College London archive; also see J. B. S. Haldane, "The Case of Dr. Alan Nunn May," *Daily Worker*, May 18, 1946.

256 **"He seems to be going":** Mitchison, *Among You Taking Notes*, 188.

257 **his presence on government committees:** Gavan Tredoux claims, in *Comrade Haldane Is Too Busy to Go on Holiday*, that the Admiralty stopped using the services of Haldane and "his cell of communist assistants" after deciding that they posed a security risk. In this case, however, Haldane was pulled from his experiments at Siebe-Gorman for other reasons: a broader move to stop using civilian scientists, a need to expand research beyond personal experimentation, and concern regarding Haldane himself. (He was also, as ever, proving difficult to work with.) The minutes of the Admiralty meeting that dealt with the termination of Haldane's appointment made no

mention of security risks or his political affiliations. The relevant file is at the National Archives, with the record number ADM 178/313.

257 **"They don't pay me anything"**: "Haldane: Let Them Sack Me," *Daily Express*, March 16, 1948.

257 **"an avowed Communist"**: Waldron Smithers, MP for Orpington, in the House of Commons, Hansard record of April 26, 1948.

257 **"I . . . said that there was a risk"**: G. R. Mitchell responding to a government query, in MI5, KV-2-1832_2, National Archives.

257 **"D. At. En. rather suspect that Haldane's motives"**: MI5, KV-2-1832_1, National Archives.

258 **"I have delayed answering"**: Haldane to Harold Himsworth, July 1950, HALDANE/5/6/3/2/12, University College London archive.

258 **"no interest to anyone"**: MI5, KV-2-1832_1, National Archives.

260 **"which in many points differs"**: Trofim Lysenko to Haldane, December 1944, HALDANE/4/9/1/1, University College London archive.

260 **the other a lengthy letter**: Haldane to M. Teich, October 1948, HALDANE/5/1/2/8/26, University College London archive.

262 **"A sharp struggle of ideas"**: "Why Bourgeois Science Is Against the Work of Soviet Scientists," *Literaturnaya Gazeta*, October 18, 1947. Two English translations of the article, differing slightly from each other, are among the Haldane papers, numbered HAL-DANE/5/1/2/8/7 and HALDANE/5/1/2/8/8, University College London archive.

262 **"An argument I do remember"**: John Maynard Smith, "J. B. S. Haldane," in Sarkar's *The Founders of Evolutionary Genetics*, 50.

262 **"He was not over-concerned"**: Kalmus, *Odyssey of a Scientist*, 82.

263 **Soviet geneticists had been "liquidated"**: Huxley would use that word in his book *Soviet Genetics and World Science*, (London: Chatto and Windus, 1949).

263 **"Prof. Haldane said he might"**: Minutes of the meeting of the *Modern Quarterly* editorial board, April 11, 1947, HAL-DANE/4/17/15, University College London archive.

264 **"Try to find out how"**: John Lewis to Haldane, October 1946, HALDANE/4/17/2/11, University College London archive.

264 **"If the central committee"**: John Maynard Smith, "J. B. S. Haldane," in *The Founders of Evolutionary Genetics*, 49.

264   **Haldane wasn't making his bloody mind up:** MI5, KV-2-1832_2, National Archives.

264   **It sounded too liberal:** MI5, KV-2-1832_2, National Archives.

266   **If the commentary wasn't withdrawn:** Ivar Montagu to Haldane, January 1949, HALDANE/5/1/3/46, University College London archive.

266   **"are so completely false":** "Note by J. B. S. Haldane on the *Daily Worker* Educational Commentary," February 1949, People's History Museum, Manchester.

266   **He marked up the draft:** HALDANE/4/9/1/7, University College London archive.

266   **He sent a letter to Maurice Cornforth:** Haldane to Cornforth, November 1948, HALDANE/4/9/1/5, University College London archive.

266   **"I cannot agree with the resolution":** Haldane to Cornforth, February 1949, HALDANE/4/9/2/9, University College London archive.

267   **"During the past few months":** Haldane to John Campbell, March 1949, HALDANE/4/8/3/18, University College London archive.

267   **calling Cornforth a bloody liar:** Gabriel Abraham Almond, *Appeals of Communism.*

267   **"From the two letters":** Haldane to John Campbell, December 1949, HALDANE/4/8/3/24, University College London archive.

268   **"Pollitt apologized":** MI5, KV-2-1832_2, National Archives.

268   **"And he and the goons":** John Maynard Smith, 2004, interviewed for Web of Stories, https://www.youtube.com/watch?v=J0OSK5m3XBk.

269   **he was back in Pollitt's office:** MI5, KV-2-1832_2, National Archives.

269   **"Professor Haldane considers that":** MI5, KV-2-1832_1, National Archives.

271   **"Well, the alternatives are probably worse":** Haldane, interviewed by Margaret Lane, Stephen Black, and Jack Morpurgo, "Frankly Speaking," May 1956.

271   **"Lysenko was too dogmatic":** Haldane to Edward Primba, September 1962, part of Acc. 9589/176, National Library of Scotland.

271   **"In my opinion":** Haldane's self-obituary, recorded for the BBC in February 1964, broadcast on December 1, 1964.

271   **"I thought, and think":** Quoted in Clark, *J. B. S.*, 160.

272   **"Yudhishthira, who, on the whole":** See Haldane's "A Passage to India," in *Science and Life*, 133.

Chapter 6: India

273   **"They will be sealed up":** Haldane to S. C. Sen, June 1957, Indian Statistical Institute archive.

273   **"I do not want to be a citizen":** "Prof. J. B. S. Haldane to Live in India," *The Times*, November 21, 1956.

273   **"I want to live in":** "Haldane, Geneticist, Quits Britain for India Which Has No G. I.'s," *New York Times*, July 25, 1957.

274   **Silver entered and absorbed:** Brian Silver, *The Ascent of Science*, 353.

274   **"paraded about like a Rajah's elephant":** Haldane to V. K. Krishna Menon, February 1952, HALDANE/5/7/5, University College London archive.

275   **"I could, I believe":** Haldane to Jawaharlal Nehru, February 1952, HALDANE/5/7/5, University College London archive.

275   **"Your help will, of course":** Nehru to Haldane, March 1952, HALDANE/5/7/5, University College London archive.

276   **"I have more or less retired":** Haldane to Stanley Kunitz, December 1952, HALDANE/5/1/4/98, University College London archive.

277   **"To take one example":** Haldane to D. W. Logan, July 1957, HALDANE/3/1/1/9/2, University College London archive.

277   **"looking after a certain number":** Haldane, interviewed by Margaret Lane, Stephen Black, and Jack Morpurgo, "Frankly Speaking," May 1956.

278   **"Dear Madam":** Haldane to Moore, Ruth, March 1952, HALDANE/5/1/4/164, University College London archive.

278   **"Your request that I should":** Haldane to Marvin Seiger, November 1953, HALDANE/5/1/4/222, University College London archive.

279   **"One of my reasons":** Quoted in Clark, *J. B. S.*, 131.

282   **"As our small group continued":** Krishna R. Dronamraju, *The Life and Work of J. B. S. Haldane with Special Reference to India*.

283   **"Wouldn't any biologist like to live":** Helen Spurway, scrap of a memoir, National Library of Scotland.

284   **"do not take the laws of gods":** See Haldane's essay "On Expecting the Unexpected," in *Science and Life*, 142.

284   **With these data, he improved:** J. B. S. Haldane, "The Mutation

Rate of the Gene for Haemophilia, and Its Segregation Ratios in Males and Females," *Annals of Eugenics* 13, no. 1 (January 1946): 262–71.

285 **Again with more fresh data:** J. B. S. Haldane, "A New Estimate for the Linkage between the Genes for Colourblindness and Haemophilia in Man," *Annals of Eugenics* 14, no. 1 (October 1947): 10–31.

285 **an offhand curl of thought:** Joshua Lederberg, "J. B. S. Haldane (1949) on Infectious Disease and Evolution," *Genetics* 153, no. 1 (September 1999): 1–3. Also see Sahotra Sarkar, "Sex, Disease, and Evolution: Variations on a Theme from J. B. S. Haldane," *BioScience* 42, no. 6, (June 1992): 448–54.

286 **"Haldane invited me up to his room":** A. C. Allison, "Genetics and Infectious Disease," in K. R. Dronamraju, ed., *Haldane and Modern Biology*, 181.

286 **A 2011 paper in *Science*:** Marek Cyrklaff et al., "Hemoglobins S and C Interfere with Actin Remodeling in *Plasmodium falciparum*–Infected Erythrocytes," *Science* 334, no. 6060 (December 2, 2011): 1283–6.

287 **This was how Maynard Smith:** John Maynard Smith, "Survival through Suicide," *New Scientist*, August 28, 1975.

288 **"Let us suppose that":** J. B. S. Haldane, "Population Genetics," *New Biology* 18 (1955): 34–51.

289 **Published in 1957 and titled:** J. B. S. Haldane, "The Cost of Natural Selection," *Journal of Genetics* 55, no. 3 (December 1957): 511.

290 **others have contested its details:** For a discussion of the criticisms of Haldane's cost of selection framework and the neutral theory of molecular evolution, see: George R. Gluesing and Fathi Abdel-Hameed, "Impact of Neutral Mutations on Evolution," in *Evolutionary Models and Studies in Human Diversity*, ed. Robert J. Meier, Charlotte M. Otten, and Fathi Abdel-Hameed (Berlin: Walter de Gruyter, 2011); Motoo Kimura, "Evolutionary Rate at the Molecular Level," *Nature* 217 (February 17, 1968): 624–6; Masatoshi Nei, "Selectionism and Neutralism in Molecular Evolution," *Molecular Biology and Evolution* 22, no. 12 (December 2005): 2318–42; Motoo Kimura, *The Neutral Theory of Molecular Evolution* (Cambridge: Cambridge University Press, 1985); Monroe W. Strickberger,, *Evolution* (Burlington, MA: Jones and Bartlett, 1990).

290 **lower than Haldane thought:** He had run his equations on the basis of an infinite population and a selective pressure that stayed the same over the years, but these weren't the conditions of the real

world. He hadn't factored the density of the population into his model, and he hadn't allowed for genetic drift, the fluctuations in the frequencies of genes that occur out of sheer randomness.

291 **What about kangaroos?:** Heather Denise Harden Botting and Gary Botting, *The Orwellian World of Jehovah's Witnesses*, xiv.

293 **"I was asked":** Haldane to John Tyler Bonner, November 1960, quoted by Bonner in "An Extract from 'Memoirs for Family and Friends,'" *Genetics* 150, no. 2 (October 1998): 519–21.

293 **"My European tail coat":** Haldane to the Director of Hospitality, Raj Bhavan, February 1961, MS 20544, National Library of Scotland.

294 **"not because Indians are stupid or lazy":** J. B. S. Haldane, "What Ails Indian Science," in *Science and Indian Culture*, 3.

295 **"a rather poor copy of British teaching":** Haldane to P. C. Mahalanobis, April 1958, Indian Statistical Institute archive.

295 **"who have a smaller salary":** Mahalanobis to Haldane, November 1959, Indian Statistical Institute archive.

296 **"arrange for its destruction":** Haldane to unknown official, September 1958, Indian Statistical Institute archive.

296 **"I find this idea of destroying":** Unknown official to Mahalanobis, Indian Statistical Institute archive.

296 **His note to Mahalanobis:** Haldane to Mahalanobis, February 1961, Indian Statistical Institute archive.

297 **"There can be no doubt":** Nehru to M. S. Thacker, May 1961, quoted in Madhavan K. Palat, ed., *Selected Works of Jawaharlal Nehru: Second Series*, vol. 69 (Kharagpur, India. National Digital Library of India, May/June 1961).

297 **"There are several hungry chaps":** Haldane to Ritchie Calder, March 1962, MS 20543, National Library of Scotland.

298 **"consists of half a table":** Haldane to Dronamraju, June 1962, MS 20544, National Library of Scotland.

298 **"I am now settling down here":** Haldane to Kurt Mothes, November 1962, MS 20545, National Library of Scotland.

299 **The roads were new:** Ravi Kalia, "Bhubaneswar: Contrasting Visions in Traditional Indian and Modern European Architecture," *Journal of Urban History* 23, no. 2 (January 1997): 164–91.

301 **"would today make me":** Haldane to Calder, March 1962, MS 20543, National Library of Scotland.

301 **Dante's foreshadowing of mutations:** Haldane, "Closing Address," *IUBS Symposium on Genetics of Population Structure*, Pavia, Italy, August 1953. Also see Ernst Mayr, "As I Knew J. B. S. Haldane," quoted in Sahotra Sarkar, "Mayr's Recollections of Haldane: A Document with Brief Commentary," *History and Philosophy of the Life Sciences* 35, no. 2 (2013): 265–75.

302 **"So have I":** Joe Felsenstein, interview with the author, April 2019.

302 **Where are you from?:** Richard Lewontin, as told to Joe Felsenstein.

302 **"I suppose that Oparin and I":** Quoted in Sydney W. Fox, ed., *The Origins of Prebiological Systems and of Their Molecular Matrices*, 89.

303 **While he was still recovering:** Jonathan Howard, interview with the author, August 2018.

304 **"I wish I had the voice of Homer":** "Cancer's a Funny Thing," *New Statesman*, February 21, 1964.

306 **"I am told that my chance":** Haldane to Arthur C. Clarke, February 1964, Royal Society archive.

306 **"He did my Greek homework":** Julian Maynard Smith, interview with the author, March 2019.

306 **"I'm afraid I'll never see this again":** John Maynard Smith, as interviewed by Kohn, in *A Reason for Everything*, 223.

307 **"I have a real snorter":** Haldane to Theodosius Dobzhansky, July 1964, MS 20545, National Library of Scotland.

309 **"For one who has taken birth":** Ramachandra Guha, "On Becoming an Indian," *Journal of Genetics* 96, no. 5 (November 2017): 805–14. Translation by K. R. Subramanian.

309 **"I don't mind dying":** Quoted in Clark, *J. B. S.*, 171.

309 **he kept in his hand the stone:** Recounted by Maynard Smith, as told to him by Helen Spurway, in Web of Stories interview, https://www.webofstories.com/play/7286?o=MS.

310 **"I hope that I have been":** See Haldane's essay "When I Am Dead," in *On Being the Right Size*, 26.

310 **His organs went to:** "Haldane's Body Given for Research," *Medical News*, April 9, 1965.

Chapter 7: Ten Thousand Years

311 **After the foundation reopened:** Gordon Wolstenholme, ed., *Man and His Future* (London: J. & A. Churchill Ltd., 1963).

312   **"so that much that is now unthinkable":** Julian Huxley, *UNESCO: Its Purpose and Its Philosophy*, 1946.

317   **Every year from 2014 to 2018:** "UCL to Investigate Eugenics Conference Secretly Held on Campus," *The Guardian*, January 11, 2018.

317   **"In the long run":** Haldane, *Heredity and Politics*, 180.

319   **"dilute ink with a little sugar":** Haldane, *What Is Life?*, 213.

Selected Bibliography

ARCHIVES

Belgium
In Flanders Field Museum, Ypres

Great Britain
Bodleian Library, Oxford
The British Library, London
The College Archives, Eton College
The Imperial War Museum, London
The John Innes Centre, Norwich
The Keep, Brighton
National Archives, Kew
National Library of Scotland, Edinburgh
People's History Museum, Manchester
The Royal Society, London
University College, London
Wellcome Library, London
Working Class Movement Library, Manchester

India
The Centre for Cellular & Molecular Biology, Hyderabad
Indian Statistical Institute, Kolkata

Ireland
National Library of Ireland, Dublin

United States
Columbia University, New York

BOOKS AND ARTICLES

Almond, Gabriel Abraham. *Appeals of Communism*. Princeton: Princeton University Press, 1954.

Ashby, Eric. *Scientist in Russia*. London: Pelican Books, 1946.

Ayala, Francisco J., and John C. Avise, ed. *Essential Readings in Evolutionary Biology*. Baltimore: Johns Hopkins University Press, 2014.

Bailes, Kendall E. *Technology and Society under Lenin and Stalin: Origins of the Soviet Technical Intelligentsia 1917–1941*. Princeton: Princeton University Press, 1978.

Ball, Simon. *The Guardsmen: Harold Macmillan, Three Friends and the World They Made*. London: HarperCollins, 2010.

Baxell, Richard. *British Volunteers in the Spanish Civil War: The British Battalion in the International Brigades, 1936–1939*. London: Routledge, 2004.

Baxell, Richard. *Unlikely Warriors: The British in the Spanish Civil War and the Struggle against Fascism*. London: Aurum Press, 2012.

Bernal, J. D. *The Social Function of Science*. London: George Routledge & Sons, 1939.

Botting, Heather, and Gary Botting. *The Orwellian World of Jehovah's Witnesses*. Toronto: University of Toronto Press, 1984.

Bowler, Peter J. *The Eclipse of Darwinism: Anti-Darwinian Evolution Theories in the Decades around 1900*. Baltimore: Johns Hopkins University Press, 1983.

Brannigan, Augustine. *The Social Basis of Scientific Discoveries*. New York: Cambridge University Press, 1981.

Brown, Andrew. *J. D. Bernal: The Sage of Science*. Oxford: Oxford University Press, 2005.

Bush, Julia. *Edwardian Ladies and Imperial Power*. Leicester, UK: Leicester University Press, 2000.

Butler, Samuel. *Life and Habit*. London: Jonathan Cape, 1878.

Calder, Jenni. *The Nine Lives of Naomi Mitchison*. London: Virago, 1997.

Clark, Ronald. *J. B. S.: The Life and Work of J. B. S. Haldane*. Oxford: Oxford University Press, 1984.

Cook, Tim. *No Place to Run: The Canadian Corps and Gas Warfare in the First World War*. Vancouver: UBC Press, 2001.

Copeman, Fred. *Reason in Revolt*. London: Blandford Press, 1948.

Cowles, Virginia. *Looking for Trouble*. London: Faber & Faber, 2014.

DeJong-Lambert, William, and Nikolai Krementsov, ed. *The Lysenko*

Controversy as a Global Phenomenon. Cham, Switzerland: Palgrave-Macmillan, 2017.

Dronamraju, Krishna R., ed. *Haldane and Modern Biology*. Baltimore: Johns Hopkins University Press, 1968.

Dronamraju, Krishna R. *The Life and Work of J. B. S. Haldane with Special Reference to India*. Aberdeen, UK: Aberdeen University Press, 1985.

Dronamraju, Krishna R. *Popularizing Science: The Life and Work of J. B. S. Haldane*. Oxford: Oxford University Press, 2017.

Dronamraju, Krishna R., ed. *Selected Genetic Papers of J. B. S. Haldane*. New York: Garland Publishing, 1990.

Dugdale, R. L. *The Jukes: A Study in Crime, Pauperism, Disease and Heredity*. New York: G. P. Putnam's Sons, 1877.

Fairbanks, Daniel J., and Scott Abbott. "Darwin's Influence on Mendel: Evidence from a New Translation of Mendel's Paper." *Genetics* 204, no. 2 (October 2016): 401–405.

Fox, Sidney W., ed. *The Origins of Prebiological Systems and of Their Molecular Matrices: Proceedings of a Conference Conducted at Wakulla Springs, Florida, on 27–30 October 1963 under the Auspices of the Institute for Space Biosciences, the Florida State University and the National Aeronautics and Space Administration*. Cambridge, MA: Academic Press, 1965.

Girard, Marion. *A Strange and Formidable Weapon: British Responses to World War I Poison Gas*. Lincoln: University of Nebraska Press, 2008.

Goodman, Martin. *Suffer and Survive: Gas Attacks, Miners' Canaries, Spacesuits and the Bends: The Extreme Life of Dr. J. S. Haldane*. London: Simon & Schuster, 2007.

Gould, Stephen Jay. *The Structure of Evolutionary Theory*. Cambridge, MA: The Belknap Press of Harvard University Press, 2002.

Guénet, Jean-Louis, Fernando Benavides, Jean-Jacques Panthier, and Xavier Montagutelli. *Genetics of the Mouse*. New York: Springer, 2014.

Haldane, Charlotte. *Truth Will Out*. London: The Right Book Club, 1949.

Haldane, Elizabeth. *From One Century to Another*. London: Alexander Maclehose & Co., 1937.

Haldane, J. B. S. *A. R. P.* London: Victor Gollancz Ltd., 1938.

Haldane, J. B. S. *A Banned Broadcast and Other Essays*. London: Chatto and Windus, 1946.

Haldane, J. B. S. *The Biochemistry of Genetics*. London: Allen & Unwin, 1954.

Haldane, J. B. S. *Callinicus*. New York: E. P. Dutton & Company, 1925.

Haldane, J. B. S. *The Causes of Evolution*. London: Longmans, Green and Co., 1932.

Haldane, J. B. S. *Daedalus; or, Science and the Future*. New York: E. P. Dutton & Company, 1924.

Haldane, J. B. S. *Enzymes*. London: Longmans, Green and Co., 1930.

Haldane. J. B. S. *Everything Has a History*. London: Allen & Unwin, 1951.

Haldane, J. B. S. *Fact and Faith*. London: Watts & Co., 1934.

Haldane, J. B. S. *Heredity and Politics*. London: Allen and Unwin, 1938.

Haldane, J. B. S. *The Inequality of Man*. London: Chatto & Windus, 1932.

Haldane, J. B. S. *Keeping Cool and Other Essays*. London: Chatto & Windus, 1944.

Haldane, J. B. S. *The Marxist Philosophy and the Sciences*. London: Allen & Unwin, 1938.

Haldane, J. B. S. *My Friend Mr. Leakey*. London: Cresset Press, 1937.

Haldane, J. B. S. *New Paths in Genetics*. London: Allen & Unwin, 1943.

Haldane, J. B. S. *On Being the Right Size and Other Essays*. Oxford: Oxford University Press, 1985.

Haldane, J. B. S. *Possible Worlds and Other Essays*. London: Chatto & Windus, 1928.

Haldane, J. B S. *Science Advances*. London: Allen & Unwin, 1947.

Haldane, J. B. S. *Science and Ethics*. London: Watts & Co., 1928.

Haldane, J. B. S. *Science and Everyday Life*. Allahabad: Kitab Mahal, 1945.

Haldane, J. B. S. *Science and Human Life*. New York: Harper and Brothers, 1933.

Haldane, J. B. S. *Science and Indian Culture*. Calcutta: New Age Publications, 1965.

Haldane, J. B. S. *Science and Life*. London: Pemberton Publishing, 1968.

Haldane, J. B. S. *Science in Peace and War*. London: Lawrence & Wishart, 1943.

Haldane, J. B. S. *The Unity and Diversity of Life*. Delhi: The Publications Division, 1958.

Haldane, J. B. S. *What I Require from Life*. Oxford: Oxford University Press, 2009.

Haldane, J. B. S. *What Is Life?* London: Lindsay Drummond, 1949.

Haldane, J. B. S. *Why I Am a Cooperator*. Unpublished autobiography.

Haldane, J. S. *Methods of Air Analysis*. London: Charles Griffin & Company, 1912.

Haldane, J. S. *The Philosophy of a Biologist.* Oxford: Oxford University Press, 1935.

Haldane, J. S. *Report to the Secretary of State for the Home Department on the Causes of Death in Colliery Explosions and Underground Fires.* London: Eyre and Spottiswoode, 1896.

Haldane, J. S. and Smith, Lorrain, "The Physiological Effects of Air Vitiated by Respiration." *Journal of Pathology and Bacteriology* 1, no. 2 (October 1892): 168–86.

Haldane, Louisa Kathleen. *Friends and Kindred.* London: Faber and Faber, 1961.

Haldane, Richard Burdon. *Richard Burdon Haldane: An Autobiography.* London: Hodder and Stoughton, 1929.

Harman, Oren Solomon. *The Man Who Invented the Chromosome: A Life of Cyril Darlington.* Cambridge, MA: Harvard University Press, 2004.

Hawks, John, Eric T. Wang, Gregory M. Cochran, Henry C. Harpending, and Robert K. Moyzis. "Recent Acceleration of Human Adaptive Evolution." *Proceedings of the National Academy of Sciences* 104, no. 52 (December 2007): 20753–58.

Henig, Robin Marantz. *The Monk in the Garden: The Lost and Found Genius of Gregor Mendel, the Father of Genetics.* New York: Houghton Mifflin Company, 2000.

Hermiston, Roger *All Behind You, Winston: Churchill's Great Coalition 1940–45.* London: Aurum Press, 2016.

Hobsbawm, Eric. *Revolutionaries.* New York: The New Press, 2001.

Howard, Victor. *The MacKenzie-Papineau Battalion: The Canadian Contingent in the Spanish Civil War.* Ottawa: Carleton University Press, 1986.

Howarth, T. E. B. *Cambridge between Two Wars.* London: Collins, 1978.

Huxley, Julian. *Memories.* London: Allen and Unwin, 1970.

Ipatieff, Vladimir. *The Life of a Chemist,* trans. Vladimir Haensel and Mrs. Ralph H. Lusher. Palo Alto, CA: Stanford University Press, 1946.

Joravsky, David. *The Lysenko Affair.* Cambridge, MA: Harvard University Press, 1970.

Kalia, Ravi. *Bhubaneswar: From Temple Town to Capital City.* Carbondale, IL: Southern Illinois University Press, 1994.

Kevles, Daniel J. *In The Name of Eugenics: Genetics and the Uses of Human Heredity.* New York: Alfred A. Knopf, 1985.

Kohn, Marek. *A Reason for Everything: Natural Selection and the English Imagination.* London: Faber and Faber, 2004.

Krementsov, Nikolai. *Revolutionary Experiments: The Quest for Immortality in Bolshevik Science*. Oxford and New York: Oxford University Press, 2013.

Langdon-Davies, John. *Russia Puts the Clock Back: A Study of Soviet Science and Some British Scientists*. London: Gollancz, 1949.

Lightman, Bernard, ed. *Victorian Science in Context*. Chicago: The University of Chicago Press, 1997.

Linklater, Andro, and Eric Linklater. *Black Watch: History of the Royal Highland Regiment*. London: Barrie & Jenkins, 1977.

Luck, James Murray. *Reminiscences*. Palo Alto: Annual Reviews, 1999.

Medawar, Peter. *Memoir of a Thinking Radish*. Oxford: Oxford University Press, 1986.

Medvedev, Zhores. *The Rise and Fall of T. D. Lysenko*, trans. I. Michael Lerner. New York: Columbia University Press, 1969.

Meyer, Jessica. *Men of War: Masculinity and the First World War in Britain*. New York: Springer, 2016.

Mitchison, Naomi. *Among You Taking Notes: The Wartime Diary of Naomi Mitchison*. Oxford: Oxford University Press, 1986.

Mitchison, Naomi. *As It Was: An Autobiography 1897–1918*. Glasgow: Richard Drew Publishing, 1988.

Mitchison, Naomi. *The Moral Basis of Politics*. London: Constable & Co., 1938.

Mitchison, Naomi. *Mucking Around*. London: Victor Gollancz, 1981.

Mitchison, Naomi. *You May Well Ask*. London: Fontana Paperbacks, 1986.

Montagu, Ivor. *The Youngest Son*. London: Lawrence and Wishart, 1970.

Morison, Patricia. *J. T. Wilson and the Fraternity of Duckmaloi*. Amsterdam: Rodopi, 1997.

Mukherjee, Siddhartha. *The Gene: An Intimate History*. New York: Scribner, 2016.

Nei, Masatoshi. *Mutation-Driven Evolution*. Oxford: Oxford University Press, 2013.

Paul, Diane B. "A War on Two Fronts: J. B. S. Haldane and the Response to Lysenkoism in Britain." *Journal of the History of Biology* 16, no. 1 (Spring 1983): 1–37.

Preston, Diana. *A Higher Form of Killing: Six Weeks in World War I That Forever Changed the Nature of Warfare*. London: Bloomsbury, 2015.

Pringle, Peter. *The Murder of Nikolai Vavilov: The Story of Stalin's Persecution of One of the Great Scientists of the Twentieth Century.* New York: Simon & Schuster, 2011.

Rhodes, Richard. *Hell and Good Company: The Spanish Civil War and the World It Made.* New York: Simon & Schuster, 2015.

Roll-Hansen, Nils. *The Lysenko Effect: The Politics of Science.* Amherst, NY: Humanity Books, 2005.

Rudra, Ashok. *Prashanta Chandra Mahalanobis: A Biography.* Oxford: Oxford University Press, 1996.

Sarkar, Sahotra, ed. *The Founders of Evolutionary Genetics: A Centenary Reappraisal.* Dordrecht: Springer Netherlands, 1992.

Schofield, Victoria. *The Black Watch: Fighting in the Front Line, 1899–2006.* London: Head of Zeus, 2017.

Segerstrale, Ullica. *Nature's Oracle: The Life and Work of W. D. Hamilton.* Oxford: Oxford University Press, 2013.

Seth, Andrew, and R. B. Haldane, eds. *Essays in Philosophical Criticism.* New York: Burt Franklin, 1883.

Silver, Brian L. *The Ascent of Science.* Oxford: Oxford University Press, 1998.

The Situation in Biological Science: Proceedings of the Lenin Academy of Agricultural Sciences of the U.S.S.R., July 31–August 7, 1948. Moscow: Foreign Languages Publishing House, 1949.

Stewart, Roderick. *Bethune in Spain.* Montreal: McGill-Queen's University Press, 2014.

Stoltzfus, Arlin, and Kele Cable. "Mendelian-Mutationism: The Forgotten Evolutionary Synthesis." *Journal of the History of Biology* 47, no. 4 (Winter 2014): 501–46.

Stone, Dan. *Breeding Superman: Nietzsche, Race and Eugenics in Edwardian and Interwar Britain.* Liverpool: Liverpool University Press, 2002.

Sturdy, Steve. "Biology as Social Theory: John Scott Haldane and Physiological Regulation." *British Journal for the History of Science* 21, no. 3 (September 1988): 315–40.

Sturdy, Steve. "A Co-ordinated Whole: The Life and Work of John Scott Haldane." PhD thesis, University of Edinburgh, 1987.

Sturdy, Steve. "The Meanings of 'Life': Biology and Biography in the Work of J. S. Haldane (186 –1936)." *Transactions of the Royal Historical Society* (Sixth Series) 21 (December 2011): 171–91.

Teich, Mikulas. "Haldane and Lysenko Revisited." *Journal of the History of Biology* 40, no. 3 (September 2007): 557–63.

Tredoux, Gavan. *Comrade Haldane Is Too Busy to Go on Holiday*. New York: Encounter Books, 2018.

Warren, C., and J. Benson. *Thetis: The Admiralty Regrets: The Disaster in Liverpool Bay*. Bebington, UK: Avid Publications, 1997.

Wauchope, A. G., ed. *A History of the Black Watch (Royal Highlanders) in the Great War 1914–18*. Uckfield, UK: Naval & Military Press, 2016.

Werskey, Gary. *The Visible College: The Collective Biography of British Scientific Socialists of the 1930s*. New York: Holt, Rinehart and Winston, 1978.

Wood, Elizabeth A. *Performing Justice: Agitation Trials in Early Soviet Russia*. Ithaca, NY: Cornell University Press, 2005.

Illustration Credits

Index

Page references after 324 refer to notes. Page references in *italics* refer to illustrations.

Burdon-Sanderson, John, 36
Burdon-Sanderson, Mary, 34
Burghes, Charlotte. *See also*
  Haldane, Charlotte
  attractiveness to Haldane, 149–50
  letters to Haldane, 149–50
  meeting with Haldane, 146–48
Burghes, Jack, 149, 152
Burghes, Ronnie (Charlotte's son),
  149, 178, 209
  Haldane's anger and, 196
  in Spain, 216
  and Spanish Civil War, 210
Butler, Samuel, 62

Calcutta, Haldane's invitation to,
  291
Cambridge, 155
  Benskin Society, 180
  Haldane in, 105
  Haldanes' house, 178
  socialists in, 166
Cambridge Five, 253
Cambridge Scientists Anti-War
  Group, 170
Cambridge Union, 166, 167
Campbell, John, 267
cancer, genetics of, 207
capitalism, 13, 236
carbon dioxide, 37
  in blood, 173
  vs. oxygen shortage, 37–38
  system to measure, 36
carbon monoxide, and mine deaths,
  48–50
Case, Martin, 179, 180, 184–85,
  196, 249
cats, inheritance study, 190–91
Catto, A. Y., 29
Cement Makers' Federation, 236

Central House of Scientists
  (Moscow), 10–11
Century, 146
Chain, Ernest, 221
Chamberlain, Neville, 220
chance, in genetics, 120
chemical warfare, in First World
  War, 144–45. *See also* chlorine
  gas
Cherwell
  convalescence at, 93
  house in, 43–44
  Naomi's guinea pigs, 77
Chesterton, G. K., 141
children, 196, 197
  Charlotte and Haldane's interest
  in, 151
chlorine gas, 91
Christian clerics, 134
Christianity, 331
Churchill, Winston, 238, 241
Ciba Foundation, 311
civilization, science and, 143
Clarke, Arthur C., 4
class, 55–56
  Haldane's exposure to those
  outside his, 99
  and military, 88
class warfare, 166
classics, Haldane's love for, 81
Cloan (family home), 55–56
Cockburn, Patricia, 162
Cold War, 257
colonial thought, Haldane and, 98
color blindness, genetic mapping
  for, 3
Communism, 13
  vs. fascism, 219
  Haldane and, 220
  science and, 169
  Stalinist, 228